Molecular Advances in Dental Pulp Tissue Engineering

Molecular Advances in Dental Pulp Tissue Engineering

Editors

Matthias Widbiller
Kerstin M. Galler

Basel • Beijing • Wuhan • Barcelona • Belgrade • Novi Sad • Cluj • Manchester

Editors
Matthias Widbiller
University Hospital
Regensburg
Regensburg, Germany

Kerstin M. Galler
Friedrich-Alexander-University
Erlangen-Nuernberg
Erlangen, Germany

Editorial Office
MDPI
St. Alban-Anlage 66
4052 Basel, Switzerland

This is a reprint of articles from the Special Issue published online in the open access journal *International Journal of Molecular Sciences* (ISSN 1422-0067) (available at: https://www.mdpi.com/journal/ijms/special_issues/DPT_engineering).

For citation purposes, cite each article independently as indicated on the article page online and as indicated below:

Lastname, A.A.; Lastname, B.B. Article Title. *Journal Name* **Year**, *Volume Number*, Page Range.

ISBN 978-3-0365-8476-8 (Hbk)
ISBN 978-3-0365-8477-5 (PDF)
doi.org/10.3390/books978-3-0365-8477-5

Cover image courtesy of Matthias Widbiller

Contents

About the Editors . . . . **vii**

Matthias Widbiller and Kerstin M. Galler
Engineering the Future of Dental Health: Exploring Molecular Advancements in Dental Pulp
Regeneration
Reprinted from: *Int. J. Mol. Sci.* **2023**, *24*, 11453, doi:10.3390/ijms241411453 . . . . **1**

Ella Ohlsson, Kerstin M. Galler and Matthias Widbiller
A Compilation of Study Models for Dental Pulp Regeneration
Reprinted from: *Int. J. Mol. Sci.* **2022**, *23*, 14361, doi:10.3390/ijms232214361 . . . . **9**

**Sandra Minic, Sibylle Vital, Catherine Chaussain, Tchilalo Boukpessi and
Francesca Mangione**
Tissue Characteristics in Endodontic Regeneration: A Systematic Review
Reprinted from: *Int. J. Mol. Sci.* **2022**, *23*, 10534, doi:10.3390/ijms231810534 . . . . **29**

**Marion Florimond, Sandra Minic, Paul Sharpe, Catherine Chaussain, Emmanuelle Renard
and Tchilalo Boukpessi**
Modulators of Wnt Signaling Pathway Implied in Dentin Pulp Complex Engineering: A
Literature Review
Reprinted from: *Int. J. Mol. Sci.* **2022**, *23*, 10582, doi:10.3390/ijms231810582 . . . . **57**

**Martyna Smeda, Kerstin M. Galler, Melanie Woelflick, Andreas Rosendahl,
Christoph Moehle, Beate Lenhardt, et al.**
Molecular Biological Comparison of Dental Pulp- and Apical Papilla-Derived Stem Cells
Reprinted from: *Int. J. Mol. Sci.* **2022**, *23*, 2615, doi:10.3390/ijms23052615 . . . . **69**

**Cristina Bucchi, Ella Ohlsson, Josep Maria de Anta, Melanie Woelflick, Kerstin Galler,
María Cristina Manzanares-Cespedes and Matthias Widbiller**
Human Amnion Epithelial Cells: A Potential Cell Source for Pulp Regeneration?
Reprinted from: *Int. J. Mol. Sci.* **2022**, *23*, 2830, doi:10.3390/ijms23052830 . . . . **87**

**Ryo Kunimatsu, Kodai Rikitake, Yuki Yoshimi, Nurul Aisyah Rizky Putranti, Yoko Hayashi
and Kotaro Tanimoto**
Bone Differentiation Ability of CD146-Positive Stem Cells from Human Exfoliated Deciduous
Teeth
Reprinted from: *Int. J. Mol. Sci.* **2023**, *24*, 4048, doi:10.3390/ijms24044048 . . . . **101**

**Yelyzaveta Razghonova, Valeriia Zymovets, Philip Wadelius, Olena Rakhimova,
Lokeshwaran Manoharan, Malin Brundin, et al.**
Transcriptome Analysis Reveals Modulation of Human Stem Cells from the Apical Papilla by
Species Associated with Dental Root Canal Infection
Reprinted from: *Int. J. Mol. Sci.* **2022**, *23*, 14420, doi:10.3390/ijms232214420 . . . . **115**

**Afzan A. Ayoub, Abdel H. Mahmoud, Juliana S. Ribeiro, Arwa Daghrery, Jinping Xu,
J. Christopher Fenno, et al.**
Electrospun Azithromycin-Laden Gelatin Methacryloyl Fibers for Endodontic Infection Control
Reprinted from: *Int. J. Mol. Sci.* **2022**, *23*, 13761, doi:10.3390/ijms232213761 . . . . **135**

**Juliana S. Ribeiro, Eliseu A. Münchow, Ester A. F. Bordini, Nathalie S. Rodrigues,
Nileshkumar Dubey, Hajime Sasaki, et al.**
Engineering of Injectable Antibiotic-Laden Fibrous Microparticles Gelatin Methacryloyl
Hydrogel for Endodontic Infection Ablation
Reprinted from: *Int. J. Mol. Sci.* **2022**, *23*, 971, doi:10.3390/ijms23020971 . . . . **151**

Konstantinos Loukelis, Foteini Machla, Athina Bakopoulou and Maria Chatzinikolaidou
Kappa-Carrageenan/Chitosan/Gelatin Scaffolds Provide a Biomimetic Microenvironment for Dentin-Pulp Regeneration
Reprinted from: *Int. J. Mol. Sci.* **2023**, *24*, 6465, doi:10.3390/ijms24076465 . . . . **167**

About the Editors

Matthias Widbiller

Matthias Widbiller started his academic career in 2013 after completing his dental studies at the University of Regensburg, Germany. He worked at the Department of Conservative Dentistry and Periodontology at the University Hospital Regensburg, where he received his doctorate in 2015. Matthias Widbiller is a lecturer in conservative dentistry and a certified university teacher. His passion for research led him to a postdoctoral position at the University of Texas Health Science Center at San Antonio (UTHSCSA), USA, from 2017 to 2018, where he worked in the renowned research laboratory of Dr. Kenneth M. Hargreaves and Dr. Anibal Diogenes. His scientific focus is on dental pulp tissue engineering, bioactive dental materials and dentin matrix proteins. As a clinical expert, Dr. Widbiller specialises in dental traumatology, vital pulp therapy and endodontic regeneration. Since 2016, he has coordinated the Centre for Dental Traumatology at the University Hospital of Regensburg, and since 2021, he has been the head of the research department at the same institution. In 2019, his expertise and contributions to the scientific community were recognised with his habilitation, and finally, in 2023, Dr Widbiller was appointed Professor of Endodontics by the University of Regensburg.

Kerstin M. Galler

Kerstin M. Galler obtained her degree in dentistry from the Ludwig Maximilian University of Munich in 2000. She worked in Private Practice until 2002 and then joined the Department of Conservative Dentistry and Periodontology at the University Hospital Regensburg, Germany. She received post-doctoral training at the University of Texas Health Science Center in Houston from 2004 to 2006 and earned her Ph.D. in Biomedical Engineering from Rice University in Houston in 2009. After her return to Regensburg, she built her own research group as an Associated Professor in Endodontology, with a focus on pulp biology and immune responses, on the clinical procedure of revitalization, and on dental pulp stem cells, tunable scaffolds and the use of dentine matrix proteins for dental pulp tissue engineering. From 2015 to 2021, she was Deputy Head and Section Leader of Endodontology at the Department of Conservative Dentistry and Periodontology at Regensburg University Hospital. Today, Dr. Galler is Director and Chair of the Department of Restorative Dentistry and Periodontology at the University Hospital Erlangen, Germany. She has published extensively, lectured in numerous national and international congresses, and received, among other honors, the IADR Distinguished Scientist Award in 2022.

International Journal of
Molecular Sciences

Editorial

Engineering the Future of Dental Health: Exploring Molecular Advancements in Dental Pulp Regeneration

Matthias Widbiller [1,*] and Kerstin M. Galler [2]

1   Department of Conservative Dentistry and Periodontology, University Hospital Regensburg, Franz-Josef-Strauß-Allee 11, D-93093 Regensburg, Germany
2   Department of Operative Dentistry and Periodontology, Friedrich-Alexander-University Erlangen-Nuernberg, D-91054 Erlangen, Germany; kerstin.galler@uk-erlangen.de
*   Correspondence: matthias.widbiller@ukr.de

Citation: Widbiller, M.; Galler, K.M. Engineering the Future of Dental Health: Exploring Molecular Advancements in Dental Pulp Regeneration. *Int. J. Mol. Sci.* **2023**, 24, 11453. https://doi.org/10.3390/ijms241411453

Received: 6 June 2023
Accepted: 2 July 2023
Published: 14 July 2023

Protected by the surrounding mineralized barriers of enamel, dentin, and cementum, dental pulp is a functionally versatile tissue that fulfills multiple roles. In addition to the perception of thermal and mechanical stimuli as a warning system and the deposition of dentin, the pulp performs a variety of immunological functions against invading microorganisms, both in terms of recognition and defense. Especially, in young patients, dental pulp is essential for the completion of root development, and early pulp necrosis results in fracture-prone teeth with fragile root walls [1–4]. Whether in young or adult patients, the loss of pulp tissue due to caries or trauma requires a therapeutic intervention by means of root canal treatment and obturation with a synthetic material.

In recent years, innovative attempts have been made to regenerate dental pulp using advanced molecular biology techniques [5,6]. Promising approaches, based on tissue engineering and regenerative medicine, have been developed for this purpose [7,8]. In this context, stem cell-based or primarily cell-free approaches use specifically tailored scaffold materials and signaling molecules to achieve pulp regeneration, both in terms of tissue microarchitecture and functionality (Figure 1). Several of these approaches already take into account the requirements of potential clinical applications [9–11].

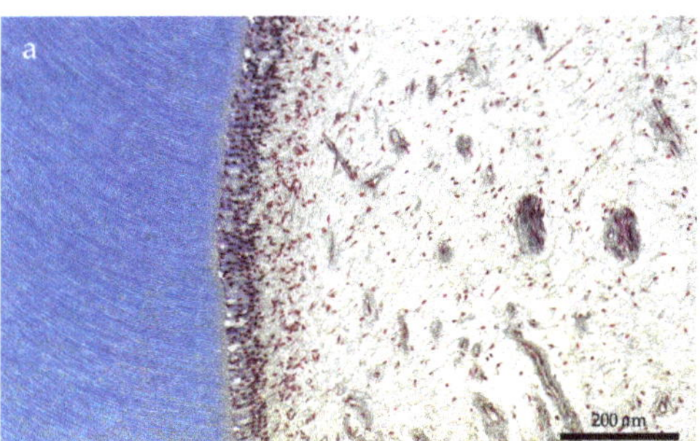
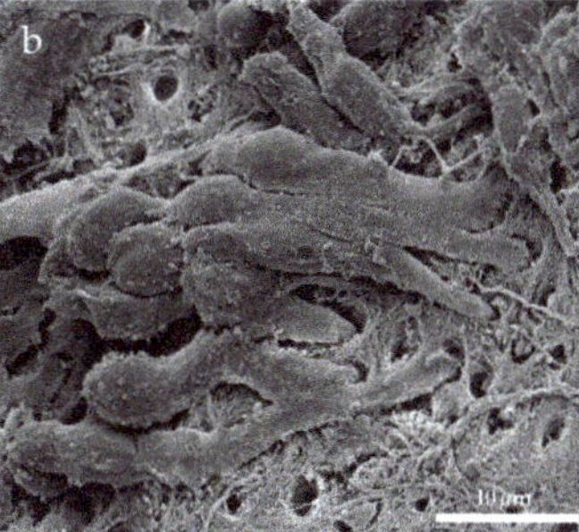

Figure 1. The goal of regenerative endodontics is the restoration of the dentin–pulp complex with all its cellular and structural elements. (**a**) Histological image of the dentin–pulp complex. The odontoblast layer lies on tubular dentin (blue). Vessels and nerve fibers cross the pulp tissue. (**b**) Odontoblasts seen through a scanning electron microscope. The cells extend their processes into the dentin tubules [12].

Essentially, all these methods follow the traditional principles of tissue engineering. They involve the incorporation of stem cells and growth factors into a scaffold material (Figure 2). Ongoing scientific research is, therefore, focusing on advances in three key areas: stem cell biology, scaffold materials, and molecular signaling. The ultimate goal is to bring us closer to the potential of biological regeneration within the root canal. Consequently, the purpose of this Special Issue is to present scientific advances in pulp regeneration, which play a crucial role in translating knowledge from the research setting to the clinic.

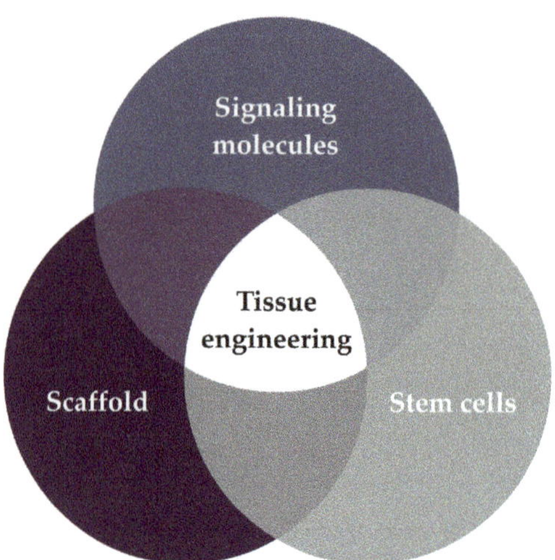

Figure 2. The tissue engineering paradigm involves delivering stem cells and signaling molecules in a scaffold material [13,14].

Accurately defining the goal of regenerative endodontic treatment is paramount [7,15–17]. Biologically, this treatment aims to fully restore all cellular and structural elements of the dentin–pulp complex, while at the same time enabling the newly formed tissue to perform all functions of the original pulp tissue. As it is difficult to assess these criteria directly in a clinical setting, scientists are more dependent than ever on appropriate experimental models to gain insights into treatment outcomes at the tissue and cell level.

Ohlsson et al. have undertaken a comprehensive compilation of preclinical study models [18]. Their review provides an overview of various ectopic, semi-orthotopic, and orthotopic in vitro and in vivo models to investigate regenerative endodontics. One key focus is the critical evaluation of the monolayer and three-dimensional cell cultures. Additionally, a semi-orthotopic transplantation model, as well as various animal models for orthotopic in-vivo-investigations, are presented and critically contrasted against each other. Nowadays, different three-dimensional cell culture techniques offer viable alternatives to animal experimentation. Certain research questions can be easily addressed in vitro, and the continued development of organoid and spheroid cultures, for example, may expand their applications in the future. However, in order to gain further insight into the results in a physiological environment, animal studies are still necessary, and the final evaluation of the research objective must be carried out using an in situ approach. Considering the diversity of in vitro and in vivo study models available, there is no single model that can answer all questions related to pulp regeneration. Depending on the research question, the appropriate model situation must be selected.

To provide an overview of the possible outcomes of regenerative endodontic procedures and to highlight the inconsistency of criteria for the successful regeneration in both animal models and human studies, Minic et al. conducted a systematic review [19]. The aim was to categorize the characteristics of the resulting tissues that could be evaluated in each study design. In addition to the tissue engineering approaches, the search also included blood clot-based procedures. With the latter, tissue formation is initiated by induced bleeding into the root canal. The search identified 75 studies in humans and 49 studies, which reported data from animal models. In humans, the assessment criteria were mainly based on clinical and radiographic examinations, with the majority of studies reporting successful clinical outcomes with relief of symptoms, healing of apical lesions, and increases in root thickness and length. In animal studies, evaluations included both histological and radiological criteria. None of the studies included in the analysis reported successful regeneration of the dentin–pulp complex based on biological criteria. Instead, both preclinical and clinical studies demonstrated the presence of connective tissue, accompanied by cementum or bone-like tissue within the root canal. Several animal studies reported the presence of vascularized and innervated regenerated pulp, however, the clinical response to these findings remained unclear. In human studies, a proportion of patients have regained sensitivity following regenerative endodontic therapy approaches. While the formation of repair tissue may meet the clinical needs of patients and clinicians, further research is needed to identify procedures that come closer to the goal of pulp regeneration.

One key is to identify the molecular–biological signaling pathways that may drive the differentiation of individual cell types. The most distinctive cell type found in dental pulp tissue is the odontoblast, which forms a closed cell layer at the interface of dentin. This unique histological feature is of great importance, as only native odontoblasts have the ability to produce new tubular dentin, known as reactionary dentin. However, other cell types predominantly deposit hard tissue without characteristic tubular morphology. Due to the complexity of regenerating the dentin–pulp complex, many studies have focused on identifying novel molecular biological strategies. One such strategy is to target Wnt proteins, as they have demonstrated their potential as highly relevant stem cell factors capable of regulating the self-renewal and proliferation of various adult stem cell populations. This ability makes Wnt proteins suitable for maintaining homeostasis, promoting tissue healing, and facilitating regeneration. With this in mind, Florimond et al. conducted a systematic review to outline different agents that target Wnt signaling [20]. The aim was to identify strategies to harness Wnt signaling via modulatory molecules and to evaluate their applicability in the dental environment. Numerous studies have highlighted the importance of Wnt signals in the formation and repair of the dentin–pulp complex. Furthermore, research groups have successfully demonstrated that a Wnt stimulus can induce tissue regeneration. As a result, small molecule drugs that stimulate the Wnt/β-catenin pathway have emerged as a promising therapy. However, it is crucial to consider the local application of these drugs to avoid potential systemic side effects. Although there is still a need for proof-of-concept, these advancements bring us closer to realizing the potential of regenerating pulp tissue.

Cell differentiation and the development of new pulp tissue rely on stem cells, which have a high potential for specialization [21,22]. These stem cells can either be transplanted into the root canal or harvested from local tissues [23–25]. This leads to two concepts in endodontic tissue engineering: cell transplantation (cell-based) and cell homing (primarily cell-free). In cell-based methods, cells are cultured ex vivo and then introduced into the root canal using scaffolds containing signaling molecules. Clinical translation of this approach poses significant challenges, as it requires donor tissue or cell banks. In contrast, cell-homing methods utilize naturally residing stem or progenitor cells within the body that are locally available, bypassing any ex vivo manipulation. Here, scaffolding materials that are primarily cell-free are placed in the root canal, along with signaling molecules to attract cells from the remaining pulp tissue or the surrounding periapical space.

Among various types of stem cells associated with teeth, dental pulp stem cells (DPSCs) and apical papilla stem cells (SCAPs) are particularly suitable for cell harvesting because they are naturally found in the root canal or apical papilla at the root tip. Therefore, Smeda et al. investigated whether both types of stem cells are equally suitable for regenerative endodontic protocols [26]. To investigate this, DPSCs and SCAPs were isolated from the same donor, and their characteristics were extensively studied, with a focus on gene expression profiling during induced differentiation. The viability, proliferation, and migration of DPSCs and SCAPs appeared very similar. However, differences in the profile of secreted molecules were observed. Transcriptome analysis identified 13 biomarkers whose regulation was critical for the differentiation of both cell types. These findings suggest that DPSCs and SCAPs share many similarities and exhibit similar patterns of differentiation. From a molecular biology point of view, both types of stem cells appear to be equally suitable for dental pulp tissue engineering.

Furthermore, the possibility of cell transplantation has been extensively studied pre-clinically and clinically in recent years [27,28]. Recent in vivo studies have shown that pulp regeneration after cell transplantation appears promising [29–31]. For cell transplantation, it is of central interest to identify cell sources in the body that are readily accessible and capable of differentiating into odontoblasts. The approach of cell transplantation with DPSCs and other tooth-derived cell types is problematic because of the need for cell expansion to obtain sufficient numbers of cells and the need for a donor tooth. A possible cell source to overcome these obstacles could be the amnion, the innermost layer of the human placenta [32,33]. It contains amniotic epithelial cells (AECs), which retain plasticity and thus are able to differentiate into cells of all three embryonic layers [21,34].

Bucchi et al. set out to investigate the suitability of amnion-derived pluripotent stem cells as a potential cell source for pulp regeneration in vitro [34]. In addition to viability and cell attachment to dentin, the expression of genes associated with mineralization and odontogenic differentiation, mineralization and epithelial–mesenchymal transition was analyzed. The viability of AECs was significantly lower than that of DPSCs, whereas both AECs and DPSCs adhered to dentin. Regulation of genes associated with odontoblast differentiation and mineralization showed that AECs did not transition to an odontoblastic phenotype, but genes associated with epithelial–mesenchymal transition were significantly upregulated. In analogy, AECs showed low levels of calcium deposition after differentiation. Although pluripotent AECs adhered to dentin and had the ability to mineralize, they showed unfavorable proliferation behavior and did not undergo odontoblastic transition.

In addition to the use of stem cells from different regions of the body, teeth can also be a source of stem cells for use elsewhere in the body, e.g., for bone regeneration. Interestingly, Kunimatsu et al. pursued the idea of using stem cells from the pulp of primary teeth for bone regeneration [35]. The scientists focused on a very specific population of stem cells from deciduous human teeth (SHEDs), namely, the fraction expressing the cluster of differentiation (CD) 146 marker. CD146 is a stem cell surface antigen that is critical for angiogenic and osteogenic differentiation. Bone regeneration has been reported to be accelerated after transplantation of CD146-positive mesenchymal stem cells derived from deciduous dental pulp. The aim of this study was, therefore, to compare the effects of CD146 in a population of SHEDs using molecular biological methods. A CD146-positive cell population (CD146+) and a CD146-negative cell population (CD146-) were obtained by cell sorting. While cell proliferation was indifferent, mineralization and marker expression were highest in the CD146+ group. CD146+ SHEDs showed the highest osteogenic differentiation potential and, therefore, may represent a valuable cell population for bone regeneration therapy.

Turning back to tissue engineering of the dental pulp, the root canal poses a number of challenges for stem cells. One of these is bacterial infection, which sometimes cannot be completely eliminated, despite intensive disinfection that knowingly affects the outcome [36,37]. If suitable stem cells are successfully introduced into the root canal, residual bacteria may influence their differentiation and thus affect the outcome of regeneration [38,39]. To explore the potential impact of oral bacteria on regenerative endodontic treatment, Razghonova et al. conducted a study, focusing on the interaction between stem cells from the apical papilla and these bacteria [40]. Using RNA-seq transcriptomic analysis, the researchers investigated the impact of Enterococcus faecalis and Fusobacterium nucleatum, along with their metabolites, on the differentiation of SCAPs towards osteogenic and dentinogenic lineages in vitro. Gene analysis revealed that bacterial metabolites had a significant influence on SCAPs, and the transcriptome profiles indicated a negative effect on the osteogenic and dentinogenic differentiation when exposed to E. faecalis or F. nucleatum. The findings demonstrated that F. nucleatum, E. faecalis, and their metabolites can alter the expression of genes involved in dentinogenesis and osteogenesis, potentially affecting the outcomes of regenerative endodontic procedures.

While disinfection is crucial, the long-term challenge lies in addressing microorganisms that may evade disinfection [39]. Therefore, the development of scaffold materials with antimicrobial properties is highly desirable for applications in dental pulp tissue engineering. In this context, Ayoub et al. utilized electrospinning techniques to create a gelatin methacryloyl (GelMA) fiber loaded with azithromycin (AZ) [41]. Gelatine itself is a natural biomaterial derived from hydrolyzed collagen, containing an alternating sequence of the arginine, glycine, and aspartate (RGD) tripeptide motif, which promotes cell adhesion and very low exhibition of immune responses. The study encompassed a comprehensive investigation of various aspects, including the morphology and diameter of the fibers, the incorporation of AZ, the mechanical properties of the fibers, the degradation profile, and antimicrobial activity. Furthermore, in vitro compatibility with DPSCs and an in vivo evaluation of the inflammatory response were conducted. The findings demonstrated that the presence of AZ did not lead to significant toxicity. In vivo results revealed increased vascularization, mild inflammation, and the presence of mature, well oriented collagen fibers that were interwoven with the engineered fibers.

The same research group carried out a study to explore the potential of enhancing the GelMA scaffold material by combining it with other antimicrobial agents. Ribeiro et al. incorporated clindamycin or metronidazole microparticles into the scaffold [42]. The experimental hydrogels were subjected to various analyses, including swelling and degradation assessments, evaluation of toxicity towards dental stem cells, and examination of antimicrobial activity against endodontic pathogens. The introduction of antibiotic-loaded fibrous microparticles into GelMA resulted in an increase in both the swelling ratio and degradation rate of the hydrogels, and the hydrogels containing antibiotic-loaded fibrous microparticles exhibited significant antibiofilm effects, although cell viability was compromised. Overall, these results suggest that the injectable hydrogels with antibiotic-loaded fibrous microparticles have promising clinical potential for the effective eradication of endodontic infections.

There is a continuous flow of new developments and innovations in the field of scaffold materials. While antimicrobial properties would be helpful, there are many requirements to be met for pulp tissue engineering applications [43], e.g., control of the mechanical properties, such as stiffness and setting time, of the material used for preparation and injection into the root canal, is critical. Furthermore, it is important that the material allows for the binding, stabilization, and controlled release of signaling molecules. The cured material should be permeable for nutrients and metabolites, as well as biocompatible and biodegradable, to allow cell migration and transformation of the material into new tissue. Here, natural biomaterials have proven to be especially suitable [44,45]. One innovative natural material is carrageenan, a polysaccharide obtained from red algae [46]. Among them, kappa- and iota-carrageenan have received much attention as scaffold materials due to their gelling properties in the food industry, but also in regenerative medicine due to their cytocompatibility. The biological properties of kappa-carrageenan are based on its three hydroxyl groups per disaccharide repeating unit, which make it hydrophilic, as well as its one negatively charged sulphate group, which enables chemical reactions. Loukelis et al. investigated the effect of kappa-carrageenan on the behavior of dental pulp stem cells. Kappa-carrageenan/chitosan/gelatin scaffolds cross-linked with KCl were compared with carrageenan/chitosan/gelatin without KCl and chitosan/gelatin. Cell viability was significantly increased over the experimental period, as was alkaline phosphatase (ALP) activity and the expression of odontogenic marker genes. The results demonstrate that incorporation of kappa-carrageenan into scaffolds significantly enhances the odontogenic potential of DPSCs and may support dentin–pulp regeneration.

In summary, the field of dental pulp regeneration has made significant progress in recent years. Preclinical and clinical studies have provided insight into the results of regenerative endodontic procedures. Although repair tissue formation has been observed, true regeneration of the dentin–pulp complex, based on biological criteria, has yet to be achieved. Further research is needed to overcome existing challenges and bridge the gap between preclinical findings and clinical applications. Continued efforts in stem cell biology, scaffold materials, molecular signaling, and experimental models will bring us closer to the goal of true pulp regeneration and restoration of pulp tissue functionality.

Author Contributions: Conceptualization, M.W. and K.M.G.; writing—original draft preparation, M.W.; writing—review and editing, M.W. and K.M.G.; visualization, M.W. All authors have read and agreed to the published version of the manuscript.

Conflicts of Interest: The authors declare no conflict of interest.

References

1. Bucchi, C.; Marcé-Nogué, J.; Galler, K.M.; Widbiller, M. Biomechanical performance of an immature maxillary central incisor after revitalization: A finite element analysis. *Int. Endod. J.* **2019**, *52*, 1508–1518. [CrossRef] [PubMed]
2. Cvek, M. Prognosis of luxated non-vital maxillary incisors treated with calcium hydroxide and filled with gutta-percha. A retrospective clinical study. *Dent. Traumatol.* **1992**, *8*, 45–55. [CrossRef] [PubMed]
3. Talati, A.; Disfani, R.; Afshar, A.; Rastegar, A.F. Finite element evaluation of stress distribution in mature and immature teeth. *Iran. Endod. J.* **2007**, *2*, 47–53. [PubMed]
4. Zhou, R.; Wang, Y.; Chen, Y.; Chen, S.; Lyu, H.; Cai, Z.; Huang, X. Radiographic, histologic, and biomechanical evaluation of combined application of platelet-rich fibrin with blood clot in regenerative endodontics. *J. Endod.* **2017**, *43*, 2034–2040. [CrossRef] [PubMed]
5. Galler, K.M.; Widbiller, M. Cell-free approaches for dental pulp tissue engineering. *J. Endod.* **2020**, *46*, S143–S149. [CrossRef]
6. Kim, S.G.; Malek, M.; Sigurdsson, A.; Lin, L.M.; Kahler, B. Regenerative endodontics: A comprehensive review. *Int. Endod. J.* **2018**, *51*, 1367–1388. [CrossRef]
7. Widbiller, M.; Schmalz, G. Endodontic regeneration: Hard shell, soft core. *Odontology* **2021**, *109*, 303–312. [CrossRef]
8. Lin, L.M.; Huang, G.T.; Sigurdsson, A.; Kahler, B. Clinical cell-based versus cell-free regenerative endodontics: Clarification of concept and term. *Int. Endod. J.* **2021**, *54*, 887–901. [CrossRef]
9. Widbiller, M.; Rosendahl, A.; Wölflick, M.; Linnebank, M.; Welzenbach, B.; Hiller, K.-A.; Buchalla, W.; Galler, K.M. Isolation of endogenous TGF-β1 from root canals for pulp tissue engineering: A translational study. *Biology* **2022**, *11*, 227. [CrossRef]
10. Xu, X.-Y.; Li, X.; Wang, J.; He, X.-T.; Sun, H.-H.; Chen, F.-M. Concise review: Periodontal tissue regeneration using stem cells: Strategies and translational considerations. *Stem Cells Transl. Med.* **2019**, *8*, 392–403. [CrossRef]
11. Brizuela, C.; Meza, G.; Urrejola, D.; Quezada, M.A.; Concha, G.; Ramírez, V.; Angelopoulos, I.; Cadiz, M.; Tapia-Limonchi, R.; Khoury, M. Cell-based regenerative endodontics for treatment of periapical lesions: A randomized, controlled phase I/II clinical trial. *J. Dent. Res.* **2020**, *99*, 523–529. [CrossRef] [PubMed]
12. Gallorini, M.; Krifka, S.; Widbiller, M.; Schröder, A.; Brochhausen, C.; Cataldi, A.; Hiller, K.-A.; Buchalla, W.; Schweikl, H. Distinguished properties of cells isolated from the dentin-pulp interface. *Ann. Anat.* **2021**, *234*, 151628. [CrossRef]
13. Langer, R.; Vacanti, J.P. Tissue engineering. *Science* **1993**, *260*, 920–926. [CrossRef]
14. Goss, R.J. Regeneration. Encyclopaedia Britannica. 2020. Available online: https://www.britannica.com/science/regeneration-biology (accessed on 15 June 2023).
15. Simon, S.R.J.; Tomson, P.L.; Berdal, A. Regenerative endodontics: Regeneration or repair? *J. Endod.* **2014**, *40*, S70–S75. [CrossRef] [PubMed]
16. Goldberg, M. Pulp healing and regeneration: More questions than answers. *Adv. Dent. Res.* **2011**, *23*, 270–274. [CrossRef]
17. Hargreaves, K.M.; Diogenes, A.R.; Teixeira, F.B. Paradigm lost: A perspective on the design and interpretation of regenerative endodontic research. *J. Endod.* **2014**, *40*, S65–S69. [CrossRef] [PubMed]
18. Ohlsson, E.; Galler, K.M.; Widbiller, M. A Compilation of study models for dental pulp regeneration. *Int. J. Mol. Sci.* **2022**, *23*, 14361. [CrossRef] [PubMed]
19. Minic, S.; Vital, S.; Chaussain, C.; Boukpessi, T.; Mangione, F. Tissue characteristics in endodontic regeneration: A systematic review. *Int. J. Mol. Sci.* **2022**, *23*, 10534. [CrossRef]
20. Florimond, M.; Minic, S.; Sharpe, P.; Chaussain, C.; Renard, E.; Boukpessi, T. Modulators of Wnt signaling pathway implied in dentin pulp complex engineering: A literature review. *Int. J. Mol. Sci.* **2022**, *23*, 10582. [CrossRef]
21. Huang, G.T.-J.; Gronthos, S.; Shi, S. Mesenchymal stem cells derived from dental tissues vs. those from other sources: Their biology and role in regenerative medicine. *J. Dent. Res.* **2009**, *88*, 792–806. [CrossRef]
22. Nuti, N.; Corallo, C.; Chan, B.M.F.; Ferrari, M.; Gerami-Naini, B. Multipotent differentiation of human dental pulp stem cells: A literature review. *Stem Cell Rev. Rep.* **2016**, *12*, 511–523. [CrossRef]
23. Sengupta, D.; Waldman, S.D.; Li, S. From in vitro to in situ tissue engineering. *Ann. Biomed. Eng.* **2014**, *42*, 1537–1545. [CrossRef]
24. Ko, I.K.; Lee, S.J.; Atala, A.; Yoo, J.J. In situ tissue regeneration through host stem cell recruitment. *Exp. Mol. Med.* **2013**, *45*, e57. [CrossRef]
25. Chen, F.-M.; Wu, L.-A.; Zhang, M.; Zhang, R.; Sun, H.-H. Homing of endogenous stem/progenitor cells for in situ tissue regeneration: Promises, strategies, and translational perspectives. *Biomaterials* **2011**, *32*, 3189–3209. [CrossRef]

26. Smeda, M.; Galler, K.M.; Woelflick, M.; Rosendahl, A.; Moehle, C.; Lenhardt, B.; Buchalla, W.; Widbiller, M. Molecular biological comparison of dental pulp- and apical papilla-derived stem cells. *Int. J. Mol. Sci.* **2022**, *23*, 2615. [CrossRef]
27. Widbiller, M.; Knüttel, H.; Meschi, N.; Terol, F.D. Effectiveness of endodontic tissue engineering in treatment of apical periodontitis: A systematic review. *Int. Endod. J.* **2022**, *online ahead of print*. [CrossRef]
28. Nakashima, M.; Iohara, K.; Murakami, M.; Nakamura, H.; Sato, Y.; Ariji, Y.; Matsushita, K. Pulp regeneration by transplantation of dental pulp stem cells in pulpitis: A pilot clinical study. *Stem Cell Res. Ther.* **2017**, *8*, 61. [CrossRef]
29. Xuan, K.; Li, B.; Guo, H.; Sun, W.; Kou, X.; He, X.; Zhang, Y.; Sun, J.; Liu, A.; Liao, L.; et al. Deciduous autologous tooth stem cells regenerate dental pulp after implantation into injured teeth. *Sci. Transl. Med.* **2018**, *10*, eaaf3227. [CrossRef]
30. Itoh, Y.; Sasaki, J.I.; Hashimoto, M.; Katata, C.; Hayashi, M.; Imazato, S. Pulp regeneration by 3-dimensional dental pulp stem cell constructs. *J. Dent. Res.* **2018**, *97*, 1137–1143. [CrossRef]
31. Iohara, K.; Imabayashi, K.; Ishizaka, R.; Watanabe, A.; Nabekura, J.; Ito, M.; Matsushita, K.; Nakamura, H.; Nakashima, M. Complete pulp regeneration after pulpectomy by transplantation of CD105+ stem cells with stromal cell-derived factor-1. *Tissue Eng. Part A* **2011**, *17*, 1911–1920. [CrossRef]
32. Gramignoli, R.; Srinivasan, R.C.; Kannisto, K.; Strom, S.C. Isolation of human amnion epithelial cells according to current good manufacturing procedures. *Curr. Protoc. Stem Cell Biol.* **2016**, *37*, 1E.10.1–1E.10.13. [CrossRef]
33. Koike, C.; Zhou, K.; Takeda, Y.; Fathy, M.; Okabe, M.; Yoshida, T.; Nakamura, Y.; Kato, Y.; Nikaido, T. Characterization of amniotic stem cells. *Cell. Reprogram.* **2014**, *16*, 298–305. [CrossRef] [PubMed]
34. Bucchi, C.; Ohlsson, E.; de Anta, J.M.; Woelflick, M.; Galler, K.; Manzanares-Cespedes, M.C.; Widbiller, M. Human amnion epithelial cells: A potential cell source for pulp regeneration? *Int. J. Mol. Sci.* **2022**, *23*, 2830. [CrossRef] [PubMed]
35. Kunimatsu, R.; Rikitake, K.; Yoshimi, Y.; Putranti, N.A.R.; Hayashi, Y.; Tanimoto, K. Bone differentiation ability of CD146-positive stem cells from human exfoliated deciduous teeth. *Int. J. Mol. Sci.* **2023**, *24*, 4048. [CrossRef]
36. Verma, P.; Nosrat, A.; Kim, J.R.; Price, J.B.; Wang, P.; Bair, E.; Xu, H.H.; Fouad, A.F. Effect of residual bacteria on the outcome of pulp regeneration in vivo. *J. Dent. Res.* **2017**, *96*, 100–106. [CrossRef]
37. Vishwanat, L.; Duong, R.; Takimoto, K.; Phillips, L.; Espitia, C.O.; Diogenes, A.R.; Ruparel, S.B.; Kolodrubetz, D.; Ruparel, N.B. Effect of bacterial biofilm on the osteogenic differentiation of stem cells of apical papilla. *J. Endod.* **2017**, *43*, 916–922. [CrossRef]
38. Almutairi, W.; Yassen, G.H.; Aminoshariae, A.; Williams, K.A.; Mickel, A. Regenerative endodontics: A systematic analysis of the failed cases. *J. Endod.* **2019**, *5*, 567–577. [CrossRef]
39. Fouad, A.F. Microbial factors and antimicrobial strategies in dental pulp regeneration. *J. Endod.* **2017**, *43*, S46–S50. [CrossRef]
40. Razghonova, Y.; Zymovets, V.; Wadelius, P.; Rakhimova, O.; Manoharan, L.; Brundin, M.; Kelk, P.; Vestman, N.R. Transcriptome analysis reveals modulation of human stem cells from the apical papilla by species associated with dental root canal infection. *Int. J. Mol. Sci.* **2022**, *23*, 14420. [CrossRef]
41. Ayoub, A.A.; Mahmoud, A.H.; Ribeiro, J.S.; Daghrery, A.; Xu, J.; Fenno, J.C.; Schwendeman, A.; Sasaki, H.; Dal-Fabbro, R.; Bottino, M.C. Electrospun azithromycin-laden gelatin methacryloyl fibers for endodontic infection control. *Int. J. Mol. Sci.* **2022**, *23*, 13761. [CrossRef]
42. Ribeiro, J.S.; Münchow, E.A.; Bordini, E.A.F.; Rodrigues, N.S.; Dubey, N.; Sasaki, H.; Fenno, J.C.; Schwendeman, S.; Bottino, M.C. Engineering of injectable antibiotic-laden fibrous microparticles gelatin methacryloyl hydrogel for endodontic infection ablation. *Int. J. Mol. Sci.* **2022**, *23*, 971. [CrossRef]
43. Galler, K.M.; Widbiller, M. Perspectives for cell-homing approaches to engineer dental pulp. *J. Endod.* **2017**, *43*, S40–S45. [CrossRef]
44. Galler, K.M.; Brandl, F.P.; Kirchhof, S.; Widbiller, M.; Eidt, A.; Buchalla, W.; Göpferich, A.; Schmalz, G. Suitability of different natural and synthetic biomaterials for dental pulp tissue engineering. *Tissue Eng. Part A* **2018**, *24*, 234–244. [CrossRef]
45. Widbiller, M.; Driesen, R.B.; Eidt, A.; Lambrichts, I.; Hiller, K.-A.; Buchalla, W.; Schmalz, G.; Galler, K.M. Cell homing for pulp tissue engineering with endogenous dentin matrix proteins. *J. Endod.* **2018**, *44*, 956–962.e2. [CrossRef]
46. Loukelis, K.; Machla, F.; Bakopoulou, A.; Chatzinikolaidou, M. Kappa-carrageenan/chitosan/gelatin scaffolds provide a biomimetic microenvironment for dentin-pulp regeneration. *Int. J. Mol. Sci.* **2023**, *24*, 6465. [CrossRef]

International Journal of
Molecular Sciences

Review

A Compilation of Study Models for Dental Pulp Regeneration

Ella Ohlsson [1], Kerstin M. Galler [1] and Matthias Widbiller [2,*]

[1] Department of Operative Dentistry and Periodontology, Friedrich-Alexander-University Erlangen-Nuernberg, D-91054 Erlangen, Germany
[2] Department of Conservative Dentistry and Periodontology, University Hospital Regensburg, D-93053 Regensburg, Germany
* Correspondence: matthias.widbiller@ukr.de

Abstract: Efforts to heal damaged pulp tissue through tissue engineering have produced positive results in pilot trials. However, the differentiation between real regeneration and mere repair is not possible through clinical measures. Therefore, preclinical study models are still of great importance, both to gain insights into treatment outcomes on tissue and cell levels and to develop further concepts for dental pulp regeneration. This review aims at compiling information about different in vitro and in vivo ectopic, semiorthotopic, and orthotopic models. In this context, the differences between monolayer and three-dimensional cell cultures are discussed, a semiorthotopic transplantation model is introduced as an in vivo model for dental pulp regeneration, and finally, different animal models used for in vivo orthotopic investigations are presented.

Keywords: regenerative endodontics; study model; dental pulp; regeneration; tissue engineering; cell culture techniques; animal models; translational research

1. Introduction

The dental pulp has important functions, and its loss can have serious consequences. A root-filled tooth may remain in the oral cavity without pulp, but it lacks the ability to react to sensory stimuli, issue an immune response, or form reparative dentin [1]. Additionally, remaining hard tissue is weakened, and as a result, root fractures occur more frequently than in vital teeth [2]. If immature teeth are affected, root development comes to a halt, leaving thin dentin walls and an open apex behind, which complicates further therapies [3,4]. To overcome the biological and mechanical drawbacks of traditional endodontic treatment, research focused on pulp regeneration has gained interest over the last years. Several approaches in the realm of endodontic tissue engineering are being explored, which can be categorized into primarily cell-free methods, where resident stem cells re-populate the root canal, and cell-based approaches, where cells are introduced by transplantation [5]. In both approaches, the three pillars of classical tissue engineering, i.e., stem cells, signaling molecules, and a scaffold material, are present [6].

Interestingly, endodontic tissue engineering has already been translated into randomized clinical trials. Patients with irreversible pulpitis have been treated with transplantation of autologous [7] or allogenic [8] mesenchymal stem cells into root canals. In these studies, all teeth that received treatment through tissue engineering have survived after 12 months, and even positive responses to sensitivity testing were evident in a considerable number of cases. Further observations, such as radiographic reduction in apical lesion size and root lengthening and thickening, demonstrated clinical success [9]. However, the question remains whether this success is associated with biological regeneration of the pulp or a repair process. Teeth that have undergone regenerative procedures and are later extracted for other reasons often show proof of repair by ectopically formed tissues instead of restitutio ad integrum [5]. Strengthening the tooth root by the apposition of any hard tissue may be of clinical value and could contribute to increased mechanical resistance, but functional

Citation: Ohlsson, E.; Galler, K.M.; Widbiller, M. A Compilation of Study Models for Dental Pulp Regeneration. *Int. J. Mol. Sci.* **2022**, *23*, 14361. https://doi.org/10.3390/ijms232214361

Academic Editor: Young-Jin Kim

Received: 20 October 2022
Accepted: 14 November 2022
Published: 18 November 2022

Publisher's Note: MDPI stays neutral with regard to jurisdictional claims in published maps and institutional affiliations.

issues, e.g., adequate biological response of the dental pulp to external stimuli, remain unresolved [3,10]. In this context, histological examination is the only way to determine the exact nature of newly generated tissues. However, this is, of course, impossible in a systematic way in clinical studies. For this reason, preclinical study models are still indispensable for the development of new pulp regeneration procedures and for the biological evaluation of outcomes.

The aim of this review is to compile different study models for both cell-based and primarily cell-free tissue engineering approaches for pulp regeneration that have emerged and developed over the past years. These can be grouped roughly into categories: in vitro; in vivo ectopic, referring to the ectopic transplantation of scaffolds and cells into immunocompromised animals; in vivo semiorthotopic, where cells are cultured in a tooth framework, which is transplanted into animals; and in vivo orthotopic, meaning the in situ simulation of clinical procedures in study animals, as shown in Figure 1. The advantages and disadvantages of the available in vitro and in vivo models are compared and discussed.

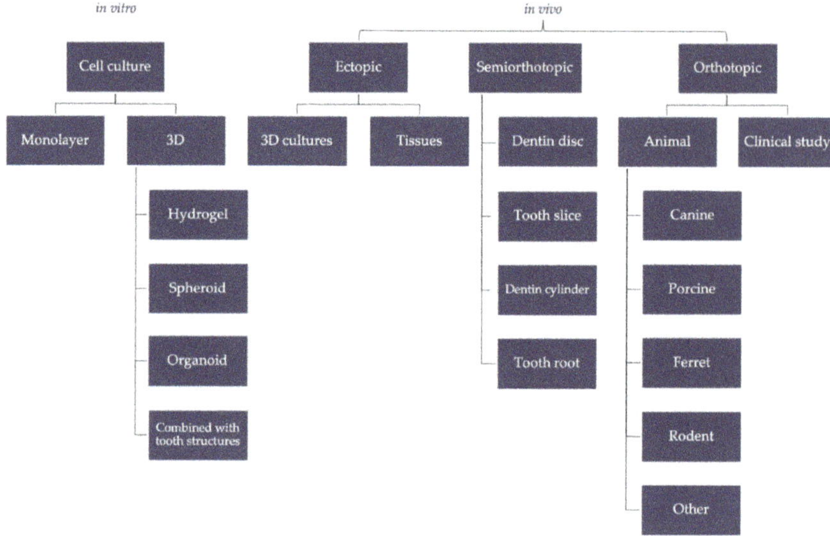

Figure 1. Compilation of study models for dental pulp tissue engineering.

2. In Vitro

2.1. Monolayer Cell Culture

The monolayer cell culture presents the most basic laboratory technique. Distinguished by the locations of their origins, several cell sources have been identified. Of particular interest for dental pulp tissue engineering are mesenchymal stem cells, such as dental pulp stem cells (DPSCs), stem cells from the apical papilla, and periodontal ligament stem cells [11–13]. Dental stem cells can be obtained from human teeth, as well as from other species [14–16]. Furthermore, the use of non-oral stem cells for dental pulp regeneration, such as umbilical cord stem cells or amniotic epithelial stem cells, has also been investigated [17,18].

In terms of dental pulp, stem cells are isolated from pulp tissue by enzyme digestion or the outgrowth method and then cultured in medium supplemented with fetal bovine serum [19]. As adherent cells, they attach to the bottom of the culture vessel and form a confluent monolayer. In this culture environment, many cell characteristics, such as viability, population doubling, senescence, gene expression, or differentiability, and their responses to signaling molecules or biomaterials can be assessed.

The strengths of this model are controllable and reproduceable experimental conditions [20]. Different aspects of a complex in vivo system can be simplified and explored mechanistically in an in vitro culture setting [21], and costs are also very low compared to other models in use [20]. However, there are some disadvantages as well. Due to its simplicity, it has limitations when it comes to reproducing physiologic conditions. This includes tissue architecture, cell–cell communication, cellular movement, and cell–matrix interaction [22]. An adequate rendering of biological processes is, therefore, difficult, and the model may produce misleading and nonpredictive data [20,23].

Still, it has been instrumental in the characterization of tooth-derived stem cells [24] and remains the basis of most research even today. Many attempts have been made to culture and analyze odontoblast-like cells in vitro by the addition of signaling molecules to stem cells [25–28]. Signaling pathways in these cells have been investigated [29,30], gene expression patterns during cell differentiation have been revealed [31–33], and mineralization has been observed through alkaline phosphatase or alizarin red staining [17,34,35]. Since the cells at the interface with dentin are an integral part of the pulp–dentin complex, this model can also be adapted to study the behavior of DPSCs seeded directly onto the surface of dentin disks [19,36]. Furthermore, dentin matrix proteins, which are rich in growth factors that modulate cell differentiation, can be isolated from human dentin and supplemented in cell culture media to study the behavior of pulp cells [28,37,38].

2.2. Three-Dimensional (3D) Cell Culture

A two-dimensional approach can be enhanced by the utilization of three-dimensional culture methods, such as scaffold cell cultures, spheroids, or organoids (Figure 2). While scaffold cultures are mostly applied in material testing, spheroids and organoids were originally developed for tumor research and personalized medicine [23]. Unfortunately, the terms are used inconsistently in the literature. The term spheroid describes a conglomerate of adult cells without any scaffold, whereas an organoid consists of self-organized stem or progenitor cells forming organ-specific constructs with the help of a scaffolding environment [20,39].

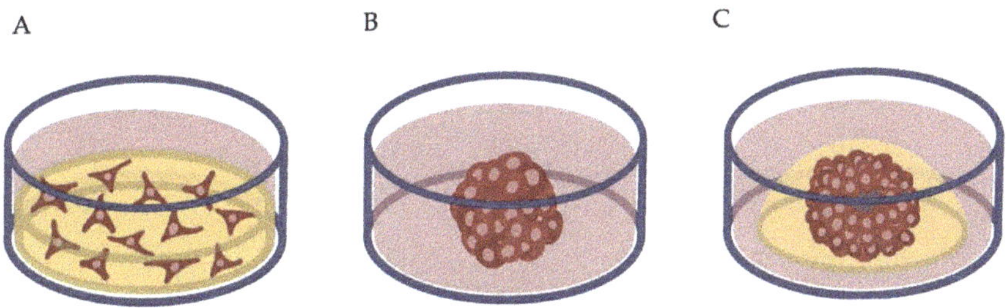

Figure 2. Three-dimensional cell culture models. (**A**) Hydrogel culture: cells are embedded in a scaffold material with a supernatant of culture medium. (**B**) Spheroid culture. (**C**) Organoid culture: cells are embedded in an extracellular matrix.

These three 3D-culturing methods have in common that the cells are spatially distributed within a supporting structure, i.e., an extracellular matrix [40–42]. 3D-cultured cells differ in morphology and physiology from two-dimensional cultures, as nonadherent cells are given the possibility to unfold their cellular shape and display greater heterogeneity, either in morphology, lineage, function, or age (Figure 3). Whereas necrotic cells in 2D cultures quickly detach from the surface of the culture flask and are rinsed out, 3D cultures consist of cells in different stages of aging. The core of agglomerates is often composed of necrotic cells, while the outermost layer consists of viable and proliferating cells, which emulates natural processes more closely [23]. The matrix itself also influences the cell behavior, e.g., by its rigidity. Stiff matrices can drive stem cell differentiation towards the osteogenic line [43].

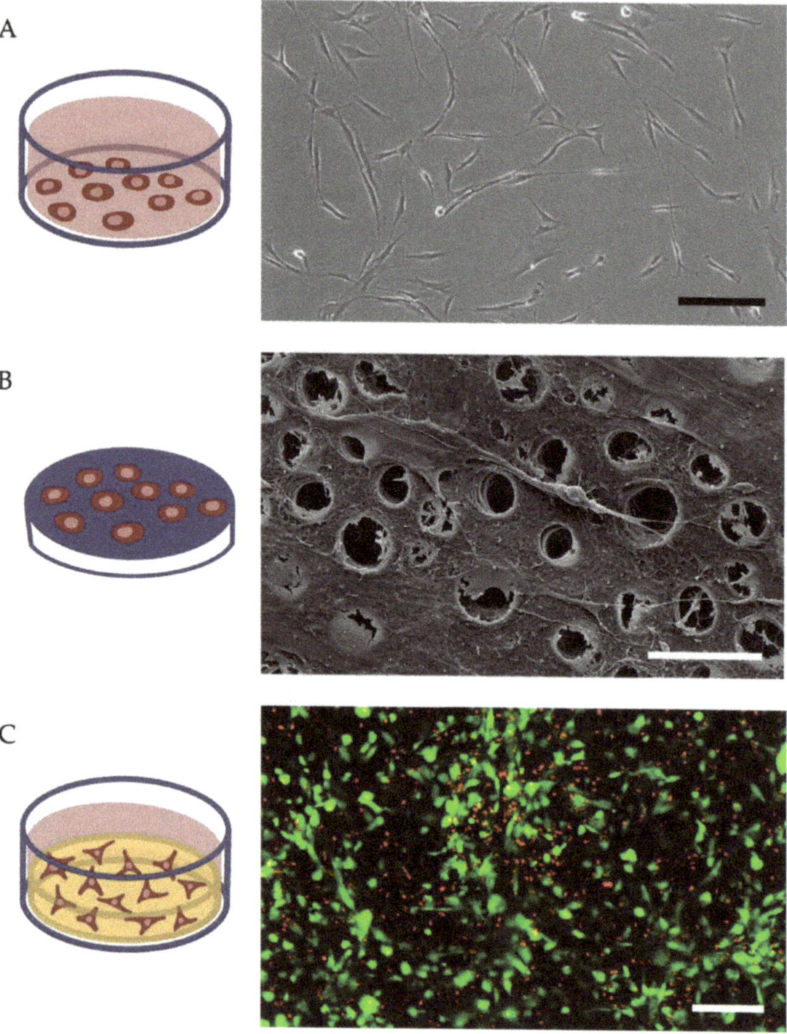

Figure 3. Comparison of morphologies of differently cultured cells. (**A**) DPSCs cultured on a tissue culture plate; scale bar of 200 μm. (**B**) REM image of DPSCs cultured on a dentin disk, where cells extend their processes into dentin tubules; scale bar of 10 μm. (**C**) Confocal laser scanning microscopy of live (green) and dead (red) DPSCs cultured in a collagen hydrogel shows the high turnover of cells; scale bar of 200 μm.

In general, 3D cell cultures are more apt to reflect in vivo mechanisms than monolayer cultures [23] and, therefore, produce more accurate insights [44]. Thus, 3D cultures have the potential to bridge the gap between simple cell cultures and in vivo experiments, which can reduce the need for ethically challenging animal models [41]. However, as of now, 3D culturing is less established than monolayer culturing and is associated with greater effort and higher costs. Furthermore, the analysis of cell cultures is more difficult since cells need to be separated from the extracellular matrix, and high variability in the produced agglomerates reduces reproducibility [22,23].

Since there is always a close connection between hard and soft tissues in the pulp–dentin complex, it is also possible to combine 3D cell cultures with tooth structures. In this case, the pulp cavity of a slice of a tooth crown or the empty canal of a root can serve as reservoir to receive 3D-cultured cells. These constructs can be maintained in culture and studied in vitro [45,46]. Nevertheless, tooth slice or root fragment models are commonly used in in vivo model situations, which is discussed in detail in a following section.

2.2.1. Hydrogels

Scaffolds can not only serve as a cell matrix in vitro, but constitute a pillar of tissue engineering, which makes them the subject of research in the realm of pulp regeneration. Essentially, these can be divided structurally into porous scaffolds, fibrous scaffolds, and hydrogels, where hydrogels are primarily used in the field of pulp biology research. They best imitate the mechanical properties of the dental pulp, and furthermore, their injectability makes them suitable for use in the root canal [47,48]. Appropriate materials should restore tissue architecture and guide cell growth but also degrade over time to provide space for new tissue formation [48,49]. The combination of scaffold materials and stem cells in vitro lays the foundation for assessing the eligibility of materials for clinical applications. Herein, cells can be cultured inside or on top of a hydrogel material to assess both the scaffold properties [50,51], such as inductivity or degradability, and the cell behavior, such as proliferation and migration, as well as cell–cell and cell–matrix interactions. Furthermore, the cytotoxicity of dental materials can be tested in a hydrogel model by adding substances to the culture medium or in direct contact with cells [52].

To emulate the mechanical and functional relationship between hard tissue and cells within the pulp–dentin complex, dentin can also be incorporated into three-dimensional culture systems. Rosa et al. filled tooth roots with stem cells from exfoliated deciduous teeth (SHED) that were encapsulated in a collagen matrix and cultured the fragments in vitro to investigate whether odontoblastic differentiation of cells was possible in full-length roots [46].

An innovative method for producing scaffolds and even hydrogels to incorporate cells is 3D bioprinting. Two different approaches are usually applied: one is the printing of acellular scaffolds, such as PCL [53], and the other is the additive manufacturing of scaffolds that already contain cells and signaling molecules [54,55]. Both types can be used in vitro and in vivo and are captivating because of their rapid fabrication, high precision, and customized production; however, the limitations of a low number of suitable materials, high costs, and possible undesirable additives could restrict applicability at present [56].

2.2.2. Spheroids

Spheroids can be defined as a conglomerate of cells that self-assemble or are forced to aggregate [22,44]. They can be produced from a single cell type or as a multicellular spheroid and can be fabricated in a myriad of ways, from the gravity-enforced hanging drop method to the layering of cell sheets, as well as low-attachment culture plates [57], pellet culturing [58], the utilization of micro molds [59], and magnetic levitation [41].

In terms of pulp regeneration, researchers are studying the possibility to use in-vitro-generated cell agglomerates to replace damaged pulp tissue directly, but the use of engineered pulp tissue replicas also provides a novel model to study the processes in dental pulp regeneration and to assess the biocompatibility of various materials used [60].

2.2.3. Organoids

Organoids commonly refer to self-organizing 3D cell structures of organ-specific cell types that arise from the differentiation of stem or progenitor cells [20]. They partially resemble the architecture and function of the target organ and are usually fabricated using decellularized extracellular matrices, such as Matrigel or collagen, to mimic a tissue's noncellular components [61]. Matrigel is a commercially available material that contains structural proteins, such as collagen, elastin, and laminin, comparable to the basal lamina in vivo. However, it is not suited for in vivo application, as it is extracted from mouse sarcoma cells [62].

Organoids are, to a certain degree, able to simulate the architecture and functionality of a native organ [20,63]. Fashioned from embryonic cells, for example, organoids can recreate both hard and soft tissues. Cells in an organoid can be cultured for an extended time and mimic signaling pathways and niche conditions more closely compared to cells in a 2D system. Compared to an animal model, the implementation of organoids provides greater accessibility and feasibility [64]. However, the creation of organoids also requires certain laboratory skills, and protocols, including which cells and signaling molecules to use, still need to be revised [64].

Thus far, intestinal, cerebral, and renal organoids have been established [65]. Research into oral organoids is also being conducted, e.g., salivary glands have been recreated that can restore nerval connections and produce saliva when implanted orthotopically [66]. Jeong et al. managed to construct dentin-pulp-like organoids that expressed odontoblast-like markers and issued a biological response to the application of hydraulic calcium silicate cements [60]. Xu et al. also established an organoid model that was recommended for the toxicity screening of dental materials used, e.g., for direct pulp capping [67].

Outside the realm of dental pulp regeneration, researchers have even attempted to engineer whole tooth germ organoids. This has been partially successful by layering a multitude of different cell types [68,69]. These constructs display odontogenic markers and are also capable of epithelial invagination into the mesenchymal layer, mimicking the tissue interactions and signaling pathways at play during human tooth development [70]. Furthermore, the vision for these organoids is to replace dental implants, but further development is necessary [64].

2.2.4. Bioreactors

One drawback of 3D cultures is that nutrients cannot efficiently penetrate the center of the 3D structures and waste accumulates, which affects cell survival. Consequently, these cultures are difficult to maintain for longer time periods [60]. However, in an in vivo environment, a steady blood supply guarantees tissue homeostasis. Bioreactors are, therefore, designed to mimic this natural phenomenon and to actively supply cells in the depth of 3D structures with nutrients and oxygen [71]. Examples for simple bioreactors are magnetic rod stirrers, rockers, rotating wall vessels, and peristaltic pumps [72–74]. What these methods have in common is that they set culture medium in motion in order to achieve deeper penetration into matrix structures. By choosing either laminar or a more turbulent flow, mechanical stimuli, such as sheer stress, flow-induced pressure, or dynamic compression, can also be applied to cells in culture vessels to further emulate an in vivo situation [71]. Naturally, each individual tissue requires specific stimuli. For example, cells differentiating towards an osteogenic cell fate have proven particularly perceptive to hydrostatic pressure and sheer stresses [75–77]. However, which stimuli best support the odontogenic differentiation of cells remains to be determined.

2.2.5. Tooth-on-a-Chip Model

The so-called "organ-on-a-chip" techniques can be viewed as an extension of bioreactors. Here, cells are seeded in a microfluid device that ensures nutrient transportation through small channels and recreates physical parameters, such as pressure or shear stress [20]. Monitoring tools can also be included in this device [20]. The first organ to be emulated in a small plastic device was the lung. For example, the alveolar–capillary barrier was simulated by combining alveolar epithelial and endothelial cells, with both blood and oxygen flow, as well as cyclic mechanical stretching, in a 3D multichannel microfluid culture vessel [78]. França et al. were the first to build a tooth-on-a-chip model [79]. It consisted of two separate, closed-circuit channels filled with medium and two reaction chambers separated by a dentin fragment. Dental stem cells were added to one side, whereas the other side of the dentin barrier mimicked a tooth cavity. This model was used to test cell reactions to biomaterials by injecting solvents of the materials into the cavity side of the chip. Morphological changes in the cells could now be observed by direct cell imaging [79].

With further development, this approach holds many opportunities to enhance research into materials and signaling molecules used in dental pulp tissue engineering.

3. In Vivo Ectopic and Semiorthotopic Models

The transplantation of biological samples into the subcutaneous space of experimental animals is another method to create a physiological environment. In this context, ectopic means that tissues or cells are transplanted into experimental animals at a nonphysiological location. Cells in scaffolds can be transplanted by themselves or with signaling molecules [12,80–83]. However, especially in the context of pulp biology, cells are often implanted together on dentin disks [84,85], in tooth slices [59,86–88], in dentin cylinders, or in tooth roots [46,50,89–93] in order to simulate their natural environment. Since the directly surrounding or adjoining tissue is not ectopic, but rather corresponds to the natural environment (orthotopic), the term semiorthotopic is often used [94]. Here, the proximity to blood vessels enables nutrient supply to cells and the removal of waste products, and the animals can, thus, be considered in vivo bioreactors [71]. Additionally, interactions with resident peripheral nerve cells, connective tissues, and the immune system can be studied. Immunodeficient animals are most often utilized to prevent unwanted immunogenic reactions.

Implantation sites can vary. Small incisions through the skin can, for example, be made on the dorsum of mice, and subcutaneous pockets created by blunt dissection. After implant placement, wounds are closed by stapling or stitching [88]. Due to its abundant blood supply, the rat renal capsule is another location for ectopic transplantation; however, it is more difficult to access, and the mortality rate of experimental animals is higher than after subcutaneous implantation [95,96]. The subcutaneous implantation of autologous dental pulp cells or scaffold constructs into the dorsal surface of rabbits was also suggested as a valid ectopic model [97]. Ruangsawasdi et al. investigated the implantation of cell-free tooth roots filled with fibrin into the calvaria of rats and found that this placement produced more tissue ingrowth in the same time period than the dorsal location. This article suggested that rat calvaria could provide a microenvironment similar to the tooth socket [98].

Favorable outcomes can be achieved with ectopic and semiorthotopic transplantation, as they offer very translational features, are reproducible, and are well-described in the literature. Compared to other preclinical in vivo models, the utilization of smaller animals, such as mice, is preferred, as breeding and housing are less expensive and murine anatomy is well-understood. The surgical procedure of implant placement is easy to perform and results in minimal distress for the animals. Nevertheless, ethical concerns still need to be considered, and especially in the early stages of research, cell cultures should be preferred. The decision to use animals should never be taken lightly. It must also be noted that newly formed tissue, blood vessels, or nerve fibers can be of human or rodent origin. These ambiguities need to be kept in mind and reviewed in order to draw the correct conclusions regarding tissue formation (Table 1).

Table 1. Strengths and weaknesses of in vitro and in vivo models.

	In Vitro		In Vivo	
	Monolayer	3D Culture	Ectopic	Semiorthotopic
high cost	+	+	++	++
ethical concerns	+	+	+++	+++
literature experience	+++	+	++	++
difficult implementation	+	++	++	++
reproducibility	+++	++	+	+
mimicry of natural situation	+	++	++	+++

Further variations of these model are presented below, and selected references for applications of both in vitro and in vivo ectopic and semiorthotopic models can be found in Table 2.

Table 2. Selected references for applications of each study model, including both in vitro and in vivo. The asterisk indicates categories that are not conceivable in the present classification.

Study Models	In Vitro	In Vivo	
		Ectopic	Semiorthotopic
Scaffold culture	Wang et al., 2010 [81] Galler et al., 2012 [50] Qu and Liu, 2013 [40] Widbiller et al., 2016 [52] Lin et al., 2021 [42]	Buurma et al., 1999 [80] Gronthos et al., 2000 [12] Wang et al., 2010 [81] Lee et al., 2011 [82] De Almeidas et al., 2014 [83]	*
Spheroid and organoid	Xiao and Tsutsui, 2013 [99] Dissanayaka et al., 2014 [59] Jeong et al., 2020 [60] Zheng et al., 2021 [100] Chan et al., 2021 [101]		*
Dentin disk	Sloan et al., 1998 [102] Huang et al., 2006 [19] Widbiller et al., 2019 [103] Atesci et al., 2020 [104]	*	Batouli et al., 2003 [84] Goncalves et al., 2007 [85]
Tooth slice	Casagrande et al., 2010 [45]	*	Cordeiro et al., 2008 [86] Prescott et al., 2009 [87] Sakai et al., 2010 [105] Casagrande et al., 2010 [45] Sakai et al., 2011 [88] Dissanayaka et al., 2014 [59]
Dentin cylinder and tooth root	Rosa et al., 2013 [46]	*	Galler et al., 2011 [89] Galler et al., 2012 [50] Rosa et al., 2013 [46] Takeuchi et al., 2015 [90] Widbiller et al., 2018 [91] With coronal plug: Huang et al., 2010 [92] Zhu et al., 2018 [93]

3.1. Dentin Disk and Tooth Slice

Despite the fact that various research applications are based on the ectopic implantation of cells alone or cells encapsulated in a scaffold material [24,83,106], pulp cannot be restored without considering the pulp–dentin complex. The close mechanical and functional connections of cells and dentin are the reasons why many researchers choose to combine pulp-derived cells with dentin disks or tooth slices in vitro and implant them subcutaneously. Therefore, dentin disks or tooth slices are usually obtained in the area of solid coronal dentin or the pulp cavity from human molars respectively. The cells can then be seeded on top of solid dentin disks or cast within a scaffold into the former pulp chamber [45,87,88] (Figure 4A,B).

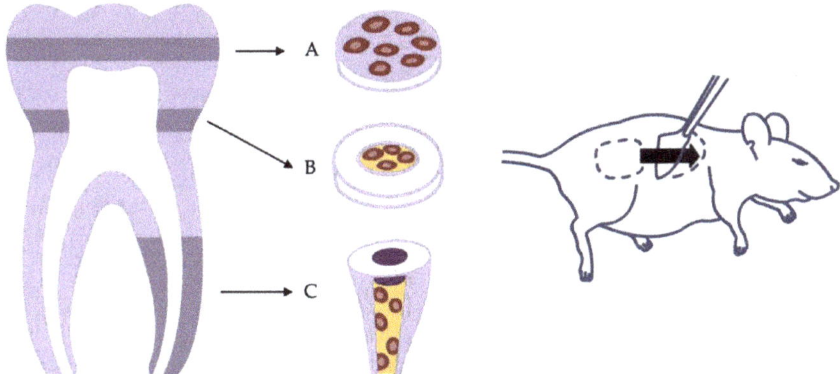

Figure 4. Variants of the ectopic transplantation model. (**A**) Dentin disk with cells seeded on top. (**B**) Tooth slice with cells and scaffold inserted into the pulp chamber. (**C**) Root fragment model with cells and scaffold inserted into the root canal.

The tooth slice model has proven to be a valid semiorthotopic approach to observe and evaluate mechanisms of differentiation, vascularization, and regeneration [45,105]. It can answer questions regarding the pretreatment of dentin surfaces, the host integration of transplants, the deposition of extracellular matrix, and tumorigenic potential. Furthermore, the assessment of various scaffold materials and their suitability for regenerative procedures is possible [107]. The research team around Nör has used it to investigate genetically modified DPSCs and to better understand signaling pathways [29,30]. It also allows for the transplantation of traceable cells to analyze cell fate in vivo [86,105].

3.2. Dentin Cylinder and Tooth Root

Sufficient vascularization is a prerequisite for cells to survive and generate new tissue [71,86,108]. In a tooth slice model, nutrients and oxygen may reach the cells easily by diffusion from neighboring tissues, as the diffusion distance is short. However, the anatomy of an actual tooth is different. Diffusion from the root tip all the way to the crown is not possible. Only the advancement of a functional vascular system allows cells to expand into the entire pulp cavity and tissue to develop even far from the apical entry [108,109]. As blood vessels have only restricted access to the root canal through the apical foramen, models using dentin cylinders or tooth roots mimic the difficulties of the clinical situation more accurately. Here, whole roots or parts of them are separated from extracted teeth, prepared, and filled with cells and a scaffold material. Sample constructs can then be implanted, for example, into a mouse dorsum to be accessed by blood vessels and nerve fibers (Figure 4C). Whereas leaving both ends of the dentin cylinder open may provide optimal blood supply from two directions, sealing of the coronal opening with a bioactive material corresponds to clinical situations, as the unilateral sprouting of vessels into the tooth root presents a challenge [46,91,92]. However, the decision of how to prepare the roots must be made depending on the application and the specific research question.

This semiorthotopic model situation allows a variety of investigations and analyses. The focus can be on qualitative factors, such as the formation of odontoblast-like cells or the expressions of certain markers, as well as quantitative factors, such as the number of blood vessels or nerve fibers or the amount of newly formed tissue. Furthermore, the model has been continuously developed and modified over the past years to answer specific questions or to counteract limitations. For example, Widbiller et al. established a customized tooth root model to test cell-homing approaches for dental pulp regeneration [91]. Here, the root canal was filled with a growth-factor-laden hydrogel with the ability to promote chemotaxis. Stem cells were then placed only at the apical opening of the root to mimic the apical papilla as the stem cell source of immature teeth. After the recovery of the tooth roots from the mouse subcutaneous space, the samples can be processed histologically, and the newly formed tissue can be analyzed by various techniques (Figure 5) [110].

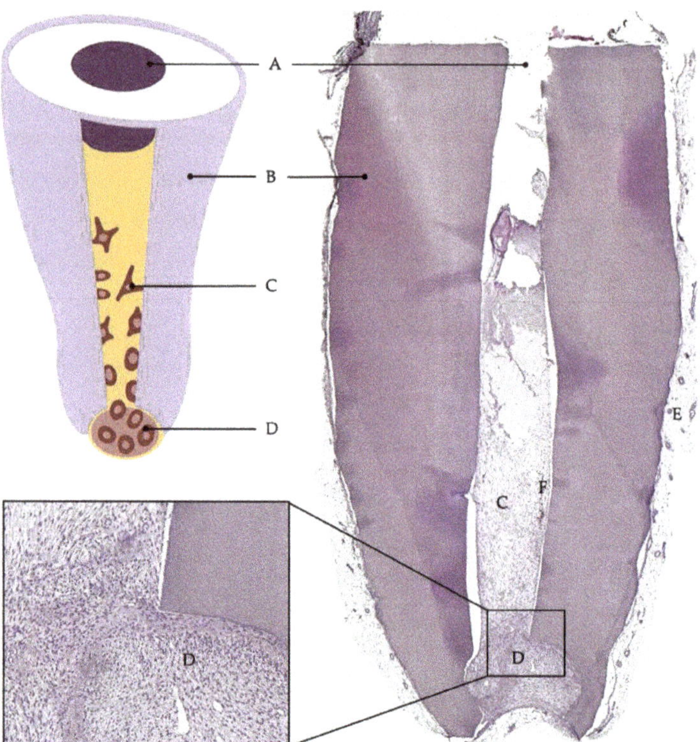

Figure 5. Cell-homing model. Tooth root recovered after 6 weeks of implantation into subcutaneous dorsal space of mice: (**A**) coronal plug, (**B**) dentin of root walls, (**C**) cells that migrated into the root canal, (**D**) apical reservoir of stem cells in collagen, (**E**) murine tissue, (**F**) blood vessel.

Another interesting variant was reported by Hilkens et al. with the aim of creating a more standardized situation. Cells were seeded not into human root fragments but into 3D-printed hydroxyapatite scaffolds shaped as tooth roots that were then implanted into mice to assess the angiogenetic potentials of different stem cells [109].

4. In Vivo Orthotopic Model

Lastly, the most translational situation to investigate stem pulp tissue engineering is the experimental animal. In this orthotopic model, signaling molecules, a scaffold, and eventually, stem cells are implanted together into an anatomically correct site, which is the root canal of a tooth in its physiological position in the oral cavity of an animal. Animal models for dental pulp tissue engineering can be grouped into small animal models and large animal models. Large animal models, such as dogs and pigs, are often preferred because it is easier to facilitate from a treatment perspective regarding tooth size (Figure 6) [95].

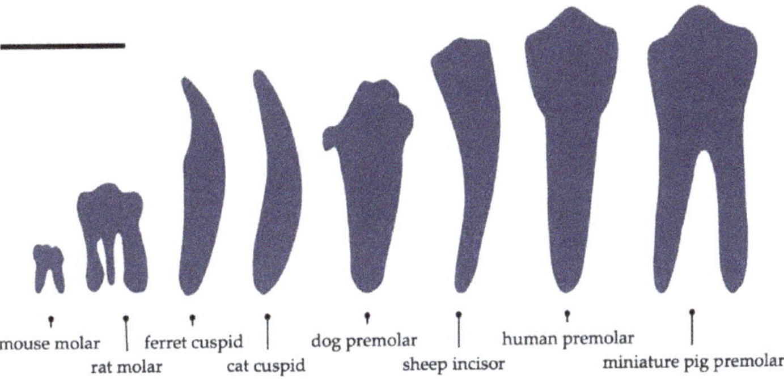

Figure 6. Relative tooth sizes of different species. Scale bar: 1 cm.

The scientific value, thereby, is that treated teeth can be examined histologically after animals have been euthanized, and the treatment outcomes of regenerative procedures can be systematically evaluated at tissue and single-cell levels. Furthermore, orthotopic models are used to evaluate the efficacy and quality of proposed regenerative strategies and to establish a data basis to design future clinical trials adequately [95]. The fields of application cover many areas, such as tumorigenesis, the testing of restorative materials, and root canal disinfection methods, as well as periodontal and endodontic regeneration [93,111].

However, it must always be kept in mind that conclusions derived from animal studies are not necessarily transferable to the clinical situation. Tooth anatomy, as well as the local microbiomes or regenerative capacities of cells or tissues, may differ. For example, autologous stem cells of animals, which are typically applied to circumvent the problem of immunocompatibility, may not behave the same way as human stem cells [111]. Furthermore, animal studies are afflicted with higher expenses than in vitro methods, which limits their availability and feasibility [111]. Most importantly, ethical concerns must always be considered when conducting research in vivo, and the step from cell culturing to animal testing should not be taken carelessly. From an ethical point of view, every single sacrifice of an animal needs to be justified by an increase in scientific knowledge. Therefore, the proposed research protocol must be reviewed before experiments can be initiated. It must align with local animal welfare laws and regulations, and it is important to ensure that researchers are educated in the handling of the animal in use. In addition, the least sentient animal should be preferred when choosing a suitable species (Table 3) [71]. When planning animal-based research, one should consider the principle of the three Rs: (1) replacement with alternative methods, such as in vitro cell cultures, whenever possible; (2) reduction in number, which may include performing multiple experiments on the same animal; and (3) refinement of the projects and techniques used in order to minimize pain and stress [112,113]. However, a reduction in number that results in invalid data and the need for repetition of the experiment is to be avoided [71]. When reporting the findings of animal studies, the ARRIVE guidelines (Animal Research: Reporting of in vivo Experiments) should be observed [114].

Table 3. Strengths and weaknesses of in vivo orthotopic models.

	In Vivo Orthotopic			
	Dog	**Pig**	**Ferret**	**Rodent**
high cost	+++	+++	++	+
ethical concerns	+++	++	++	++
literature experience	+++	++	+	+
housing requirements	++	+++	+	+
animal handling	+++	+	++	+++
similarity of tooth anatomy	++	++	+	+
similarity of tooth size	+++	+++	++	+
access to teeth	++	++	++	+

4.1. Large Animal Models

4.1.1. Dogs

Dogs have often been used as a model in dental pulp regeneration research [115]. In general, canine teeth are similar to human teeth in anatomy, growth patterns, and pathophysiology [116]. Canine premolars are preferred, as they present the greatest similarity to human molars [71]. However, even other teeth, such as incisors or canines, are suitable [117].

Differently from human teeth, the root canal system of an adult dog ends in a highly branched apical delta with multiple ramifications, which makes disinfection by irrigation difficult [116]. The premolars of younger dogs, however, have not yet formed the complex delta and still present a more stringent apical anatomy [71]. Still, the enlargement of the apical opening is a necessary step during the operation procedure [118]. Of course, there are ethical concerns and public criticism associated with the use of the canine model, which are justified and understandable, as dogs are considered companions to humans and are usually kept as pets.

Overall, there is extensive knowledge of this study model and, thus, its predictability in outcome can be seen as an advantage. Groups around Nakashima and Iohara have established and refined canine models using the beagle dog breed due to its friendly temperament and small size [118,119], which is advantageous for the housing and handling of the animals. They have worked for many years on the development of clinically applicable protocols for pulp regeneration by cell transplantation, making extensive use of the canine model, thereby, for example, proving the successful regeneration of pulp tissue using cell transplantation approaches by transplanting autologous canine stem cells into canine teeth after partial or total removal of the pulp [108,118]. Furthermore, this experimental set-up has been used to investigate different subclasses of canine dental stem cells [120], the influence of various signaling molecules [58,120], and the impacts of age [121] and inflammation [122] on endodontic regeneration.

4.1.2. Pigs

Pigs are used in various areas of research, especially as surgical models. This is because their growth patterns, physiology, and head size come close to humans [123,124]. Endodontic procedures were performed on porcine premolars, which were easy to access and were deemed suitable for experimentation in dental pulp tissue engineering [93]. Another advantage is that euthanizing pigs, which are regarded as livestock, is considered less critically [71,117]. However, there are also shortcomings to this model, such as the heavy weight of the animals compared to dogs, the small size of pig pulp chambers, and the challenges of adequate housing and high demands for feeding and care [93]. Additionally, their posterior teeth were described as difficult to access, and the root canal morphology was irregular and not ideal [93]. Interestingly, pigs were reported to possess a "disobedient temperament" or "uncooperative behavior", which deemed them difficult to manage [117]. However, Zhu et al. isolated porcine dental pulp stem cells and could prove the formation of vascularized pulp-like tissue in pig teeth and reparative dentin formation [93]. On the other hand, the implantation of porcine dental pulp stem cells in induced pulp defects did not result in regenerated pulp or reparative dentinogenesis in other studies [124]. Therefore, there are still challenges that need to be overcome before this model can find widespread use.

4.2. Small Animal Models

4.2.1. Rodents

Small animal models are often automatically excluded from use as orthotopic study models because of the diminutive size of the teeth. In addition, rodent incisors grow continuously throughout their lives and are only shortened due to attritive wear and tear. In contrast, their molars are brachydont and can, therefore, be considered for endodontic treatments. However, small mouth size limits access, and teeth are minute compared to human teeth. When using standard endodontic instruments, there is a high risk of perforating the soft dentin walls, especially in curved roots. Nowadays, the use of magnification by, e.g., operative microscopes and small instruments enables the use of rodents for endodontic applications [71,125,126]. Despite the difficulties in treating teeth, the animals' small sizes are beneficial when it comes to housing. Another advantage is that rodents possess faster biological responses to treatment; one month for rats is equal to 30 months in humans [125].

Thus, the rat model was reported as a suitable model to study novel methods of root canal treatment after apical periodontitis [126]. On various occasions, rodent models have also been used for the study of pulpal healing in direct pulp capping [127–129]. Furthermore, Almushayt et al. used rats to test the functionality of DMP1 as a signaling molecule for dental pulp tissue engineering [130].

4.2.2. Ferrets

The ferret is a medium sized carnivore that is much smaller than dogs or pigs. Their teeth exhibit anatomical, physiologic, histologic, and pathologic characteristics that resemble human teeth [111]. In particular, their single-rooted canine is suited for endodontic procedures [71,131,132].

Ferrets have the advantage of being less expensive to house and easier to maintain and breed in the laboratory than larger animals and are typically not considered as pets [111]. Because their root apices are wide open, ferret teeth lend themselves to the study of regenerative endodontic procedures where the pulp tissue is removed and bleeding is induced in order to facilitate the formation of new tissue in the root canal [133,134]. In addition, periapical infections can predictably be induced, and ferret canines can be used to investigate irrigation and medication protocols [131].

4.3. Untypical or Inappropriate Models

4.3.1. Feline Model

Other animals have been considered for stem-cell-based oral tissue engineering, as well. Cats are easy to anesthetize and have four single-rooted cuspid teeth that are suitable for endodontic procedures. However, they are more expensive to accommodate than small animals, and in analogy to dogs, they are commonly considered as pets, which induces emotional problems and public objections [111]. Although they have been described historically as a possible model, e.g., for the study of periapical lesions [135], they have not been used as such for a long time.

4.3.2. Ovine Model

Sheep present a less-developed study model in dental research but were reported to be very promising [117]. Because they are ruminants, the salivary pH of sheep is higher than humans [117]. Furthermore, ovine teeth are different from those of humans, although there are similarities in anatomy and size [136,137]. The permanent first incisors of 12-to-18-month sheep are suited for regenerative endodontic studies, as they possess an open apex and thin dentinal walls. Further advantages can be seen in the low ethical concerns regarding farm animals and the easy upkeep, as they can be released on fields [137]. Even if sheep have been used in other research areas, such as periodontology [138] or bone regeneration [139], further investigations need to be conducted before sheep can be utilized as a study model for dental pulp tissue engineering [136].

4.3.3. Primate Model

Because of their sentient character, long life span, and expensive acquisition and care, non-human primates are not an adequate model for research in dental regeneration [140,141]. Furthermore, despite presenting great anatomical similarities to humans, non-human primates are not ideal for endodontic research, as they have far better recovering abilities than humans. The artificial induction of pulpitis was hindered by the strong resistance of primate pulp to oral contamination [71]. For various ethical, legal, and physiological reasons, primates may not be used in this context, and other animal models must be preferred.

5. Conclusions

Today, various 3D cell culture models offer good alternatives to animal studies. Certain questions can easily be resolved in vitro, and the ongoing development of organoid and spheroid cultures, for example, could expand this area of application in the future. In order to gain further insight into outcomes in a physiological environment, there is, of course, also a necessity for animal studies. In consideration of the 3 Rs, study designs based on the semiorthotopic approach are of great benefit here. However, the final investigation of the research goal must be carried out in an in situ approach. Small animal studies should also be considered in this context in order to reduce the number of currently used large and more sentient animals.

Looking at the variety of in vitro and in vivo study models, there is not a single model that is suitable to answer all questions related to dental pulp regeneration. In each case, the appropriate model situation must be selected to correspond with the specific research question and the current state of development on the way to clinical application. Requirements, costs, and above all, ethical considerations should be included in the decision-making process.

Author Contributions: Conceptualization, M.W. and E.O.; writing—original draft preparation, E.O.; writing—review and editing, E.O., M.W., and K.M.G.; visualization, E.O. and M.W. All authors have read and agreed to the published version of the manuscript.

Funding: This research received no external funding.

Institutional Review Board Statement: Not applicable.

Informed Consent Statement: Not applicable.

Data Availability Statement: Not applicable.

Conflicts of Interest: The authors declare no conflict of interest.

References

1. Caplan, D.J.; Cai, J.; Yin, G.; White, B.A. Root canal filled versus non-root canal filled teeth: A retrospective comparison of survival times. *J. Public Health Dent.* **2005**, *65*, 90–96. [CrossRef] [PubMed]
2. Yan, W.; Montoya, C.; Øilo, M.; Ossa, A.; Paranjpe, A.; Zhang, H.; Arola, D.D. Contribution of root canal treatment to the fracture resistance of dentin. *J. Endod.* **2019**, *45*, 189–193. [CrossRef] [PubMed]
3. Bucchi, C.; Marcé-Nogué, J.; Galler, K.M.; Widbiller, M. Biomechanical performance of an immature maxillary central incisor after revitalization: A finite element analysis. *Int. Endod. J.* **2019**, *52*, 1508–1518. [CrossRef] [PubMed]
4. Cvek, M. Prognosis of luxated non-vital maxillary incisors treated with calcium hydroxide and filled with guttapercha: A retrospective clinical study. *Dent. Traumatol.* **1992**, *8*, 45–55. [CrossRef]
5. Widbiller, M.; Schmalz, G. Endodontic regeneration: Hard shell, soft core. *Odontology* **2021**, *109*, 303–312. [CrossRef] [PubMed]
6. Langer, R.; Vacanti, J.P. Tissue engineering. *Science* **1993**, *260*, 920–926. [CrossRef]

7. Xuan, K.; Li, B.; Guo, H.; Sun, W.; Kou, X.; He, X.; Zhang, Y.; Sun, J.; Liu, A.; Liao, L.; et al. Deciduous autologous tooth stem cells regenerate dental pulp after implantation into injured teeth. *Sci. Transl. Med.* **2018**, *10*, eaaf3227. [CrossRef]
8. Brizuela, C.; Meza, G.; Urrejola, D.; Quezada, M.A.; Concha, G.; Ramírez, V.; Angelopoulos, I.; Cadiz, M.I.; Tapia-Limonchi, R.; Khoury, M. Cell-based regenerative endodontics for treatment of periapical lesions: A randomized, controlled phase I/II clinical trial. *J. Dent. Res.* **2020**, *99*, 523–529. [CrossRef]
9. Widbiller, M.; Knüttel, H.; Meschi, N.; Durán-Sindreu Terol, F. Effectiveness of endodontic tissue engineering in treatment of apical periodontitis: A systematic review. *Int. Endod. J.* **2022**. *Online ahead of print.* [CrossRef]
10. Nakashima, M.; Iohara, K.; Murakami, M.; Nakamura, H.; Sato, Y.; Ariji, Y.; Matsushita, K. Pulp regeneration by transplantation of dental pulp stem cells in pulpitis: A pilot clinical study. *Stem. Cell Res. Ther.* **2017**, *8*, 61. [CrossRef]
11. Gay, I.; Chen, S.; MacDougall, M. Isolation and characterization of multipotent human periodontal ligament stem cells. *Orthod. Craniofac. Res.* **2007**, *10*, 149–160. [CrossRef]
12. Gronthos, S.; Mankani, M.; Brahim, J.; Robey, P.G.; Shi, S. Postnatal human dental pulp stem cells (DPSCs) in vitro and in vivo. *Proc. Natl. Acad. Sci. USA* **2000**, *97*, 13625–13630. [CrossRef]
13. Miura, M.; Gronthos, S.; Zhao, M.; Lu, B.; Fisher, L.W.; Robey, P.G.; Shi, S. SHED: Stem cells from human exfoliated deciduous teeth. *Proc. Natl. Acad. Sci. USA* **2003**, *100*, 5807–5812. [CrossRef]
14. Iohara, K.; Zheng, L.; Ito, M.; Tomokiyo, A.; Matsushita, K.; Nakashima, M. Side population cells isolated from porcine dental pulp tissue with self-renewal and multipotency for dentinogenesis, chondrogenesis, adipogenesis, and neurogenesis. *Stem Cells* **2006**, *24*, 2493–2503. [CrossRef]
15. Nakashima, M. Establishment of primary cultures of pulp cells from bovine permanent incisors. *Arch. Oral Biol.* **1991**, *36*, 655–663. [CrossRef]
16. Yang, X.; van den Dolder, J.; Walboomers, X.F.; Zhang, W.; Bian, Z.; Fan, M.; Jansen, J.A. The odontogenic potential of stro-1 sorted rat dental pulp stem cells in vitro. *J. Tissue Eng. Regen. Med.* **2007**, *1*, 66–73. [CrossRef]
17. Bucchi, C.; Ohlsson, E.; de Anta, J.M.; Woelflick, M.; Galler, K.; Manzanares-Cespedes, M.C.; Widbiller, M. Human amnion epithelial cells: A potential cell source for pulp regeneration? *Int. J. Mol. Sci.* **2022**, *23*, 2830. [CrossRef]
18. Huang, G.T.-J.; Gronthos, S.; Shi, S. Mesenchymal stem cells derived from dental tissues vs. those from other sources: Their biology and role in regenerative medicine. *J. Dent. Res.* **2009**, *88*, 792–806. [CrossRef]
19. Huang, G.T.-J.; Sonoyama, W.; Chen, J.; Park, S.H. In vitro characterization of human dental pulp cells: Various isolation methods and culturing environments. *Cell Tissue Res.* **2006**, *324*, 225–236. [CrossRef]
20. Kang, S.; Kim, D.; Lee, J.; Takayama, S.; Park, J.Y. Engineered microsystems for spheroid and organoid studies. *Adv. Healthc. Mater.* **2021**, *10*, 2001284. [CrossRef]
21. Schmalz, G. Use of cell cultures for toxicity testing of dental materials—Advantages and limitations. *J. Dent.* **1994**, *22*, S6–S11. [CrossRef]
22. Pampaloni, F.; Reynaud, E.G.; Stelzer, E.H.K. The third dimension bridges the gap between cell culture and live tissue. *Nat. Rev. Mol. Cell Biol.* **2007**, *8*, 839–845. [CrossRef] [PubMed]
23. Edmondson, R.; Broglie, J.J.; Adcock, A.F.; Yang, L. Three-dimensional cell culture systems and their applications in drug discovery and cell-based biosensors. *ASSAY Drug Dev. Technol.* **2014**, *12*, 207–218. [CrossRef] [PubMed]
24. Gronthos, S.; Brahim, J.; Li, W.; Fisher, L.W.; Cherman, N.; Boyde, A.; DenBesten, P.; Robey, P.G.; Shi, S. Stem cell properties of human dental pulp stem cells. *J. Dent. Res.* **2002**, *81*, 531–535. [CrossRef] [PubMed]
25. Couble, M.-L.; Farges, J.-C.; Bleicher, F.; Perrat-Mabillon, B.; Boudeulle, M.; Magloire, H. Odontoblast differentiation of human dental pulp cells in explant cultures. *Calcif. Tissue Int.* **2000**, *66*, 129–138. [CrossRef]
26. Narayanan, K.; Srinivas, R.; Ramachandran, A.; Hao, J.; Quinn, B.; George, A. Differentiation of embryonic mesenchymal cells to odontoblast-like cells by overexpression of dentin matrix protein 1. *Proc. Natl. Acad. Sci. USA* **2001**, *98*, 4516–4521. [CrossRef]
27. Seki, D.; Takeshita, N.; Oyanagi, T.; Sasaki, S.; Takano, I.; Hasegawa, M.; Takano-Yamamoto, T. Differentiation of odontoblast-like cells from mouse induced pluripotent stem cells by Pax9 and Bmp4 transfection. *Stem Cells Transl. Med.* **2015**, *4*, 993–997. [CrossRef]
28. Widbiller, M.; Eidt, A.; Lindner, S.R.; Hiller, K.-A.; Schweikl, H.; Buchalla, W.; Galler, K.M. Dentine matrix proteins: Isolation and effects on human pulp cells. *Int. Endod. J.* **2018**, *51*, e278–e290. [CrossRef]
29. Bento, L.W.; Zhang, Z.; Imai, A.; Nör, F.; Dong, Z.; Shi, S.; Araujo, F.B.; Nör, J.E. Endothelial differentiation of SHED requires MEK1/ERK signaling. *J. Dent. Res.* **2013**, *92*, 51–57. [CrossRef]
30. Zhang, Z.; Nör, F.; Oh, M.; Cucco, C.; Shi, S.; Nör, J.E. Wnt/β-catenin signaling determines the vasculogenic fate of postnatal mesenchymal stem cells. *Stem Cells* **2016**, *34*, 1576–1587. [CrossRef]
31. Liu, L.; Wei, X.; Ling, J.; Wu, L.; Xiao, Y. Expression pattern of Oct-4, Sox2, and c-Myc in the primary culture of human dental pulp derived cells. *J. Endod.* **2011**, *37*, 466–472. [CrossRef]
32. Smeda, M.; Galler, K.M.; Woelflick, M.; Rosendahl, A.; Moehle, C.; Lenhardt, B.; Buchalla, W.; Widbiller, M. Molecular biological comparison of dental pulp- and apical papilla-derived stem cells. *Int. J. Mol. Sci.* **2022**, *23*, 2615. [CrossRef]
33. Wei, X.; Ling, J.; Wu, L.; Liu, L.; Xiao, Y. Expression of Mineralization Markers in Dental Pulp Cells. *J. Endod.* **2007**, *33*, 703–708. [CrossRef]
34. Magne, D.; Bluteau, G.; Lopez-Cazaux, S.; Weiss, P.; Pilet, P.; Ritchie, H.H.; Daculsi, G.; Guicheux, J. Development of an odontoblast in vitro model to study dentin mineralization. *Connect. Tissue Res.* **2004**, *45*, 101–110. [CrossRef]

35. Riccio, M.; Resca, E.; Maraldi, T.; Pisciotta, A.; Ferrari, A.; Bruzzesi, G.; De Pol, A. Human dental pulp stem cells produce mineralized matrix in 2D and 3D cultures. *Eur. J. Histochem.* **2010**, *54*, 46. [CrossRef]

36. Galler, K.M.; Widbiller, M.; Buchalla, W.; Eidt, A.; Hiller, K.-A.; Hoffer, P.C.; Schmalz, G. EDTA conditioning of dentine promotes adhesion, migration and differentiation of dental pulp stem cells. *Int. Endod. J.* **2016**, *49*, 581–590. [CrossRef]

37. Smith, A.J.; Scheven, B.A.; Takahashi, Y.; Ferracane, J.L.; Shelton, R.M.; Cooper, P.R. Dentine as a bioactive extracellular matrix. *Arch. Oral Biol.* **2012**, *57*, 109–121. [CrossRef]

38. Widbiller, M.; Schweikl, H.; Bruckmann, A.; Rosendahl, A.; Hochmuth, E.; Lindner, S.R.; Buchalla, W.; Galler, K.M. Shotgun proteomics of human dentin with different prefractionation methods. *Sci. Rep.* **2019**, *9*, 4457. [CrossRef]

39. Knowlton, S.; Cho, Y.; Li, X.-J.; Khademhosseini, A.; Tasoglu, S. Utilizing stem cells for three-dimensional neural tissue engineering. *Biomater. Sci.* **2016**, *4*, 768–784. [CrossRef]

40. Qu, T.; Liu, X. Nano-structured gelatin/bioactive glass hybrid scaffolds for the enhancement of odontogenic differentiation of human dental pulp stem cells. *J. Mater. Chem. B* **2013**, *1*, 4764. [CrossRef]

41. Lv, D.; Hu, Z.; Lu, L.; Lu, H.; Xu, X. Three-dimensional cell culture: A powerful tool in tumor research and drug discovery (review). *Oncol. Lett.* **2017**, *14*, 6999–7010. [CrossRef] [PubMed]

42. Lin, Y.-T.; Hsu, T.-T.; Liu, Y.-W.; Kao, C.-T.; Huang, T.-H. Bidirectional differentiation of human-derived stem cells induced by biomimetic calcium silicate-reinforced gelatin methacrylate bioink for odontogenic regeneration. *Biomedicines* **2021**, *9*, 929. [CrossRef] [PubMed]

43. Lv, H.; Wang, H.; Zhang, Z.; Yang, W.; Liu, W.; Li, Y.; Li, L. Biomaterial Stiffness Determines Stem Cell Fate. *Life Sci.* **2017**, *178*, 42–48. [CrossRef] [PubMed]

44. Zanoni, M.; Cortesi, M.; Zamagni, A.; Arienti, C.; Pignatta, S.; Tesei, A. Modeling neoplastic disease with spheroids and organoids. *J. Hematol. Oncol.* **2020**, *13*, 97. [CrossRef] [PubMed]

45. Casagrande, L.; Demarco, F.F.; Zhang, Z.; Araujo, F.B.; Shi, S.; Nör, J.E. Dentin-derived BMP-2 and odontoblast differentiation. *J. Dent. Res.* **2010**, *89*, 603–608. [CrossRef]

46. Rosa, V.; Zhang, Z.; Grande, R.H.M.; Nör, J.E. Dental pulp tissue engineering in full-length human root canals. *J. Dent. Res.* **2013**, *92*, 970–997. [CrossRef]

47. Chiu, L.L.Y.; Chu, Z.; Radisic, M. Tissue Engineering. In *Comprehensive Nanoscience and Technology*; Elsevier: Amsterdam, The Netherlands, 2011; pp. 175–211. ISBN 978-0-12-374396-1.

48. Galler, K.M.; Widbiller, M. Perspectives for cell-homing approaches to engineer dental pulp. *J. Endod.* **2017**, *43*, S40–S45. [CrossRef]

49. Yildirimer, L.; Seifalian, A.M. Three-dimensional biomaterial degradation—Material choice, design and extrinsic factor considerations. *Biotechnol. Adv.* **2014**, *32*, 984–999. [CrossRef]

50. Galler, K.M.; Hartgerink, J.D.; Cavender, A.C.; Schmalz, G.; D'Souza, R.N. A customized self-assembling peptide hydrogel for dental pulp tissue engineering. *Tissue Eng. Part A* **2012**, *18*, 176–184. [CrossRef]

51. Soares, D.G.; Anovazzi, G.; Bordini, E.A.F.; Zuta, U.O.; Silva Leite, M.L.A.; Basso, F.G.; Hebling, J.; de Souza Costa, C.A. Biological analysis of simvastatin-releasing chitosan scaffold as a cell-free system for pulp-dentin regeneration. *J. Endod.* **2018**, *44*, 971–976.e1. [CrossRef]

52. Widbiller, M.; Lindner, S.R.; Buchalla, W.; Eidt, A.; Hiller, K.-A.; Schmalz, G.; Galler, K.M. Three-dimensional culture of dental pulp stem cells in direct contact to tricalcium silicate cements. *Clin. Oral Investig.* **2016**, *20*, 237–246. [CrossRef]

53. Li, X.; Zhang, Q.; Ye, D.; Zhang, J.; Guo, Y.; You, R.; Yan, S.; Li, M.; Qu, J. Fabrication and characterization of electrospun PCL/ *antheraea pernyi* silk fibroin nanofibrous scaffolds. *Polym. Eng. Sci.* **2017**, *57*, 206–213. [CrossRef]

54. Han, J.; Kim, D.S.; Jang, H.; Kim, H.-R.; Kang, H.-W. Bioprinting of Three-Dimensional Dentin–Pulp Complex with Local Differentiation of Human Dental Pulp Stem Cells. *J. Tissue Eng.* **2019**, *10*, 204173141984584. [CrossRef]

55. Athirasala, A.; Tahayeri, A.; Thrivikraman, G.; França, C.M.; Monteiro, N.; Tran, V.; Ferracane, J.; Bertassoni, L.E. A Dentin-Derived Hydrogel Bioink for 3D Bioprinting of Cell Laden Scaffolds for Regenerative Dentistry. *Biofabrication* **2018**, *10*, 024101. [CrossRef]

56. Gu, B.K.; Choi, D.J.; Park, S.J.; Kim, M.S.; Kang, C.M.; Kim, C.-H. 3-Dimensional Bioprinting for Tissue Engineering Applications. *Biomater. Res.* **2016**, *20*, 12. [CrossRef]

57. Yamamoto, M.; Kawashima, N.; Takashino, N.; Koizumi, Y.; Takimoto, K.; Suzuki, N.; Saito, M.; Suda, H. Three-dimensional spheroid culture promotes odonto/osteoblastic differentiation of dental pulp cells. *Arch. Oral Biol.* **2014**, *59*, 310–331. [CrossRef]

58. Iohara, K.; Nakashima, M.; Ito, M.; Ishikawa, M.; Nakasima, A.; Akamine, A. Dentin regeneration by dental pulp stem cell therapy with recombinant human bone morphogenetic protein 2. *J. Dent. Res.* **2004**, *83*, 590–595. [CrossRef]

59. Dissanayaka, W.L.; Zhu, L.; Hargreaves, K.M.; Jin, L.; Zhang, C. Scaffold-free prevascularized microtissue spheroids for pulp regeneration. *J. Dent. Res.* **2014**, *93*, 1296–1303. [CrossRef]

60. Jeong, S.Y.; Lee, S.; Choi, W.H.; Jee, J.H.; Kim, H.-R.; Yoo, J. Fabrication of dentin-pulp-like organoids using dental-pulp stem cells. *Cells* **2020**, *9*, 642. [CrossRef]

61. Gunti, S.; Hoke, A.T.K.; Vu, K.P.; London, N.R. Organoid and spheroid tumor models: Techniques and applications. *Cancers* **2021**, *13*, 874. [CrossRef]

62. Abbott, A. Biology's New Dimension. *Nature* **2003**, *424*, 870–872. [CrossRef] [PubMed]

63. Li, M.; Izpisua Belmonte, J.C. Organoids—Preclinical models of human disease. *N. Engl. J. Med.* **2019**, *380*, 569–579. [CrossRef] [PubMed]

64. Gao, X.; Wu, Y.; Liao, L.; Tian, W. Oral organoids: Progress and challenges. *J. Dent. Res.* **2021**, *100*, 454–463. [CrossRef] [PubMed]
65. Prasad, M.; Kumar, R.; Buragohain, L.; Kumari, A.; Ghosh, M. Organoid technology: A reliable developmental biology tool for organ-specific nanotoxicity evaluation. *Front. Cell Dev. Biol.* **2021**, *9*, 696668. [CrossRef] [PubMed]
66. Adine, C.; Ng, K.K.; Rungarunlert, S.; Souza, G.R.; Ferreira, J.N. Engineering innervated secretory epithelial organoids by magnetic three-dimensional bioprinting for stimulating epithelial growth in salivary glands. *Biomaterials* **2018**, *180*, 52–66. [CrossRef] [PubMed]
67. Xu, X.; Li, Z.; Ai, X.; Tang, Y.; Yang, D.; Dou, L. Human three-dimensional dental pulp organoid model for toxicity screening of dental materials on dental pulp cells and tissue. *Int. Endod. J.* **2022**, *55*, 79–88. [CrossRef]
68. Modino, S.A.C.; Sharpe, P.T. Tissue engineering of teeth using adult stem cells. *Arch. Oral Biol.* **2005**, *50*, 255–258. [CrossRef]
69. Smith, E.E.; Zhang, W.; Schiele, N.R.; Khademhosseini, A.; Kuo, C.K.; Yelick, P.C. Developing a biomimetic tooth bud model: Developing a biomimetic tooth bud model. *J. Tissue Eng. Regen. Med.* **2017**, *11*, 3326–3336. [CrossRef]
70. Rosowski, J.; Bräunig, J.; Amler, A.-K.; Strietzel, F.P.; Lauster, R.; Rosowski, M. Emulating the early phases of human tooth development in vitro. *Sci. Rep.* **2019**, *9*, 7057. [CrossRef]
71. Tayebi, L. *Applications of Biomedical Engineering in Dentistry*; Springer International Publishing: Cham, Switzerland, 2020; ISBN 978-3-030-21582-8.
72. Botchwey, E.A.; Pollack, S.R.; Levine, E.M.; Laurencin, C.T. Bone tissue engineering in a rotating bioreactor using a microcarrier matrix system. *J. Biomed. Mater. Res.* **2001**, *55*, 242–253. [CrossRef]
73. Teixeira, G.Q.; Barrias, C.C.; Lourenço, A.H.; Gonçalves, R.M. A multicompartment holder for spinner flasks improves expansion and osteogenic differentiation of mesenchymal stem cells in three-dimensional scaffolds. *Tissue Eng. Part C Methods* **2014**, *20*, 984–993. [CrossRef]
74. Hofmann, A.; Konrad, L.; Gotzen, L.; Printz, H.; Ramaswamy, A.; Hofmann, C. Bioengineered human bone tissue using autogenous osteoblasts cultured on different biomatrices. *J. Biomed. Mater. Res.* **2003**, *67*, 191–199. [CrossRef]
75. Reinwald, Y.; El Haj, A.J. Hydrostatic pressure in combination with topographical cues affects the fate of bone marrow-derived human mesenchymal stem cells for bone tissue regeneration. *J. Biomed. Mater. Res. Part A* **2018**, *106*, 629–640. [CrossRef]
76. Song, K.-D.; Liu, T.-Q.; Li, X.-Q.; Cui, Z.-F.; Sun, X.-Y.; Ma, X.-H. Three-dimensional expansion: In suspension culture of SD rat's osteoblasts in a rotating wall vessel bioreactor. *Biomed. Environ. Sci. BES* **2007**, *20*, 91–98.
77. Burdick, J.A.; Vunjak-Novakovic, G. Engineered microenvironments for controlled stem cell differentiation. *Tissue Eng. Part A* **2009**, *15*, 205–219. [CrossRef]
78. Huh, D.; Matthews, B.D.; Mammoto, A.; Montoya-Zavala, M.; Hsin, H.Y.; Ingber, D.E. Reconstituting organ-level lung functions on a chip. *Science* **2010**, *328*, 1662–1668. [CrossRef]
79. França, C.M.; Tahayeri, A.; Rodrigues, N.S.; Ferdosian, S.; Puppin Rontani, R.M.; Sereda, G.; Ferracane, J.L.; Bertassoni, L.E. The tooth on-a-chip: A microphysiologic model system mimicking the biologic interface of the tooth with biomaterials. *Lab Chip* **2020**, *20*, 405–413. [CrossRef]
80. Buurma, B.; Gu, K.; Rutherford, R.B. Transplantation of human pulpal and gingival fibroblasts attached to synthetic scaffolds: Transplantation of human pulpal and gingival fibroblasts attached to synthetic scaffolds. *Eur. J. Oral Sci.* **1999**, *107*, 282–289. [CrossRef]
81. Wang, J.; Liu, X.; Jin, X.; Ma, H.; Hu, J.; Ni, L.; Ma, P.X. The odontogenic differentiation of human dental pulp stem cells on nanofibrous poly(l-lactic acid) scaffolds in vitro and in vivo. *Acta Biomater.* **2010**, *6*, 3856–3863. [CrossRef]
82. Lee, J.-H.; Lee, D.-S.; Choung, H.-W.; Shon, W.-J.; Seo, B.-M.; Lee, E.-H.; Cho, J.-Y.; Park, J.-C. Odontogenic differentiation of human dental pulp stem cells induced by preameloblast-derived factors. *Biomaterials* **2011**, *32*, 9696–9706. [CrossRef]
83. de Almeida, J.F.A.; Chen, P.; Henry, M.A.; Diogenes, A. Stem cells of the apical papilla regulate trigeminal neurite outgrowth and targeting through a BDNF-dependent mechanism. *Tissue Eng. Part A* **2014**, *20*, 3089–3100. [CrossRef]
84. Batouli, S.; Miura, M.; Brahim, J.; Tsutsui, T.W.; Fisher, L.W.; Gronthos, S.; Robey, P.G.; Shi, S. Comparison of stem-cell-mediated osteogenesis and dentinogenesis. *J. Dent. Res.* **2003**, *82*, 976–981. [CrossRef] [PubMed]
85. Goncalves, S.; Dong, Z.; Bramante, C.; Holland, G.; Smith, A.; Nor, J. Tooth slice–based models for the study of human dental pulp angiogenesis. *J. Endod.* **2007**, *33*, 811–814. [CrossRef] [PubMed]
86. Cordeiro, M.M.; Dong, Z.; Kaneko, T.; Zhang, Z.; Miyazawa, M.; Shi, S.; Smith, A.J.; Nör, J.E. Dental pulp tissue engineering with stem cells from exfoliated deciduous teeth. *J. Endod.* **2008**, *34*, 962–969. [CrossRef] [PubMed]
87. Prescott, R.S.; Alsanea, R.; Fayad, M.I.; Johnson, B.R.; Wenckus, C.S.; Hao, J.; John, A.S.; George, A. In-vivo generation of dental pulp-like tissue using human pulpal stem cells, a collagen scaffold and dentin matrix protein 1 following subcutaneous transplantation in mice. *J. Endod.* **2009**, *34*, 421–426. [CrossRef]
88. Sakai, V.T.; Cordeiro, M.M.; Dong, Z.; Zhang, Z.; Zeitlin, B.D.; Nör, J.E. Tooth slice/scaffold model of dental pulp tissue engineering. *Adv. Dent. Res.* **2011**, *23*, 325–332. [CrossRef]
89. Galler, K.M.; Cavender, A.C.; Koeklue, U.; Suggs, L.J.; Schmalz, G.; D'Souza, R.N. Bioengineering of dental stem cells in a PEGylated fibrin gel. *Regen. Med.* **2011**, *6*, 191–200. [CrossRef]
90. Takeuchi, N.; Hayashi, Y.; Murakami, M.; Alvarez, F.; Horibe, H.; Iohara, K.; Nakata, K.; Nakamura, H.; Nakashima, M. Similar *in vitro* effects and pulp regeneration in ectopic tooth transplantation by basic fibroblast growth factor and granulocyte-colony stimulating factor. *Oral Dis.* **2015**, *21*, 113–122. [CrossRef]

91. Widbiller, M.; Driesen, R.B.; Eidt, A.; Lambrichts, I.; Hiller, K.-A.; Buchalla, W.; Schmalz, G.; Galler, K.M. Cell homing for pulp tissue engineering with endogenous dentin matrix proteins. *J. Endod.* **2018**, *44*, 956–962.e2. [CrossRef]

92. Huang, G.T.-J.; Yamaza, T.; She, A.L.D.; Djouad, F.; Kuhn, N.Z.; Tuan, R.S.; Shi, S. Stem/progenitor cell–mediated de novo regeneration of dental pulp with newly deposited continuous layer of dentin in an in vivo model. *Tissue Eng. Part A* **2010**, *16*, 605–615. [CrossRef]

93. Zhu, X.; Liu, J.; Yu, Z.; Chen, C.-A.; Aksel, H.; Azim, A.A.; Huang, G.T.-J. A miniature swine model for stem cell-based *de novo* regeneration of dental pulp and dentin-like tissue. *Tissue Eng. Part C Methods* **2018**, *24*, 108–120. [CrossRef] [PubMed]

94. Nakashima, M.; Iohara, K.; Bottino, M.C.; Fouad, A.F.; Nör, J.E.; Huang, G.T.-J. Animal Models for Stem Cell-Based Pulp Regeneration: Foundation for Human Clinical Applications. *Tissue Eng. Part B Rev.* **2019**, *25*, 100–113. [CrossRef] [PubMed]

95. Kim, S.; Shin, S.-J.; Song, Y.; Kim, E. In vivo experiments with dental pulp stem cells for pulp-dentin complex regeneration. *Mediat. Inflamm.* **2015**, *2015*, 409347. [CrossRef] [PubMed]

96. Yu, J.; Deng, Z.; Shi, J.; Zhai, H.; Nie, X.; Zhuang, H.; Li, Y.; Jin, Y. Differentiation of dental pulp stem cells into regular-shaped dentin-pulp complex induced by tooth germ cell conditioned medium. *Tissue Eng.* **2006**, *12*, 3097–3105. [CrossRef] [PubMed]

97. El-Backly, R.M.; Massoud, A.G.; El-Badry, A.M.; Sherif, R.A.; Marei, M.K. Regeneration of dentine/pulp-like tissue using a dental pulp stem cell/poly(lactic-co-glycolic) acid scaffold construct in new zealand white rabbits. *Aust. Endod. J.* **2008**, *34*, 52–67. [CrossRef] [PubMed]

98. Ruangsawasdi, N.; Zehnder, M.; Patcas, R.; Ghayor, C.; Weber, F.E. Regenerative Dentistry: Animal model for regenerative endodontology. *Transfus. Med. Hemotherapy* **2016**, *43*, 359–364. [CrossRef]

99. Xiao, L.; Tsutsui, T. Characterization of human dental pulp cells-derived spheroids in serum-free medium: Stem cells in the core: Stem cell distribution in spheroid. *J. Cell Biochem.* **2013**, *114*, 2624–2636. [CrossRef]

100. Zheng, Y.; Jiang, L.I.; Yan, M.; Gosau, M.; Smeets, R.; Kluwe, L.; Friedrich, R.E. Optimizing conditions for spheroid formation of dental pulp cells in cell culture. *In Vivo* **2021**, *35*, 1965–1972. [CrossRef]

101. Chan, Y.-H.; Lee, Y.-C.; Hung, C.-Y.; Yang, P.-J.; Lai, P.-C.; Feng, S.-W. Three-dimensional spheroid culture enhances multipotent differentiation and stemness capacities of human dental pulpderived mesenchymal stem cells by modulating MAPK and NF-ĸB signaling pathways. *Stem. Cell Rev. Rep.* **2021**, *17*, 1810–1826. [CrossRef]

102. Sloan, A.J.; Shelton, R.M.; Hann, A.C.; Moxham, B.J.; Smith, A.J. An in vitro approach for the study of dentinogenesis by organ culture of the dentine–pulp complex from rat incisor teeth. *Arch. Oral Biol.* **1998**, *43*, 421–430. [CrossRef]

103. Widbiller, M.; Althumairy, R.I.; Diogenes, A. Direct and indirect effect of chlorhexidine on survival of stem cells from the apical papilla and its neutralization. *J. Endod.* **2019**, *45*, 156–160. [CrossRef]

104. Atesci, A.A.; Avci, C.B.; Tuglu, M.I.; Ozates Ay, N.P.; Eronat, A.C. Effect of different dentin conditioning agents on growth factor release, mesenchymal stem cell attachment and morphology. *J. Endod.* **2020**, *46*, 200–208. [CrossRef]

105. Sakai, V.T.; Zhang, Z.; Dong, Z.; Neiva, K.G.; Machado, M.A.A.M.; Shi, S.; Santos, C.F.; Nör, J.E. SHED differentiate into functional odontoblasts and endothelium. *J. Dent. Res.* **2010**, *89*, 791–796. [CrossRef]

106. Nakao, K.; Morita, R.; Saji, Y.; Ishida, K.; Tomita, Y.; Ogawa, M.; Saitoh, M.; Tomooka, Y.; Tsuji, T. The development of a bioengineered organ germ method. *Nat. Methods* **2007**, *4*, 227–230. [CrossRef]

107. Demarco, F.F.; Casagrande, L.; Zhang, Z.; Dong, Z.; Tarquinio, S.B.; Zeitlin, B.D.; Shi, S.; Smith, A.J.; Nör, J.E. Effects of morphogen and scaffold porogen on the differentiation of dental pulp stem cells. *J. Endod.* **2010**, *36*, 1805–1811. [CrossRef]

108. Nakashima, M.; Iohara, K. Regeneration of dental pulp by stem cells. *Adv. Dent. Res.* **2011**, *23*, 313–319. [CrossRef]

109. Hilkens, P.; Bronckaers, A.; Ratajczak, J.; Gervois, P.; Wolfs, E.; Lambrichts, I. The angiogenic potential of dpscs and scaps in an *in vivo* model of dental pulp regeneration. *Stem. Cells Int.* **2017**, *2017*, 2582080. [CrossRef]

110. Widbiller, M.; Rothmaier, C.; Saliter, D.; Wölflick, M.; Rosendahl, A.; Buchalla, W.; Schmalz, G.; Spruss, T.; Galler, K.M. Histology of human teeth: Standard and specific staining methods revisited. *Arch. Oral Biol.* **2021**, *127*, 105136. [CrossRef]

111. Torabinejad, M.; Corr, R.; Buhrley, M.; Wright, K.; Shabahang, S. An animal model to study regenerative endodontics. *J. Endod.* **2011**, *37*, 197–202. [CrossRef]

112. Muthanandam, S.; Muthu, J.; Mony, V.; Rl, P.; Lal K, P. Animal models in dental research—A Review. *Int. Dent. J. Stud. Res.* **2020**, *8*, 44–47. [CrossRef]

113. Russell, W.M.S.; Burch, R.L. *The Principles of Humane Experimental Technique*; Universities Federation for Animal Welfare: Potters Bar, Herts, 1992; ISBN 978-0-900767-78-4.

114. Percie du Sert, N.; Hurst, V.; Ahluwalia, A.; Alam, S.; Avey, M.T.; Baker, M.; Browne, W.J.; Clark, A.; Cuthill, I.C.; Dirnagl, U.; et al. The ARRIVE guidelines 2.0: Updated guidelines for reporting animal research. *PLoS Biol.* **2020**, *18*, e3000410.

115. Wang, Y.; Zhao, Y.; Jia, W.; Yang, J.; Ge, L. Preliminary study on dental pulp stem cell-mediated pulp regeneration in canine immature permanent teeth. *J. Endod.* **2013**, *39*, 195–201. [CrossRef] [PubMed]

116. Watanabe, K.; Kikuchi, M.; Barroga, E.F.; Okumura, M.; Kadosawa, T.; Fujinaga, T. The formation of apical delta of the permanent teeth in dogs. *J. Vet. Med. Sci.* **2001**, *63*, 789–795. [CrossRef] [PubMed]

117. Mangione, F.; Salmon, B.; EzEldeen, M.; Jacobs, R.; Chaussain, C.; Vital, S. Characteristics of large animal models for current cell-based oral tissue regeneration. *Tissue Eng. Part B Rev.* **2021**, *28*, 489–505. [CrossRef] [PubMed]

118. Iohara, K.; Imabayashi, K.; Ishizaka, R.; Watanabe, A.; Nabekura, J.; Ito, M.; Matsushita, K.; Nakamura, H.; Nakashima, M. Complete pulp regeneration after pulpectomy by transplantation of cd105+ stem cells with stromal cell-derived factor-1. *Tissue Eng. Part A* **2011**, *17*, 1911–1920. [CrossRef]

119. Iohara, K.; Zheng, L.; Ito, M.; Ishizaka, R.; Nakamura, H.; Into, T.; Matsushita, K.; Nakashima, M. Regeneration of dental pulp after pulpotomy by transplantation of cd31/cd146 side population cells from a canine tooth. *Regen. Med.* **2009**, *4*, 377–385. [CrossRef]

120. Iohara, K.; Murakami, M.; Takeuchi, N.; Osako, Y.; Ito, M.; Ishizaka, R.; Utunomiya, S.; Nakamura, H.; Matsushita, K.; Nakashima, M. A novel combinatorial therapy with pulp stem cells and granulocyte colony-stimulating factor for total pulp regeneration. *STEM CELLS Transl. Med.* **2013**, *2*, 521–533. [CrossRef]

121. Iohara, K.; Zayed, M.; Takei, Y.; Watanabe, H.; Nakashima, M. Treatment of pulpectomized teeth with trypsin prior to transplantation of mobilized dental pulp stem cells enhances pulp regeneration in aged dogs. *Front. Bioeng. Biotechnol.* **2020**, *8*, 983. [CrossRef]

122. Ling, L.; Zhao, Y.M.; Wang, X.T.; Wen, Q.; Ge, L.H. Regeneration of dental pulp tissue by autologous grafting stem cells derived from inflammatory dental pulp tissue in immature premolars in a beagle dog. *Chin. J. Dent. Res. J. Sci. Sect. Chin. Stomatol. Assoc. CSA* **2020**, *23*, 143–150.

123. Kodonas, K.; Gogos, C.; Papadimitriou, S.; Kouzi-Koliakou, K.; Tziafas, D. Experimental formation of dentin-like structure in the root canal implant model using cryopreserved swine dental pulp progenitor cells. *J. Endod.* **2012**, *38*, 913–919. [CrossRef]

124. Mangione, F.; EzEldeen, M.; Bardet, C.; Lesieur, J.; Bonneau, M.; Decup, F.; Salmon, B.; Jacobs, R.; Chaussain, C.; Opsahl-Vital, S. Implanted dental pulp cells fail to induce regeneration in partial pulpotomies. *J. Dent. Res.* **2017**, *96*, 1406–1413. [CrossRef]

125. Dammaschke, T. Rat molar teeth as a study model for direct pulp capping research in dentistry. *Lab. Anim.* **2010**, *44*, 1–6. [CrossRef]

126. Yoneda, N.; Noiri, Y.; Matsui, S.; Kuremoto, K.; Maezono, H.; Ishimoto, T.; Nakano, T.; Ebisu, S.; Hayashi, M. Development of a root canal treatment model in the rat. *Sci. Rep.* **2017**, *7*, 3315. [CrossRef]

127. Chicarelli, L.P.G.; Webber, M.B.F.; Amorim, J.P.A.; Rangel, A.L.C.A.; Camilotti, V.; Sinhoreti, M.A.C.; Mendonça, M.J. Effect of tricalcium silicate on direct pulp capping: Experimental study in rats. *Eur. J. Dent.* **2021**, *15*, 101–108. [CrossRef]

128. Dammaschke, T.; Stratmann, U.; Fischer, R.-J.; Sagheri, D.; Schäfer, E. A histologic investigation of direct pulp capping in rodents with dentin adhesives and calcium hydroxide. *Quintessence Int. Berl. Ger.* **2010**, *41*, e62–e71.

129. Simon, S.; Cooper, P.; Smith, A.; Picard, B.; Ifi, C.N.; Berdal, A. Evaluation of a new laboratory model for pulp healing: Preliminary study. *Int. Endod. J.* **2008**, *41*, 781–790. [CrossRef]

130. Almushayt, A.; Narayanan, K.; Zaki, A.E.; George, A. dentin matrix protein 1 induces cytodifferentiation of dental pulp stem cells into odontoblasts. *Gene Ther.* **2006**, *13*, 611–620. [CrossRef]

131. Fouad, A.F.; Walton, R.E.; Rittman, B.R. Healing of induced periapical lesions in ferret canines. *J. Endod.* **1993**, *19*, 123–129. [CrossRef]

132. He, T.; Friede, H.; Kiliaridis, S. Dental eruption and exfoliation chronology in the ferret (mustela putorius furo). *Arch. Oral Biol.* **2002**, *47*, 619–623. [CrossRef]

133. Alexander, A.; Torabinejad, M.; Vahdati, S.A.; Nosrat, A.; Verma, P.; Grandhi, A.; Shabahang, S. Regenerative endodontic treatment in immature noninfected ferret teeth using blood clot or synoss putty as scaffolds. *J. Endod.* **2020**, *46*, 209–215. [CrossRef]

134. Bucchi, C.; Gimeno-Sandig, Á.; Valdivia-Gandur, I.; Manzanares-Céspedes, C.; De Anta, J.M. A Regenerative endodontic approach in mature ferret teeth using rodent preameloblast-conditioned medium. *Vivo* **2019**, *33*, 1143–1150. [CrossRef] [PubMed]

135. Torabinejad, M.; Bakland, L.K. An animal model for the study of immunopathogenesis of periapical lesions. *J. Endod.* **1978**, *4*, 273–277. [CrossRef]

136. Altaii, M.; Cathro, P.; Broberg, M.; Richards, L. Endodontic regeneration and tooth revitalization in immature infected sheep teeth. *Int. Endod. J.* **2017**, *50*, 480–491. [CrossRef] [PubMed]

137. Altaii, M.; Broberg, M.; Cathro, P.; Richards, L. Standardisation of sheep model for endodontic regeneration/revitalisation research. *Arch. Oral Biol.* **2016**, *65*, 87–94. [CrossRef]

138. Danesh-Meyer, M.J.; Pack, A.R.C.; McMillan, M.D. A Comparison of 2 Polytetrafluoroethylene Membranes in Guided Tissue Regeneration in Sheep. *J. Periodontal. Res.* **1997**, *32*, 20–30. [CrossRef]

139. Schliephake, H.; Knebel, J.W.; Aufderheide, M.; Tauscher, M. Use of cultivated osteoprogenitor cells to increase bone formation in segmental mandibular defects: An experimental pilot study in sheep. *Int. J. Oral. Maxillofac. Surg.* **2001**, *30*, 531–537. [CrossRef]

140. Carlsson, H.-E.; Schapiro, S.J.; Farah, I.; Hau, J. Use of primates in research: A global overview. *Am. J. Primatol.* **2004**, *63*, 225–237. [CrossRef]

141. Percie du Sert, N.; Ahluwalia, A.; Alam, S.; Avey, M.T.; Baker, M.; Browne, W.J.; Clark, A.; Cuthill, I.C.; Dirnagl, U.; Emerson, M.; et al. Reporting animal research: Explanation and elaboration for the ARRIVE guidelines 2.0. *PLoS Biol.* **2020**, *18*, e3000411. [CrossRef]

 International Journal of *Molecular Sciences*

Review

Tissue Characteristics in Endodontic Regeneration: A Systematic Review

Sandra Minic [1], Sibylle Vital [1,2], Catherine Chaussain [1,3], Tchilalo Boukpessi [1,4] and Francesca Mangione [1,5,*]

[1] URP 2496 Laboratory of Orofacial Pathologies, Imaging and Biotherapies, Life Imaging Platform (PIV), Laboratoire d'excellence INFLAMEX, UFR Odontology, Université Paris Cité, 92120 Montrouge, France
[2] Louis Mourier Hospital, AP-HP, DMU ESPRIT, 92700 Colombes, France
[3] Bretonneau Hospital Dental Department and Reference Center for Rare Diseases of Calcium and Phosphorus Metabolism, AP-HP, 75018 Paris, France
[4] Pitié Salpétrière Hospital, DMU CHIR, AP-HP, 75013 Paris, France
[5] Henri Mondor Hospital, AP-HP, 94000 Créteil, France
* Correspondence: francesca.mangione@u-paris.fr

Abstract: The regenerative endodontic procedure (REP) represents a treatment option for immature necrotic teeth with a periapical lesion. Currently, this therapy has a wide field of pre-clinical and clinical applications, but no standardization exists regarding successful criteria. Thus, by analysis of animal and human studies, the aim of this systematic review was to highlight the main characteristics of the tissue generated by REP. A customized search of PubMed, EMBASE, Scopus, and Web of Science databases from January 2000 to January 2022 was conducted. Seventy-five human and forty-nine animal studies were selected. In humans, the evaluation criteria were clinical 2D and 3D radiographic examinations. Most of the studies identified a successful REP with an asymptomatic tooth, apical lesion healing, and increased root thickness and length. In animals, histological and radiological criteria were considered. Newly formed tissues in the canals were fibrous, cementum, or bone-like tissues along the dentine walls depending on the area of the root. REP assured tooth development and viability. However, further studies are needed to identify procedures to successfully reproduce the physiological structure and function of the dentin–pulp complex.

Keywords: regenerative endodontics; dentin-pulp complex regeneration; pulp injury; pulp necrosis; animal model

Citation: Minic, S.; Vital, S.; Chaussain, C.; Boukpessi, T.; Mangione, F. Tissue Characteristics in Endodontic Regeneration: A Systematic Review. *Int. J. Mol. Sci.* **2022**, *23*, 10534. https://doi.org/10.3390/ijms231810534

Academic Editors: Kerstin M. Galler and Matthias Widbiller

Received: 3 August 2022
Accepted: 7 September 2022
Published: 11 September 2022

Publisher's Note: MDPI stays neutral with regard to jurisdictional claims in published maps and institutional affiliations.

1. Introduction

Tissue regeneration in dentistry has found clinical application in everyday practice with the regeneration endodontic procedure (REP). This treatment, relying on the triad of stem cells, scaffolds, and growth factors in a sterile environment (Figure 1), has the purpose of ideally replacing damaged structures, such as dentin and root, as well as cells of the pulp–dentin complex, in addition to the resolution of possible apical periodontitis [1]. In the case of pulp necrosis in immature teeth due to caries or developmental abnormalities, as well as dental trauma, it is recommended by several endodontic and pediatric dentistry associations to implement this therapy [2–4]. In fact, despite an important variability in the resolution of the clinical table, root edification, and neurogenesis, it is recognized as a viable management technique for immature permanent teeth with necrotic pulp [2,3,5], especially as it overcomes the challenges of conventional root canal treatment in the presence of short roots, thin fracture-prone dentine walls, and wide apices [6–8].

Indeed, REP has been a research topic for decades now, both in animals and in patients, with a wide range of protocols, materials, and success parameters evaluation.

However, at the clinical stage, while high survival rates for REPs have been reported by numerous case reports, case series, and comparative clinical trials, the few existing systematic reviews highlight the weakness of these clinical trials [9,10]. Moreover, long-term follow-up prospective studies are necessary to better identify reliable success rates and outcomes of REP [11]. In fact, one of the major drawbacks is the level of accuracy in the assessment of regenerated tissues in REP-treated teeth [11], resulting in a lack of standardization regarding successful criteria.

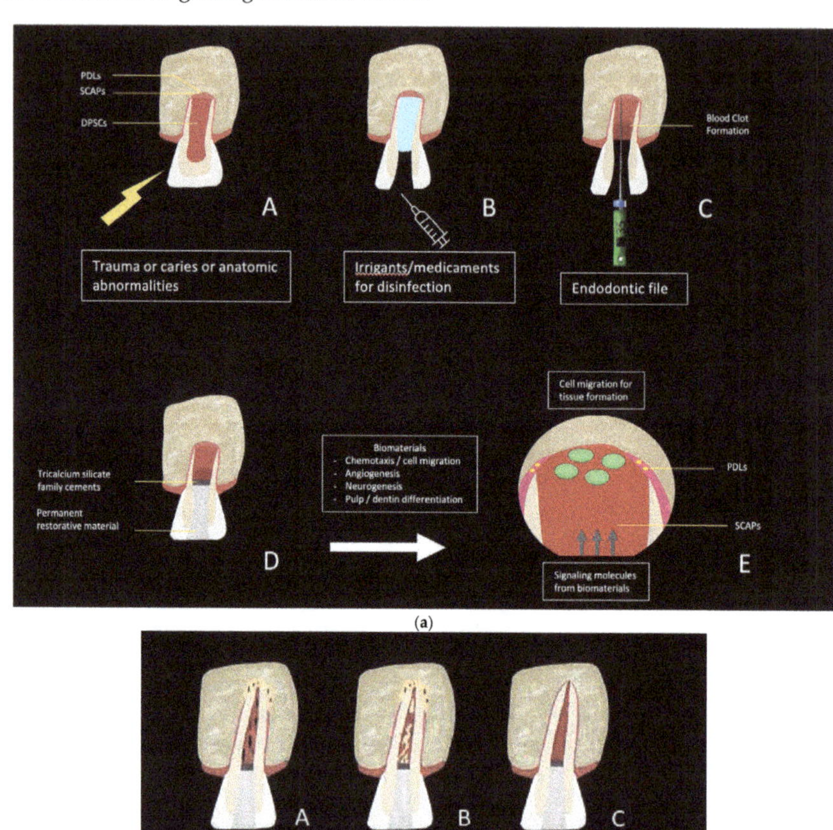

Figure 1. (**a**) Schema illustrating REP procedure. (**A**) Immature permanent incisor tooth exposed to trauma or caries. (**B**) Access cavity preparation and chemical debridement by using of irrigants and/or medicaments. (**C**) Bleeding induced by dental endodontic File to create a blood clot. (**D**) Restoration of the access cavity with a permanent restorative material covering the biomaterial in contact with blood clot or with collagen plug, PRF, PRP, etc. (**E**) Release of signaling molecules from biomaterial (growth factors, calcium silicate, depending on the material). They influence SCAPs (from apical papilla) and PDLs, including chemotaxis/cell migration, angiogenesis, neurogenesis, and differentiation into pulp/dentin complex. DPSCs (Dental Pulp Stem Cells); SCAPs (Stem Cells from Apical Papilla); PDLs (Periodontal Ligament cells) (Document of URP2496). (**b**) Different type of achievement of tissue regeneration determined by continued root development, increased dentinal wall thickness by cementum-like deposition, and apical closure. (**A**) Immature permanent incisor tooth after REP technique with "pulp-like" tissue with ligamentous, fibrous aspect and apical closure with mineralized tissue: Cement-like or bone-like tissue; (**B**) "pulp-like" tissue with ligamentous aspect and apical closure with mineralized tissue: Cement-like or bone-like tissue. Cementum island in the intracanal tissue; (**C**) the expectation of the pulp-like tissue with an increased root thickness and length, a decreasing apex diameter. Document of URP2496.

Thus, this systematic review aimed to highlight the main tissue characteristics related to the therapeutic success of REP in human and animal studies.

2. Methods

2.1. Search Strategy

The review process followed the Preferred Reporting Items for Systematic Reviews and Meta-Analyses (PRISMA) guidelines [12], and the protocol was registered in PROSPERO (International Prospective Register of Systematic Reviews) under the numbers CRD42022303001 for humans and CRD42022322610 for animals.

In order to highlight the main characteristics of dental tissues treated by REP, based on the PICO strategy, this study systematically searched the following databases: PubMed/Embase/Scopus/Web of Science. The results were limited to the English language with a date range of January 2010 to February 2022, with full text available. The following combination of keywords was used: (Revitalisation OR Revitalization OR regenerative Endodontics OR revascularization OR regenerative Endodontics procedure OR regenerative Endodontics therapeutics) AND (dental pulp).

2.2. Eligibility Criteria

Inclusion criteria were as follows: (1) All animal and human studies focusing on regenerative endodontic procedures; (2) orthotopic, semi-orthotopic, and ectopic procedures that attempted to revascularize or regenerate new pulp-like tissue; (3) English-language full text available; and (4) publication between 2010 and 2022.

The exclusion criteria were as follows: (1) In vitro studies, (2) ex vivo studies, (3) in silico studies, (4) studies on transgenic animals, (5) publications solely in a non-English language, and (6) reviews and meta-analyses. The titles and abstracts were screened by two independent reviewers (S.M. and F.M.) against eligibility criteria.

Full texts were analyzed whenever the abstract was not informative enough. A third reviewer (T.B.) was involved to resolve disagreements.

2.3. Data Extraction and Analysis

All selected articles were assigned depending on human or animal studies. Each article was subsequently classified according to (1) procedure, (2) follow-up, and (3) evaluation criteria.

2.4. Quality Analysis and Level of Evidence

The risk of bias for animal studies was evaluated using the Systematic Review Centre Laboratory animal Experimentation (SYRCLE) risk of bias tool with the following criteria: (1) Selection bias, (2) performance bias, (3) detection bias, (4) attrition, and (5) reporting bias. Studies were scored with a "yes" for low risk of bias, "no" for high risk of bias, and "?"for unclear risk of bias.

For human observational studies (cohort studies), the Newcastle–Ottawa Scale tool was used. Selections and outcomes were rated. Six binary responses contributed to an aggregate score corresponding to high risk (0–2 points), mild risk (3–4 points), and low risk of bias (5–6 points).

The risk of bias tool R.O.B 2.0 [13] was used to assess the quality of randomized clinical studies. Studies were scored with a "yes" for low risk of bias, "no" for high risk of bias, and "?" for some concerns of risk of bias.

ROBINS-I tools [14] were involved in assessing non-randomized clinical studies. By means of (1) randomization, (2) deviation from the intended intervention, (3) missing outcome data, (4) measurement of the outcome, and (5) selection of the reported results criteria, studies were scored with a "yes" for low risk of bias, "no" for high risk of bias, and "?" for medium risk of bias.

Finally, an adapted Newcastle–Ottawa Scale was used for case reports [15] with the following criteria: (1) Selection, (2) ascertainment, (3) causality, and (4) reporting. Studies were scored by eight binary responses and compiled into an aggregate score. The risk of bias was high (0–1 points), mild (2–3 points), or low (4–5 points).

The two reviewers analyzed all articles independently. Disagreement was resolved by discussion with TB and SV.

3. Results

3.1. Study Design and Characteristics of Included Studies

Details of the study selection process are outlined in Figure 2. The research retrieved 931 articles of which 774 were excluded at the title screening stage and 4 articles were excluded upon abstract screening. The full-text articles have been read in their entirety to see if they are relevant to our research. In total, 29 articles were excluded, and the reasons were as follows: 9 were not in vivo studies, 11 were on transgenic animals, and 9 studies focused on pulpotomy procedures.

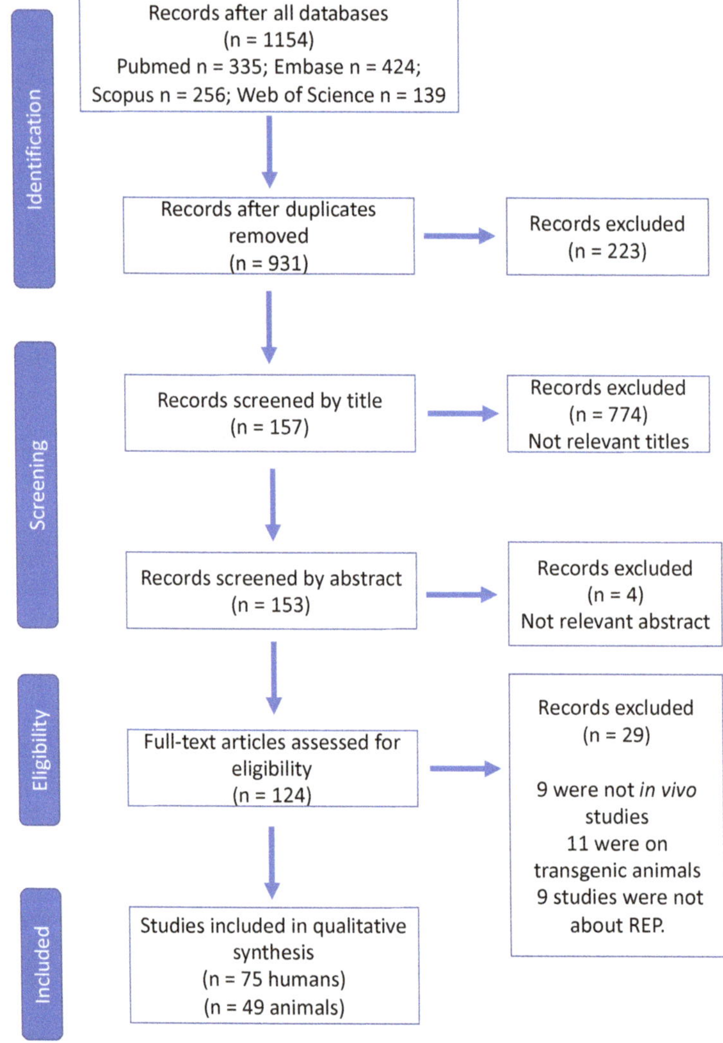

Figure 2. Flowchart of the article selection process.

Out of the 124 articles that met the inclusion criteria, 75 were clinical and 49 were animal studies. The results are summarized in Tables 1–3. All abbreviations are explained in abbreviation table. The details of each included study are provided in the Tables S1 and S2.

3.2. Animals Studies

In this review, 49 animal models are represented. Small animals including rats, mice, rabbits, and ferrets were used for ectopic [16–21] and orthotopic REP models [22–31]. Large animals such as dogs, pigs, and sheep are used for orthotopic REP models [32–64].

3.2.1. Ectopic REP

Procedure

Materials used for ectopic studies consisted of:

- Dentine slices or entire tooth roots:

    (1) With growth factors such as VEGF with or without cells as DPSCs [19,20].
    (2) With cells in the presence of a collagen scaffold and calcium silicate cement [21].
    (3) With Fibrin gel. [23].

- Polymers such as poly(lactic-co-glycolic acid) and rabbit DPSCs.

One study analyzed different disinfection agents in polyethylene tubes with triple antibiotic paste vs. calcium hydroxide paste [22] (Tables 1 and S1).

Follow-up

Studies using ectopic procedures described follow-up periods from 12 days to 3 months [19–24] (Tables 1 and S1).

Evaluation Criteria

Evaluation criteria were based on histological and immunohistochemical analyses. The goal was to analyze the type of tissue formed after implantation. Signs of inflammation (the presence of inflammatory cells) [19,22,24], signs of fibrosis [22,24], vascularization [20,21,24], and neurofilament [21] were evaluated. The nature of the new tissues that were synthetized inside the canal space was analyzed, and the presence or not of calcification was sought [21,23–25]. For IHC, different markers were used to characterize the pulp: DSPP or Nestin, vVW or CD34 for new vascularization, and PGP9,5 markers for nerve-like cell characterization (Tables 1 and S1).

Table 1. Ectopic REP techniques in animal studies.

Animal Models: Ectopic Procedure				
Assessment	**Main Results**	**Procedure**	**Follow-Up**	**Model**
Histology	Soft tissue formation	Root Tooth: VEGF + DPSCs + MTA Dentin slice: rBMSC + collagen scaffold + iRoot BP Human teeth roots + fibrin gel DPSCs + polymers scaffold	12 days–3 months	Mice [20] Rats [16,18] Rabbit [17]
	Presence of odontoblasts cells	Root Tooth: VEGF + DPSCs + MTA Dentin slice: rBMSC + collagen scaffold + iRoot BP Human teeth roots + fibrin gel DPSCs + polymers scaffold	12 days–3 months	Mice [20] Rats [16,18] Rabbit [17]
	Presence of Inflammation	Polyethylene tubes: TAP vs. CHP calcium VEGF-loaded fiber + root fragment + MTA	1.5–3 months	Mice [21] Rats [19]

Table 1. *Cont.*

Animal Models: Ectopic Procedure				
Assessment	Main Results	Procedure	Follow-Up	Model
Histology	Presence of Vessels	Polyethylene tubes: TAP vs. CHP calcium Dentin slice: rBMSC + collagen scaffold + iRoot BP Human teeth roots + fibrin gel DPSCs + polymers scaffold	12 days–3 months	Mice [20,21] Rats [16,18] Rabbit [17]
	Presence of Nerves	Dentin slice: rBMSC + collagen scaffold + iRoot BP Root Tooth: VEGF + DPSCs + MTA	2–3 months	Mice [20] Rats [16]
	Presence of mineralization	Polyethylene tubes: TAP vs. CHP calcium	3 months	Rats [19]

3.2.2. Orthotopic REP

Procedure

Both small and large animals were used for REP. This therapy relied on the triad of tissue engineering: Stem cells, scaffolds, and growth factors, in a sterile environment. For that purpose, different materials were used: The gold standard MTA [26–44], different types of hydrogels [21,23,26,29,43,45–49], PRF or PRP alone or with cement or a blood clot (29,36,38,45,50–55,) or natural products, such as propolis [31,34,56]. Autologous pulp or cells such as DPSC or buccal fat were also used [20,21,32,50,51,57–59]. BMSCs with LPS vesicles or peptide angiogenic or dentinogenic or amelogenic [30,60] were also found. Growth factors such as VEGF were the most used [19,20,61]. (Tables 2 and S1)

Follow-up

For small animal models, follow-up from 3 weeks to 12 weeks was achieved [22–31,65]. Regarding large animal models, the follow-up ranged from 1 to 28 weeks [28,32–59,61–64,66]. (Tables 2 and S1).

Evaluation Criteria

To determine the success criteria, different techniques were used. Histology and radiology were dependent on the animal model used. Histological criteria consisted of periapical inflammation [34,36,37,39,42,46,62], inflammatory cell infiltration [32,34,35,37–39,43,48,56,61,66], the presence of pulp-like/vital tissue [33–37,40–46,48,49,51–53,55–57,62–64,66], the new formation of mineralized tissue [32,34,36–39,41–45,47–54,56,57,59,61–64,66], closure of the apex [22,24,25,34,36,38,41,42,44,47–49,51,52,57,62,66], the presence of odontoblastic palisade [33,35,36,42,45,46,54], and the presence of blood vessels [35,37,42,43,47,52,55,56,64], nerve fibers [35,42], or resorptions.

The criteria for the identification of different types of mineralized tissue were:

- Dentin: Presence/absence of dentinal tubules.
- Cementum: Absence of dentinal tubules and adherence onto dentin, and the presence of cementocyte-like cells.
- Bone: Presence of Haversian canals with uniformly distributed osteocyte-like cells.
- PDL: Presence of Sharpey's fibers and fibers bridging cementum and bone.

For deeper characterization, IHC allowed specific protein localization. Specific markers such as DMP4, DLX1, GLI2, CEMP, CAP, NF, CD31, DSPP, SOX2, CGRP, and peripherin were identified in pigs, S100+ for neurofilaments, and DSSP, tenascin C, laminin, and fibronectin in dogs.

Bidimensional and/or tridimensional radiographies were performed to assess regeneration. The following radiological criteria were considered: Periapical radiolucency, root resorption, root thickening, root lengthening, and apex closure [22,23,26,27,31,32,34,41–43,46,49,51,57,59,63,64,66].

Based on histology and radiology, a scoring system was created to characterize the tissue and define the success of REP [23,37,38,48–50,52,53,61] (Tables 2 and S1).

Others

qPCR was used to quantify the gene expression of different genes (DSPP, Col1A1, DMP1, and ALP) in order to determine whether tissue regeneration was triggered or not during revitalization [38]. (Tables 2 and S1).

Table 2. Orthotopic REP techniques in animal studies.

Animal Models: Orthotopic REP Procedure				
Assessment	Main Results	Procedure	Follow-Up	Model
Histology	Presence of pulp-like / vital tissue	Gelatin and fibrin- based matrix BC alone + MTA PRP or PRF with cement and BC DPSCs and Buccal fat with BC and MTA. Nanosphere w/o BMSCs	3 months 3–7 months 3 months 1–2 months	Mini pig [34] Dogs [32,35–57,61–63] Ferrets [23–26] Rats [28–31]
	New formation of mineralized tissue	Gelatin and fibrine-based matrix BC alone + MTA PRP or PRF with cement and BC DPSCs and Buccal fat with BC and MTA.	3 months 3–7 months 3 months 3 months	Mini pig [34] Dogs [32,34,36–39,41–45,47–54,56,57,59,61–64,66] Sheep [64] Ferrets [23–26]
	Presence of odontoblastic palisade	Gelatin and fibrine-based matrix Autologous stem cells BC alone + MTA PRP or PRF with cement and BC DPSCs and Buccal fat with BC and MTA.	3–6 months 3–6 months	Mini pig [33,67] Dogs [33,35,36,42,45,46,54]
	Inflammatory cell infiltration	Gelatin and fibrine-based matrix Autologous stem cells BC alone + MTA PRP or PRF with cement and BC DPSCs and Buccal fat with BC and MTA. TAP + silver amalgam	3 months 3–6 months 3 months 1.5 months	Mini pig [34] Dogs [32,35,37–39,43,48,56,61,66] Ferrets [22,25] Rats [27]
	Presence of blood vessels	Gelatin and fibrine-based matrix Autologous stem cells BC alone + MTA PRP or PRF with cement and BC DPSCs and Buccal fat with BC and MTA.	3–6 months 3–7 months 3 months 1–1.5 months	Mini pig [33,34] Dogs [35,37,42,43,47,52,55,56,64] Ferrets [24] Rats [28,29]
	Presence of nerve fibers	SLan angiogenictarget peptide vs. SLed dentinogenic control peptide Autologous pulp + BC + MTA	3 months	Dogs [35,42]
	Presence of resorption	Gelatin and fibrine-based matrix Collagen sponge vs. PRF vs. MTA	3 months	Mini pig [34] Dogs [39]
	No intraradicular mineralized tissue deposition	Gelatin and fibrine-based matrix	3 months	Mini pig [34]
	Root maturation	BC + MTA	3 months	Sheep [64]
	Apex maturation	Gelatin and fibrine-based matrix Autologous stem cells BC alone + MTA PRP or PRF with cement and BC DPSCs and Buccal fat with BC and MTA.	3 months	Mini pig [34] Dogs [34,36,38,41,42,44,47–49,51,52,57,62,66] Sheep [64] Ferrets [22,24,25]
	Cementum cells/ tissue	Gelatin and fibrine-based matrix BC + Gelfoam BC + PRP BC + MTA	3–7 months	Mini pig [34] Dogs [41,48,50,52,56] Ferrets [24] Rats [28]
	Dentin tissue	BC + Gelfoam BC + PRP Propolis vs. MTA Autologous stem cells	1–3 months	Dogs [41,42,52] Rats [31]
	Osteodentin	(Buccal fat) vs. (BC + Buccal fat) + MTA BC + PRP BC + MTA	3–6 months	Dogs [38] Ferrets [26] Rats [31]
	Bone tissue	Autologous stem cells BC alone + MTA PRP or PRF with cement and BC DPSCs and Buccal fat with BC and MTA.	6 months	Dogs [39,48–50,52,53,59,60,63]
	Mineralized tissue deposition	Autologous stem cells BC alone + MTA PRP or PRF with cement and BC DPSCs and Buccal fat with BC and MTA.	3–6 months	Dogs [32,41–43,45,49,57,58] Ferrets [23,26]

Table 2. *Cont.*

Animal Models: Orthotopic REP Procedure				
Assessment	Main Results	Procedure	Follow-Up	Model
Radiology	Presence of pulp-like / vital tissue	Gelatin and fibrine-based matrix BC alone + MTA PRP or PRF with cement and BC DPSCs and Buccal fat with BC and MTA. Nanosphere w/o BMSCs	3 months 3–7 months 3 months 1–2 months	Mini pig [34] Dogs [32,35–57,61–63] Ferrets [23–26] Rats [28–31]
	Apex closure	Autologous stem cells BC alone + MTA PRP or PRF with cement and BC DPSCs and Buccal fat with BC and MTA. TAP + silver amalgam	3–6 months	Dogs [32,41–43,46,49,51,57–59,66] Sheep [64] Ferrets [23,26] Rats [27]
	Increase root length	Autologous stem cells BC alone + MTA PRP or PRF with cement and BC DPSCs and Buccal fat with BC and MTA. TAP + silver amalgam	3–6 months	Dogs [32,41,42,46,49,51,57–59,66] Sheep [64] Ferrets [23,26] Rats [27]
	Increase dentin thickness	Autologous stem cells BC alone + MTA PRP or PRF with cement and BC DPSCs and Buccal fat with BC and MTA. TAP + silver amalgam	3–6 months	Dogs [32,41,42,46,49,51,57–59,66] Sheep [64] Ferrets [23,26] Rats [27]
	Periapical healing	Autologous stem cells BC alone + MTA PRP or PRF with cement and BC DPSCs and Buccal fat with BC and MTA. TAP + silver amalgam	3–6 months	Dogs [32,42,46,57–59] Sheep [64] Ferrets [23,26] Rats [27]
qPCR	DSPP, COL1A1, ALP, DMP1 expression	(Buccal fat) vs. (BC + Buccal fat) + MTA	3 months	Dogs [38]

3.3. Human Studies

3.3.1. REP protocol

All human studies consisted of randomized, non-randomized, case reports, and retrospective studies. The evaluation criteria were clinical examination, 2D and 3D radiography, and/or MRI.

Many protocols have been tested to achieve pulp-like tissue regeneration. Regarding the cement used, among all included studies, MTA was the most frequently used cement [7,8,60,67–113]. Seven studies were performed with Biodentine [75,85,114–118], one study used a Calcium-Enriched Mixture [119], two studies used Synoss Putty [120,121], one used calcium hydroxide [6,90], and five used Glass Ionomer Cement [79,112,122–124] to create a mineralized bridge to close the pulp chamber. Different types of materials were used to regenerate pulp tissue. Most of the studies only used blood clots as a scaffold. Some used other supplemental scaffolding components such as collagen [72,74–76,81,83,88,90,93,115,117,124,125], PRF or iPRF (+ MTA [8,87,91,126]; + Biodentine [116–118] + BC [87,91] + Portland cement [124]); and PRP [7,65,69,71,89,91,95,105,122,126] in addition to the calcium silicate cements. Blood clots have, at times, been used alone as a control [65,91,121,122,127] in comparative studies. Two studies have used collagen scaffolds with two types of cells, such as MDSPCs or umbilical cells MSCs [115,123].

Four studies have investigated the best canal disinfection technique in use and identified bi-antibiotics and triple-antibiotics of calcium hydroxide [128–131]. However, it is also important to identify whether the procedure of REP is more successful when performed in one visit or two [132] (Tables 3 and S1).

3.3.2. Follow-Up

According to the studies, different follow-up timepoints were reported. For retrospective studies, follow-up periods of 1 month to 8.2 years were observed [73–85,99,126,132–134]; randomized studies employed follow-up periods of 21 days to 18 months [6,7,65,69–71,85–91,114,115,121,126,128,129,134]; non-randomized studies used periods of 2 weeks to 36 months [66,116,121,123,124,132,135]; and cases series employed periods of 1 week to 6 years [8,67,92–97,99–113,117–120,127,130,136] (Tables 3 and S1).

3.3.3. Clinical and Radiographic Evaluation of REP

For clinical examination, teeth could be symptomatic or not with different sensitivity tests (thermal or electric), percussion, pain on percussion, and palpation. Dyschromia, swelling, and tenderness of the surrounding tissues were also assessed. The mobility of teeth, as well as pocket probing, were evaluated.

Radiographically, the teeth should have shown the resolution of the apical lesion with PAI scoring, apical closure, root length growth, thickening of the root walls, and the formation of calcific barrier. On tridimensional radiographic observation, the lesion size, bone density, root length, and pulp area were evaluated. Rarely, MRI was used in order to identify more organized tissue in the canal, dentin deposits, or mineralization (Tables 3 and S1).

3.3.4. Others

Other techniques such as qPCR were used to quantify how many bacteria were present on the canal dentin walls after different interappointment medications [128,129] and identify stem cells in the canal [135]. Histology was used for analysis in cases of a crown fracture [107,127] or tooth extraction as a result of orthodontic reasons [106,107,121,136] (Tables 3 and S1).

Table 3. Human studies of REP.

Human Model: Regenerative Endodontic Procedure				
Assessment	**Main Results**	**Procedure**	**Follow-Up**	**Articles**
Clinical tests	Asymptomatic teeth	BC + Biodentine or MTA BC +PRF + MTA or Biodentine or GIC BC + PRP + MTA or Biodentine or GIC BC + Collagen + MTA or BC+ UC-MSCs + collagen + MTA BC + PRF + Collagen + Biodentine or Portland mDPSCs + G-CSF + Collagen + MTA Medication on different Appointment TAP vs. CaOH$_2$ vs. formocresol Bi antibiotic + GIC	21 days–79 months	[61,66,68–72,86,87,90,91,94–97,101–104,106,107,110,111,113,116–126,128,130–132]
	PAI	BC + MTA Sealbio vs. obturation	12–24 months	[80,98,134]
	Dyschromia	BC + collaplug MTA vs. Biodentine vs. GIC BC + MTA vs. Biodentine BC + PRF vs. PRP + MTA Bi-antibiotic paste + BC + GIC	12–96 months	[75,79,85,86,98,126,130]
	Mobility	BC + Synoss Putty	72 months	[120]
Radiographic observation	Apical lesion	BC + hydrogel with FGF+ MTA BC + DPSC In hydrogel + MTA or GIC BC + MTA BC + PRF + Biodentine BC + PRP + MTA BC + PRF vs. PRP + Collagen + GIC BC + Synoss putty + MTA BC + Collagen + Portland + MTA BC + LPRF + Portland cement	21 days–72 months	[8,66,70,71,79–83,90,92,93,95,97,99,100,103,105,107,112,113,116,118,120,122,124,125,128,133]
	Root length	BC + hydrogel with FGF+ MTA BC + DPSC In hydrogel + MTA or GIC BC + MTA BC + PRF + Biodentine BC + PRP + MTA BC + PRF vs. PRP + Collagen + GIC BC + Synoss putty + MTA BC + Collagen + Portland + MTA BC + LPRF + Portland cement	21 days–78 months	[66,67,69,73,80,82,83,87,90,91,94,96,97,99–102,104,105,108,110,112,119,122,124,127,128,130,132]

Table 3. *Cont.*

Human Model:
Regenerative Endodontic Procedure

Assessment	Main Results	Procedure	Follow-Up	Articles
Radiographic observation	Root thickness	BC + hydrogel with FGF+ MTA BC + DPSC In hydrogel + MTA or GIC BC + MTA BC + PRF + Biodentine BC + PRP + MTA BC + PRF vs. PRP + Collagen + GIC BC + Synoss putty + MTA BC + Collagen + Portland + MTA BC + LPRF + Portland cement	21 days–60 months	[65–67,71,80,83,87,90,91,96,97,99–102,104,105,108,111,112,118,119,122,124,127,128,130,132]
	Apical closure	BC + hydrogel with FGF+ MTA BC + DPSC In hydrogel + MTA or GIC BC + Collagen + coltosol BC + MTA BC + PRF + Biodentine BC + PRP + MTA BC + PRF vs. PRP + Collagen + GIC BC + Synoss putty + MTA BC + Collagen + Portland + MTA BC + LPRF + Portland cement	21 days–78 months	[8,67,69–74,76–78,84,88,92,97–100,102,104,105,110–112,118–120,122,124,128,132]
	Radiolucy	BC + PRF + MTA BC + MTA or CEM	6 months–78 months	[8,73,97,102,103,107–110,119]
	Bone density	BC + hydrogel with FGF+ MTA BC + DPSC in hydrogel + MTA or GIC BC + MTA BC + PRP + MTA BC + Synoss putty + MTA	24–70 months	[70,96,97,109,111,112,120]
	Resorption	BC + PRF + Collagen + Biodentine Medication on different appointment	21 days–30 months	[117,128,133]
	Calcification in the pulp	BC + MTA or CEM BC + Collagen + MTA BC + PRP + MTA BC + iPRF + Biodentine BC + Amelogen Plus	12–60 months	[66,73,78,85,88,93,96,99,110,118,119,127,133]
	Ligament repair	BC + PRF or PRP + MTA Vs BC + MTA BC + PRP + MTA BC + PRF + Collagen + Biodentine	50 months	[91,95,117]
qPCR	Quantify bacteria	Different appointment medication TAP vs. calcium hydroxide medication	21 days–19 months	[128,129]
	Cells identification in the canal	Intracanal blood sample after BC	1 month	[135]
Histology	Regenerate tissue observation	BC + MTA BC + Synoss Putty BC + Amelogen Plus BC + Collagen / MTA	7.5–36 months	[106,107,121,127,136]

3.4. Risk of Bias

For animal studies, a low risk of selection bias (baseline characteristics) was found. In all studies, the performance bias was unclear, because no information about random housing was given [16–24,26–64]. Random outcome assessment was scored as low risk for 52% of the studies and unclear for the rest of them. In none of the studies was blinding described, and the risk was rated as unclear. Low risk of attrition and reporting bias was estimated for all studies (Figures 3 and 4).

For randomized clinical studies, randomization was scored as low risk for 65% and medium for the rest. Deviation from the intended intervention was scored as low risk for 55% and the rest was medium. A low risk of missing outcome data was estimated for all studies. Measurement of the outcome was scored as low risk for 25%, while 70% showed medium risk and 5% showed high risk. Moreover, 95% of the studies present a low risk for selection of the reported results. In addition, 40% of studies presented a low risk of bias, 55% medium risk and 5% presented a high risk of bias (Figures 5 and 6).

For human non-randomized studies, all studies presented a low risk for confounding, classification of the intervention, deviation from the intended intention, missing data, and selection of the reported results. However, for the selection of participants, medium risk was noted for all of them. (Figures 7 and 8)

For case reports studies, all the studies presented a low risk of bias (Table 4). In addition, for observational studies, 17 studies presented a low risk, but one study present a mild risk of bias (Table 5).

	SELECTION BIAS *Baseline characteristics*	PERFORMANCE BIAS *Random housing*	DETECTION BIAS *Random outcome assessment*	*Blinding*	ATTRITION BIAS *Incomplete outcome data*	REPORTING BIAS *Selective outcome reporting*
Wen et al., 2020	+	?	?	?	+	+
El-Backly et al., 2008	+	?	?	?	+	+
Ruangsawasdi et al., 2014	+	?	?	?	+	+
Gomes-Filho et al., 2012	+	?	?	?	+	+
Li et al., 2016	+	?	+	?	+	+
Yadlapati et al., 2017	+	?	?	?	+	+
Alexander et al., 2020	+	?	?	?	+	+
Torabinejad et al., 2018	+	?	+	?	+	+
Torabinejad et al., 2015	+	?	+	?	+	+
Torabinejad et al., 2014	+	?	+	?	+	+
Verma et al., 2017	+	?	+	?	+	+
Scarparo et al., 2011	+	?	+	?	+	+
Moreira et al., 2017	+	?	+	?	+	+
Nabeshima et al., 2018	+	?	?	?	+	+
Chen et al., 2021	+	?	?	?	+	+

Figure 3. *Cont.*

Figure 3. *Cont.*

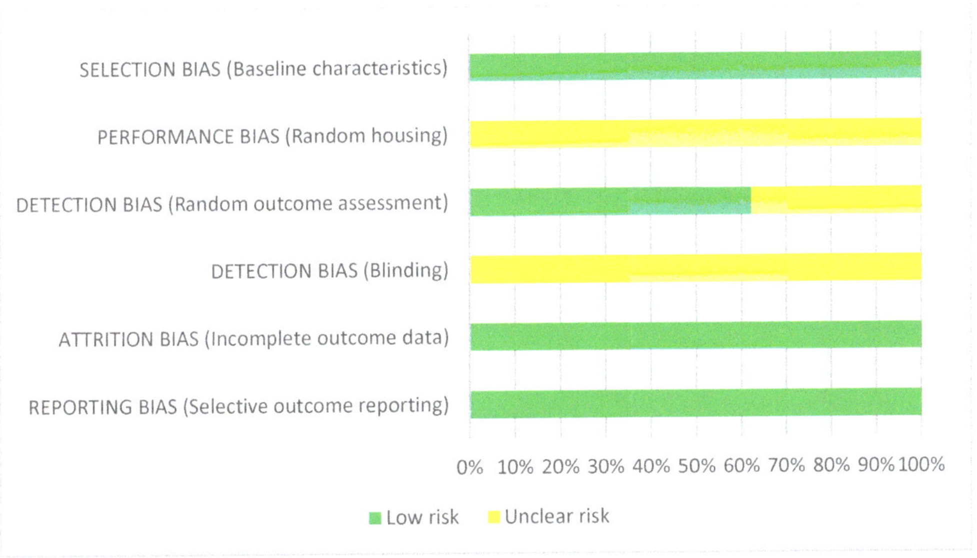

Figure 3. Risk of bias assessment of REP in animal studies according to the Systematic Review Centre for Laboratory Animal Experimentation (SYRCLE): Authors' judgment about each risk of bias item (green = low, yellow = unclear) [16–34,37–59,61–65].

Figure 4. Risk of bias assessment of REP in animal studies according to the Systematic Review Centre for Laboratory Animal Experimentation (SYRCLE) (green = low risk, yellow = unclear risk).

Study	Randomization	Deviation from intended intervention	Missing outcome data	Measurement of the outcome	Selection of the reported results	Overall risk of bias
Sallam et al. 2020	+	?	+	?	+	?
ElSheshtawy et al. 2020	+	?	+	?	+	?
Bezgin et al. 2015	?	?	+	?	+	?
Nagy et al. 2014	+	+	+	+	+	+
Alagl et al 2017	?	?	+	+	+	?
Aly et al. 2019	+	+	+	+	+	+
Arslan et al. 2019	+	+	+	+	+	+
Ragab et al. 2019	+	+	+	+	+	+
Jiang et al. 2017	?	+	+	+	+	?
Rizk et al. 2019	+	+	+	+	+	+
Botero et al. 2017	+	+	+	+	+	+
Shivashankar et al. 2017	+	?	+	?	+	?
El-Kateb et al. 2020	+	+	+	?	?	?
Brizuela et al. 2020	+	+	+	−	+	−
Narang et al., 2015	?	?	+	+	+	?
Rizk et al. 2020	+	+	+	+	+	+
De-Jesus-Soares et al.2020	?	?	+	+	+	?
Nagata et al. 2014	?	+	+	+	+	?
Jha et al. 2019	?	?	+	+	+	?

Figure 5. Risk of bias assessment evaluated according to the R.O.B 2.0. Authors' judgment about the following items: Randomization, deviation from intended intervention, missing outcome data, measurement of the outcome, selection of the reported results, and overall risk of bias (green = low, yellow = unclear, red = high). [6,7,69,70,84–90,114,115,123,126,128,129,135].

Figure 6. Risk of bias assessment evaluated according to the R.O.B 2.0 tool. Randomization, deviation from intended intervention, missing outcome data, measurement of the outcome, selection of the reported results, and overall risk of bias (green = low, yellow = moderate, red = high).

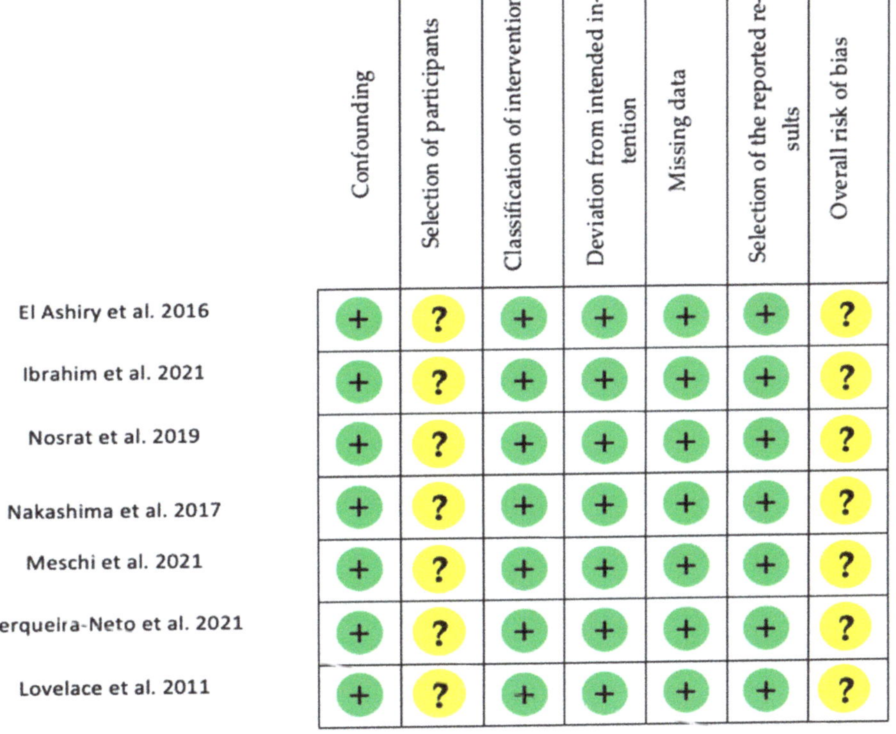

	Confounding	Selection of participants	Classification of intervention	Deviation from intended intention	Missing data	Selection of the reported results	Overall risk of bias
El Ashiry et al. 2016	+	?	+	+	+	+	?
Ibrahim et al. 2021	+	?	+	+	+	+	?
Nosrat et al. 2019	+	?	+	+	+	+	?
Nakashima et al. 2017	+	?	+	+	+	+	?
Meschi et al. 2021	+	?	+	+	+	+	?
Cerqueira-Neto et al. 2021	+	?	+	+	+	+	?
Lovelace et al. 2011	+	?	+	+	+	+	?

Figure 7. Risk of bias summary: Authors' judgement about each risk of bias item for each included non-randomized study (Robins I tool). [60,116,121,123,124,132,135].

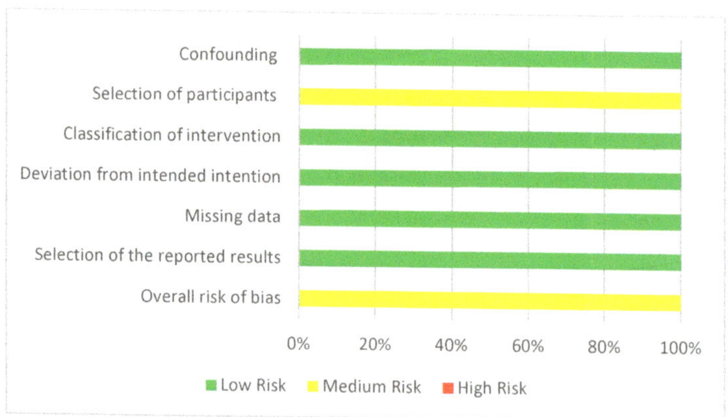

Figure 8. Risk of bias assessment evaluated according to the ROBINS I tool. Confounding, selection of participants, classification of intervention, deviation from intended intervention, missing data, selection of the reported results, overall risk of bias (green = low, yellow = moderate, red = high).

Table 4. Risk of bias assessment of case reports according to an adapted Newcastle–Ottawa Scale with the following criteria: Selection, ascertainment, causality, and reporting.

Author/Year	Selection	Ascertainment	Causality	Reporting	Results	Finality
Yoshpe et al., 2021 [8]	1	1	1	1	1	5 Low
Jiang et al., 2020 [93]	1	1	1	1	1	5 Low
Sabeti et al., 2021 [94]	1	1	1	1	1	5 Low
Gaviño et al., 2017 [95]	1	1	1	1	1	5 Low
Terauchi et al., 2021 [96]	1	1	1	1	1	5 Low
Jung et al., 2008 [97]	1	1	1	1	1	5 Low
McTigue et al., 2013 [67]	1	1	1	1	1	5 Low
Li et al., 2017 [99]	1	1	1	1	1	5 Low
Saoud et al., 2014 [100]	1	1	1	1	1	5 Low
Dabbagh et al., 2012 [101]	1	1	1	1	1	5 Low
Dudeja et al., 2015 [102]	1	1	1	1	1	5 Low
Ulusoy et al., 2017 [103]	1	1	1	1	1	5 Low
Cehreli et al., 2011 [104]	1	1	1	1	1	5 Low
Sachdeva et al., 2015 [105]	1	1	1	1	1	5 Low
Lin et al., 2014 [106]	1	1	1	1	1	5 Low
Becerra et al., 2014 [107]	1	1	1	1	1	5 Low
Chen et al., 2013 [108]	1	1	1	1	1	5 Low
Chang et al., 2013 [109]	1	1	1	1	1	5 Low
Lenzi et al., 2012 [110]	1	1	1	1	1	5 Low
Shin et al., 2009 [111]	1	1	1	1	1	5 Low
Shiehzadeh et al., 2014 [112]	1	1	1	1	1	5 Low
Plascencia et al., 2016 [113]	1	1	1	1	1	5 Low
Yoshpe et al., 2020 [117]	1	1	1	1	1	5 Low
Bakhtian et al., 2017 [118]	1	1	1	1	1	5 Low
Mehrvarzfar et al., 2017 [119]	1	1	1	1	1	5 Low
Cymerman et al., 2020 [120]	1	1	1	1	1	5 Low
Shimizu et al., 2013 [127]	1	1	1	1	1	5 Low
Nazzal et al., 2018 [130]	1	1	1	1	1	5 Low
Meschi et al., 2016 [136]	1	1	1	1	1	5 Low

Table 5. Risk of bias assessment of observational studies (cohort studies) according to the Newcastle–Ottawa Scale with the following criteria: Selection and outcome.

Author/Year	Selection				Outcome		Results	Finality
Meschi et al., 2018 [72]	1	1	1	1	1	1	6	Low
Elfrink et al., 2021 [73]	1	1	1	1	1	1	6	Low
Pereira et al., 2020 [74]	1	1	1	1	1	1	6	Low
Chrepa et al., 2020 [75]	1	1	1	1	1	1	6	Low
Mittman et al., 2020 [76]	1	1	1	1	1	1	6	Low
Linsuwanont et al., 2017 [77]	1	1	1	1	1	1	6	Low
Estefan et al., 2016 [78]	1	1	1	1	1	1	6	Low
Peng et al., 2017 [79]	1	1	1	1	1	1	6	Low
Chen et al., 2016 [80]	1	1	1	1	1	1	6	Low
Jeeruphan et al., 2012 [81]	1	1	1	1	1	1	6	Low
Bukhari et al., 2016 [82]	1	1	1	1	1	1	6	Low
Chan et al., 2017 [83]	1	1	1	1	1	1	6	Low
Song et al., 2017 [84]	1	1	1	1	1	1	6	Low
Zizka et al., 2021 [98]	1	1	1	1	1	1	6	Low
Meschi et al., 2019 [125]	1	1	1	1	1	1	6	Low
Bose et al., 2009 [131]	1	1	1	1	1	1	6	Low
Shah et al., 2012 [137]	1	1	1	1	1	1	6	Low
Sutam et al., 2018 [133]	1	1	1	1	0	0	4	Mild

4. Discussion

4.1. Success Criteria Assessment of REP

REP has been the subject of numerous studies, both in animals and humans. Its objective is to regenerate intra-canal dentin-pulp tissue that is able to promote root growth in terms of length and thickness and apical closure, while restoring the sensitivity of the tooth and leading to periapical tissue healing [23,138–140]. To achieve all these objectives, the ideal protocol has not yet been defined, and there is no clear consensus regarding the scaffolds, disinfection methods, and sealing materials that emerged from the different studies. As research continues, the need for precise, repetitive parameters for assessing success is of prime importance. This systematic review sought to elucidate the different criteria of evaluation in both animals and humans, associated with the risk of bias in the studies.

4.2. Ectopic Model

Our research identified that several rodent ectopic models of REP, involving subcutaneous implantation of DPSCs and/or growth factors such as VEGF into the tooth slice or tooth root, have been developed to assess the biocompatibility and regenerative potential of biomaterials with a follow-up time of 12 days to 3 months [16–21].

The histological criteria for success in these models are the regeneration of pulp-like tissue with the presence of the odontoblastic palisade and neo-vessels along the root or the tooth edge [16,17,20]. Regenerated tissue should fill the entire root space. Often, calcifications can be detected if cement tissue is present in the apical region or if pulp-like tissue is formed in the middle third. Furthermore, the presence of macrophages along the vessels to control the inflammation was assessed [77].

In this review, pulp-like tissue was obtained in almost all ectopic studies. The observations made in histological sections were essentially based on the presence of vascularization at different levels of the root or tooth slice. The presence of DSPP labelling was also found very often, which can confirm the presence of odontoblast-like cells. In a rabbit study, there was the presence of osteodentine-like tissue. Some authors showed that the formation of an atubular fibrodentine or osteodentine matrix is a precursor to the formation of a more organized tubular matrix [141–143].

Obviously, ectopic models represent a primary step in research focusing on REP. On the other hand, thee subcutaneous area sensibly differs from the oral cavity and periapical region mostly in terms of blood supply. Indeed, the vascularization of the subcutaneous tissue is very different from the vascularization of the normal pulp in the intra-canal area [29]. It has also been demonstrated that the transplantation of long dental fragments with a relatively closed apex (less than one millimeter) leads to inconsistent results due to the difficulty for the vascular and nervous system of the mouse to reach the interior part of the dental fragment [20,144,145]. Moreover, ectopic studies are also not very reliable due to the mixed cell population, with both animal and human regenerated tissue obtained. In addition to all these drawbacks, the technique used for REP is very different from those performed on humans, because the authors strongly suggest that, after the results are obtained by the ectopic models, tests on an orthoscopic model be performed [146].

4.3. Animal Models

Regarding small-animal models of REP, only rats and ferrets were involved. The procedure was performed on the incisors, which are continuously growing teeth, while in ferrets, REP was performed on canine teeth. The limited size of small animals and the substantial anatomical differences can impact the complete removal of the pulp and, therefore, the regeneration process [26].

Excluding studies on sheep and mini-pigs using mono-rooted or continuously growing teeth, the molar-premolar were the most used teeth. Multirooted teeth are good models allowing for the reproduction of most common clinical tables, such as deep caries inducing pulp necrosis [147].

The most used biomaterials were trisilicate cement, such as MTA in association with hydrogels seeded or not with DPSC, PRF or PRP, or a blood clot.

Indeed, DPSCs are the precursors of odontoblast-like cells [148] and, as previously demonstrated, are capable of regenerating the pulp–dentin complex in vivo [143]. These cells are often of human origin, harvested from third molars or teeth extracted for orthodontic reasons. Moreover, their association with hydrogel allows one to keep the cells in a matrix that will disintegrate with time [35,97]. PRP and PRF are widely used in regenerative dentistry, since they are known to optimize healing pathways by stimulating the scar stem cells present in the injured area.

The use of a blood clot allows the formation of even more entanglement for optimal regeneration. Thus, the recovery and regeneration of the pulp structure and function of the pulp tissues were achieved. In addition, the follow-up was long enough to observe a complete formation of the tissue and an increase in the root walls for small animals as well as large animals.

However, these procedures requiring cell cultures or blood derivates are hardly reproducible in common clinical practice [147,149].

The histological success criteria in animals were the resolution of the apical lesion, the presence of vital pulp-like tissue, the formation of mineralized neo-tissue, the closure of the apex with the possibility of newly formed blood vessels, and the presence of nerve fibers.

It can be observed that histological studies were almost systematically performed by distinguishing three regions along the root and characterizing the tissue.

The coronal third was very often at an early stage of hard tissue development on the canal walls. It is assumed that the migrating stem cells differentiate into cementoblast-like cells and deposit a matrix of collagen fibers. These fibers calcify, forming cementum islands. The further down the canal, the more these islands will fuse and form a thin layer of cementum in the second region, which was more medial. At the apical level of this region, which is closer to the apical region, a thicker, acellular matrix, such as immature acellular cementum, was found. The pulp cavity was filled with loose fibrovascular tissue. Finally, in the third region, there was mature hard tissue covering the canal walls and loose connective tissue covering the pulp canal. The hard tissue is also cementum.

Radiologically, the same characteristics were observed: Periapical lesion resolution, an increase in root length and the thickness of the root length, and a decrease in the apical diameter. In all cases, REP allowed for the healing of the periapical lesion, to decrease the diameter of the apex and the lengthening of the roots.

Inflammation is essential for tissue regeneration. The goal is to have sufficient but controlled inflammation [150]. In many animal studies, the presence of persistent periapical inflammation or intracanal and periapical inflammation may be due to remnants of intracanal medicaments or attempted healing in the newly formed tissue [24,25,50,61]. Closure of the apex is a sign of success that assures that root building is somehow complete [2,23,27,32,41–43,46,49,51,57–59,66]. However, the degree of mineralization must also be assessed, since dentin-type mineralization is not systematically found in small and large animals. Most of the time, a bone-like or cementum-like mineralized extension at the apex is found. Cement islands are also very often found in the newformed tissue. Despite the use of DPSC, there is no formation of pulp tissue with an intact cell layer similar to an odontoblast. However, in some samples, connective tissue could be observed inside the root canal with cells. The presence of ligament-like newformed tissue is also noticed. Based on large and small animal studies, stem/progenitor cells introduced into the root canal spaces of immature teeth with necrotic pulp after revascularization procedures appear to be able to differentiate into cementoblast- or osteoblast-like cells rather than odontoblasts. These stem/progenitor cells are likely derived from the periodontal ligament or periapical alveolar bone marrow, because the newly formed tissues in the canals of revascularized teeth are cementum or bone-like tissue [50,57,107,149].

Some studies have found pulp-like tissue, but it was often due to the previous presence of healthy pulp tissue [23]. This does not represent the clinical reality where REP treatment is performed in cases of necrotic pulp.

To summarize, in small animal ectopic model studies, regenerated tissues are very similar to the dentine pulp.

In small animals, REP is performed on healthy pulp. The fact that the tooth is small, resulting in infected pulp, makes the tooth even more fragile, which makes REP difficult to perform. However, some studies have been successful in performing REP on infected pulp.

In large animals, the pulp is most often infected, which is simpler to perform and closer to the clinic. To assess the inflammation, the use of radiography allows one to observe whether a preapical lesion is present or not.

4.4. Clinical Studies

In human studies, the primary outcomes were the absence of pain, inflammation, or swelling. The resolution of the periapical lesion was observable in 2D or 3D radiography.

To assess regeneration, different clinical tests were performed. In this review, a few studies succeeded in having positive responses in sensitivity tests [71,74,80,92,95,96,101,102,106,109,121–123,126,129,130,132,133], confirming that neurogenesis is variable in REP [4,11].

Radiographically, almost all the studies reported the resolution of apical lesions, very often associated with high apex closure. Periapical radiographs with a parallel technique also allow one to verify the increase in radicular length, the thickening of dentinal walls, and periapical tissue repair. It is important to consider the normal growth of the patients, as most of them are in their growing phase, which may create visual variations in the measurements [85].

In all studies, the first radiographic finding is complete apical closure with a decrease in the apical diameter, while the send radiographic outcome was thickening of the canal walls with average wall thickening and continued root elongation.

Moreover, intrapulpal calcifications were often found close to the blood clot used alone, or with scaffold materials.

When performed, histological examination showed the formation of intracanal mineralized tissue around the scaffold particles solidifying with newly formed cementum tissue along the dentinal walls. Quantitative PCR confirmed the absence of odontoblasts able to make pulp-like tissues.

Regenerated tissue, whether for REP in animals or humans, is a fibrous pulp-like tissue with cementum-based islands in the middle and mineralization along the dentin walls. Sometimes, cementoblasts are found by immunohistochemistry. One could assume that internal calcification plays a protective role, but the evolution of such calcified tissues in the long term are still not known [92].

With the wide use of the REP procedure, several questions have arisen, mostly concerning the viability of the treated teeth. A tooth is considered viable once the root is structured and the tooth is clinically and radiographically asymptomatic with no signs of failure.

In the clinical studies included in this review, success of REP therapy was represented by the persistence of the tooth remaining asymptomatic. Only few patients showed positive responses to pulp sensitivity tests after revitalization.

Data arising from prolonged follow-up periods are still lacking. In fact, all randomized studies report follow-up periods that do not exceed 2 years. In the case of retrospective studies or case reports, we found two cases in which the follow-up period was between 6 and 8 years.

4.5. Risk of Bias

The included studies presented a low risk of bias in terms of animal selection (ARRIVE guidelines were respected), attrition, and reporting. On the other hand, weak reporting in terms of performance and detection affected evaluations and the synthesis of results. Thus, SYRCLE guidelines should be followed, especially for randomization protocols, animal housing facilities, and blinding, which could improve homogeneity of small and large animal model trials focusing on REP.

Regarding randomized studies, in only 40% of the cases was there an overall low risk of bias. Indeed, there was a lack of information in terms of randomization, deviation from the intended intervention, and measurement of the outcome, complicating comparisons between studies.

For non-randomized studies, a medium risk of bias in patient selection was found.

Indeed, more structured protocols for patient enrollment should be applied.

An overall low risk of bias was found in retrospective and case reports studies, since the criteria of selection, ascertainment, causality, and reporting were respected.

5. Conclusions

Currently, regenerative endodontics is legitimately considered part of the spectrum of endodontic therapies. In fact, REP on immature or mature teeth is a reliable approach that creates a new tissue, ensuring tooth development and viability.

However, no study has succeeded in regenerating pulp-like tissue; instead, in both preclinical and clinical studies, ligamentous tissue with cementum or bone-like mineralization replaced the necrotic pulp. Intra pulpal calcification can play a protective role for the regenerated tissue, but the evolution of this calcification in the long term is not known.

Some animal studies reported a vascularized and innervated regenerated pulp, but the response to the clinical test was not verified. In human studies, only few patients have regained sensitivity after revitalization.

Preclinical and clinical studies identify the success of REP therapy as the persistence of the tooth without signs or symptoms of failure.

Even if endodontic reparation can clinically satisfy the needs of dental and alveolar bone development and preservation, further studies are still necessary to identify procedures to successfully reproduce the physiological structure and function of the dentin–pulp complex.

Supplementary Materials: The following supporting information can be downloaded at: https://www.mdpi.com/article/10.3390/ijms231810534/s1.

Author Contributions: Conceptualization, S.M. and F.M.; writing/original draft preparation, S.M.; writing review and editing, S.M., F.M., S.V., T.B. and C.C. All authors have read and agreed to the published version of the manuscript.

Funding: This research received no external funding.

Institutional Review Board Statement: Not applicable.

Informed Consent Statement: Not applicable.

Data Availability Statement: The datasets generated for this study can be obtained upon reasonable request by email to the corresponding author.

Conflicts of Interest: The authors declare no conflict of interest.

Abbreviation

ALP	Alkaline Phosphatase
BC	Blood Clot
BMSCs	Bone Marrow Stem Cells
$CaOH_2$	Calcium Hydroxide
CAP	Catabolite Activator Protein
CD31	Platelet endothelial cell adhesion molecule
CEM	Calcium Enriched Mixture
CEMP	Cathelicidin Antimicrobial Peptide
CHP	Calcium Hydroxide Paste
CGRP	Calcitonin Gene-Related Peptide
GIC	Glass Ionomer Cement
Col1A1	Collagen type I alpha 1
DMP1	Dentin Matrix acidic Phosphoprotein 1
DMP4	Dentin matrix protein 4
(m) DPSCs	(mobilized) Dental Pulp Stem Cells
DSPP	Dentin Sialophosphoprotein
DXL1	Distal-Less Homeobox 1
FGF	Fibroblast growth factor
G-CSF	Granulocyte colony-stimulating factor
GLI2	GLI Family Zinc Finger 2
LPS	Lipopolysaccharide
MTA	Mineral Trioxide Aggregate
MRI	Magnetic Resonance Imaging
NF	Neurofilament
PAI	Periapical Index
PDLs	Periodontal ligament cells
PGP 9,5	Neuronal marker
(i)- (L)- PRF	(Injection-) (Leucocyte-) Platelet Rich Fibrin
PRP	Platelet Rich Plasma
qPCR	quantitative Polymerase Chain Reaction
rBMSC	rabbit Bone Marrow Stem Cells
REP	Regenerative Endodontic Procedure
SOX2	Sex determining region Y)-box 2
TAP	Tri-Antibiotic Paste
UC-MSCs	Umbilical Cord Mesenchymental Stem Cells
VEGF	Vascular Endothelial Growth Factor
vVW	Von Willebrand
2D radiography	X-ray / Panoramic radiography
3D radiography	CBCT (Cone Beam Computed Tomography)

References

1. Diogenes, A.; Ruparel, N.B. Regenerative Endodontic Procedures: Clinical Outcomes. *Dent. Clin. N. Am.* **2017**, *61*, 111–125. [CrossRef] [PubMed]
2. Galler, K.M.; Krastl, G.; Simon, S.; Van Gorp, G.; Meschi, N.; Vahedi, B.; Lambrechts, P. European Society of Endodontology position statement: Revitalization procedures. *Int. Endod. J.* **2016**, *49*, 717–723. [CrossRef] [PubMed]

3. American Association of Endodontists. Endodontic Diagnosis. 2013. Available online: https://www.aae.org/specialty/newsletter/endodontic-diagnosis/ (accessed on 28 July 2022).
4. Kim, S.G.; Malek, M.; Sigurdsson, A.; Lin, L.M.; Kahler, B. Regenerative endodontics: A comprehensive review. *Int. Endod. J.* **2018**, *51*, 1367–1388. [CrossRef] [PubMed]
5. Duggal, M.; Tong, H.J.; Al-Ansary, M.; Twati, W.; Day, P.F.; Nazzal, H. Interventions for the endodontic management of non-vital traumatised immature permanent anterior teeth in children and adolescents: A systematic review of the evidence and guidelines of the European Academy of Paediatric Dentistry. *Eur. Arch. Paediatr. Dent.* **2017**, *18*, 139–151. [CrossRef] [PubMed]
6. Sallam, N.M.; El Kalla, I.H.; Wahba, A.H.; Salama, N.M. Clinical and radiographic evaluation of platelet-rich fibrin for revascularization of necrotic immature permanent teeth: A controlled clinical trial. *Pediatr. Dent. J.* **2020**, *30*, 182–190. [CrossRef]
7. ElSheshtawy, A.S.; Nazzal, H.; El Shahawy, O.I.; El Baz, A.A.; Ismail, S.M.; Kang, J.; Ezzat, K.M. The effect of platelet-rich plasma as a scaffold in regeneration/revitalization endodontics of immature permanent teeth assessed using 2-dimensional radiographs and cone beam computed tomography: A randomized controlled trial. *Int. Endod. J.* **2020**, *53*, 905–921. [CrossRef]
8. Yoshpe, M.; Kaufman, A.Y.; Lin, S.; Ashkenazi, M. Regenerative endodontics: A promising tool to promote periapical healing and root maturation of necrotic immature permanent molars with apical periodontitis using platelet-rich fibrin (PRF). *Eur. Arch. Paediatr. Dent.* **2021**, *22*, 527–534. [CrossRef]
9. He, L.; Zhong, J.; Gong, Q.; Cheng, B.; Kim, S.G.; Ling, J.; Mao, J.J. Regenerative Endodontics by Cell Homing. *Dent. Clin.* **2017**, *61*, 143–159. [CrossRef]
10. Kim, S.G.; Solomon, C.; Zheng, Y.; Suzuki, T.; Mo, C.; Song, S.; Jiang, N.; Cho, S.; Zhou, J.; Mao, J.J. Effects of Growth Factors on Dental Stem/ProgenitorCells. *Dent. Clin. N. Am.* **2012**, *56*, 563–575. [CrossRef]
11. Saoud, T.M.A.; Ricucci, D.; Lin, L.M.; Gaengler, P. Regeneration and Repair in Endodontics—A Special Issue of the Regenerative Endodontics—A New Era in Clinical Endodontics. *Dent. J.* **2016**, *4*, 3. [CrossRef]
12. Moher, D.; Liberati, A.; Tetzlaff, J.; Altman, D.G.; Group, T.P. Preferred Reporting Items for Systematic Reviews and Meta-Analyses: The PRISMA Statement. *PLoS Med.* **2009**, *6*, e1000097. [CrossRef] [PubMed]
13. Higgins, J.P.T.; Altman, D.G.; Gøtzsche, P.C.; Jüni, P.; Moher, D.; Oxman, A.D.; Savović, J.; Schulz, K.F.; Weeks, L.; Sterne, J.A.C. The Cochrane Collaboration's tool for assessing risk of bias in randomised trials. *BMJ* **2011**, *343*, d5928. [CrossRef] [PubMed]
14. Sterne, J.A.; Hernán, M.A.; Reeves, B.C.; Savović, J.; Berkman, N.D.; Viswanathan, M.; Henry, D.; Altman, D.G.; Ansari, M.T.; Boutron, I.; et al. ROBINS-I: A tool for assessing risk of bias in non-randomised studies of interventions. *BMJ* **2016**, *355*, i4919. [CrossRef] [PubMed]
15. Murad, M.H.; Sultan, S.; Haffar, S.; Bazerbachi, F. Methodological quality and synthesis of case series and case reports. *BMJ EBM* **2018**, *23*, 60–63. [CrossRef]
16. Wen, R.; Wang, X.; Lu, Y.; Du, Y.; Yu, X. The combined application of rat bone marrow mesenchymal stem cells and bioceramic materials in the regeneration of dental pulp-like tissues. *Int. J. Clin. Exp. Pathol.* **2020**, *13*, 1492–1499. [PubMed]
17. El-Backly, R.M.; Massoud, A.G.; El-Badry, A.M.; Sherif, R.A.; Marei, M.K. Regeneration of dentine/pulp-like tissue using a dental pulp stem cell/poly(lactic-co-glycolic) acid scaffold construct in New Zealand white rabbits. *Aust. Endod. J.* **2008**, *34*, 52–67. [CrossRef]
18. Ruangsawasdi, N.; Zehnder, M.; Weber, F.E. Fibrin Gel Improves Tissue Ingrowth and Cell Differentiation in Human Immature Premolars Implanted in Rats. *J. Endod.* **2014**, *40*, 246–250. [CrossRef]
19. Gomes-Filho, J.E.; Duarte, P.C.T.; de Oliveira, C.B.; Watanabe, S.; Lodi, C.S.; Cintra, L.T.Â.; Bernabé, P.F.E. Tissue Reaction to a Triantibiotic Paste Used for Endodontic Tissue Self-regeneration of Nonvital Immature Permanent Teeth. *J. Endod.* **2012**, *38*, 91–94. [CrossRef]
20. Li, X.; Ma, C.; Xie, X.; Sun, H.; Liu, X. Pulp regeneration in a full-length human tooth root using a hierarchical nanofibrous microsphere system. *Acta Biomater.* **2016**, *35*, 57–67. [CrossRef]
21. Yadlapati, M.; Biguetti, C.; Cavalla, F.; Nieves, F.; Bessey, C.; Bohluli, P.; Garlet, G.P.; Letra, A.; Fakhouri, W.D.; Silva, R.M. Characterization of a Vascular Endothelial Growth Factor–loaded Bioresorbable Delivery System for Pulp Regeneration. *J. Endod.* **2017**, *43*, 77–83. [CrossRef]
22. Alexander, A.; Torabinejad, M.; Vahdati, S.A.; Nosrat, A.; Verma, P.; Grandhi, A.; Shabahang, S. Regenerative Endodontic Treatment in Immature Noninfected Ferret Teeth Using Blood Clot or SynOss Putty as Scaffolds. *J. Endod.* **2020**, *46*, 209–215. [CrossRef] [PubMed]
23. Torabinejad, M.; Alexander, A.; Vahdati, S.A.; Grandhi, A.; Baylink, D.; Shabahang, S. Effect of Residual Dental Pulp Tissue on Regeneration of Dentin-pulp Complex: An In Vivo Investigation. *J. Endod.* **2018**, *44*, 1796–1801. [CrossRef] [PubMed]
24. Torabinejad, M.; Milan, M.; Shabahang, S.; Wright, K.R.; Faras, H. Histologic Examination of Teeth with Necrotic Pulps and Periapical Lesions Treated with 2 Scaffolds: An Animal Investigation. *J. Endod.* **2015**, *41*, 846–852. [CrossRef] [PubMed]
25. Torabinejad, M.; Faras, H.; Corr, R.; Wright, K.R.; Shabahang, S. Histologic Examinations of Teeth Treated with 2 Scaffolds: A Pilot Animal Investigation. *J. Endod.* **2014**, *40*, 515–520. [CrossRef] [PubMed]
26. Verma, P.; Nosrat, A.; Kim, J.R.; Price, J.B.; Wang, P.; Bair, E.; Xu, H.H.; Fouad, A.F. Effect of Residual Bacteria on the Outcome of Pulp Regeneration In Vivo. *J. Dent. Res.* **2017**, *96*, 100–106. [CrossRef]
27. Scarparo, R.K.; Dondoni, L.; Böttcher, D.E.; Grecca, F.S.; Rockenbach, M.I.B.; Batista, E.L. Response to intracanal medication in immature teeth with pulp necrosis: An experimental model in rat molars. *J. Endod.* **2011**, *37*, 1069–1073. [CrossRef]

28. Moreira, M.S.; Diniz, I.M.; Rodrigues, M.F.S.D.; de Carvalho, R.A.; de Almeida Carrer, F.C.; Neves, I.I.; Gavini, G.; Marques, M.M. In vivo experimental model of orthotopic dental pulp regeneration under the influence of photobiomodulation therapy. *J. Photochem. Photobiol. B Biol.* **2017**, *166*, 180–186. [CrossRef]

29. Nabeshima, C.K.; Valdivia, J.E.; Caballero-Flores, H.; Arana-Chavez, V.E.; de Lima Machado Machado, M.E. Immunohistological study of the effect of vascular Endothelial Growth Factor on the angiogenesis of mature root canals in rat molars. *J. Appl. Oral Sci.* **2018**, *26*, e20170437. [CrossRef]

30. Chen, W.-J.; Xie, J.; Lin, X.; Ou, M.-H.; Zhou, J.; Wei, X.-L.; Chen, W.-X. The Role of Small Extracellular Vesicles Derived from Lipopolysaccharide-preconditioned Human Dental Pulp Stem Cells in Dental Pulp Regeneration. *J. Endod.* **2021**, *47*, 961–969. [CrossRef]

31. Edanami, N.; Yoshiba, K.; Shirakashi, M.; Ibn Belal, R.S.; Yoshiba, N.; Ohkura, N.; Tohma, A.; Takeuchi, R.; Okiji, T.; Noiri, Y. Impact of remnant healthy pulp and apical tissue on outcomes after simulated regenerative endodontic procedure in rat molars. *Sci. Rep.* **2020**, *10*, 20967. [CrossRef]

32. Khademi, A.A.; Dianat, O.; Mahjour, F.; Razavi, S.M.; Younessian, F. Outcomes of revascularization treatment in immature dog's teeth. *Dent. Traumatol.* **2014**, *30*, 374–379. [CrossRef] [PubMed]

33. Guo, H.; Li, B.; Wu, M.; Zhao, W.; He, X.; Sui, B.; Dong, Z.; Wang, L.; Shi, S.; Huang, X.; et al. Odontogenesis-related developmental microenvironment facilitates deciduous dental pulp stem cell aggregates to revitalize an avulsed tooth. *Biomaterials* **2021**, *279*, 121223. [CrossRef]

34. Jang, J.-H.; Moon, J.-H.; Kim, S.G.; Kim, S.-Y. Pulp regeneration with hemostatic matrices as a scaffold in an immature tooth minipig model. *Sci. Rep.* **2020**, *10*, 12536. [CrossRef] [PubMed]

35. Siddiqui, Z.; Sarkar, B.; Kim, K.-K.; Kadincesme, N.; Paul, R.; Kumar, A.; Kobayashi, Y.; Roy, A.; Choudhury, M.; Yang, J.; et al. Angiogenic hydrogels for dental pulp revascularization. *Acta Biomater.* **2021**, *126*, 109–118. [CrossRef] [PubMed]

36. Kim, S.G.; Solomon, C.S. Regenerative Endodontic Therapy in Mature Teeth Using Human-Derived Composite Amnion-Chorion Membrane as a Bioactive Scaffold: A Pilot Animal Investigation. *J. Endod.* **2021**, *47*, 1101–1109. [CrossRef]

37. Abbas, K.F.; Tawfik, H.; Hashem, A.A.R.; Ahmed, H.M.A.; Abu-Seida, A.M.; Refai, H.M. Histopathological evaluation of different regenerative protocols using Chitosan-based formulations for management of immature non-vital teeth with apical periodontitis: In vivo study. *Aust. Endod. J.* **2020**, *46*, 405–414. [CrossRef]

38. Khazaei, S.; Khademi, A.; Torabinejad, M.; Nasr Esfahani, M.H.; Khazaei, M.; Razavi, S.M. Improving pulp revascularization outcomes with buccal fat autotransplantation. *J. Tissue Eng. Regen. Med.* **2020**, *14*, 1227–1235. [CrossRef]

39. Zaky, S.H.; AlQahtani, Q.; Chen, J.; Patil, A.; Taboas, J.; Beniash, E.; Ray, H.; Sfeir, C. Effect of the Periapical "Inflammatory Plug" on Dental Pulp Regeneration: A Histologic In Vivo Study. *J. Endod.* **2020**, *46*, 51–56. [CrossRef]

40. Mounir, M.M.F.; Rashed, F.M.; Bukhary, S.M. Regeneration of Neural Networks in Immature Teeth with Non-Vital Pulp Following a Novel Regenerative Procedure. *Int. J. Stem Cells* **2019**, *12*, 410–418. [CrossRef]

41. El-Tayeb, M.M.; Abu-Seida, A.M.; El Ashry, S.H.; El-Hady, S.A. Evaluation of antibacterial activity of propolis on regenerative potential of necrotic immature permanent teeth in dogs. *BMC Oral Health* **2019**, *19*, 174. [CrossRef]

42. Huang, Y.; Tang, X.; Cehreli, Z.C.; Dai, X.; Xu, J.; Zhu, H. Autologous transplantation of deciduous tooth pulp into necrotic young permanent teeth for pulp regeneration in a dog model. *J. Int. Med. Res.* **2019**, *47*, 5094–5105. [CrossRef] [PubMed]

43. El Halaby, H.M.; Abu-Seida, A.M.; Fawzy, M.I.; Farid, M.H.; Bastawy, H.A. Evaluation of the regenerative potential of dentin conditioning and naturally derived scaffold for necrotic immature permanent teeth in a dog model. *Int. J. Exp. Pathol.* **2020**, *101*, 264–276. [CrossRef] [PubMed]

44. El Kalla, I.H.; Salama, N.M.; Wahba, A.H.; Sallam, N.M. Histological evaluation of platelet-rich fibrin for revascularization of immature permanent teeth in dogs. *Pediatr. Dent. J.* **2019**, *29*, 72–77. [CrossRef]

45. Alqahtani, Q.; Zaky, S.H.; Patil, A.; Beniash, E.; Ray, H.; Sfeir, C. Decellularized Swine Dental Pulp Tissue for Regenerative Root Canal Therapy. *J. Dent. Res.* **2018**, *97*, 1460–1467. [CrossRef] [PubMed]

46. Yoo, Y.-J.; Perinpanayagam, H.; Choi, Y.; Gu, Y.; Chang, S.-W.; Baek, S.-H.; Zhu, Q.; Fouad, A.F.; Kum, K.-Y. Characterization of Histopathology and Microbiota in Contemporary Regenerative Endodontic Procedures: Still Coming up Short. *J. Endod.* **2021**, *47*, 1285–1293.e1. [CrossRef]

47. Zarei, M.; Jafarian, A.H.; Harandi, A.; Javidi, M.; Gharechahi, M. Evaluation of the expression of VIII factor and VEGF in the regeneration of non-vital teeth in dogs using propolis. *Iran. J. Basic Med. Sci.* **2017**, *20*, 172–177. [CrossRef]

48. De Londero, C.L.D.; Pagliarin, C.M.L.; Felippe, M.C.S.; Felippe, W.T.; Danesi, C.C.; Barletta, F.B. Histologic Analysis of the Influence of a Gelatin-based Scaffold in the Repair of Immature Dog Teeth Subjected to Regenerative Endodontic Treatment. *J. Endod.* **2015**, *41*, 1619–1625. [CrossRef]

49. Stambolsky, C.; Rodríguez-Benítez, S.; Gutiérrez-Pérez, J.L.; Torres-Lagares, D.; Martín-González, J.; Segura-Egea, J.J. Histologic characterization of regenerated tissues after pulp revascularization of immature dog teeth with apical periodontitis using tri-antibiotic paste and platelet-rich plasma. *Arch. Oral Biol.* **2016**, *71*, 122–128. [CrossRef]

50. Wang, X.; Thibodeau, B.; Trope, M.; Lin, L.M.; Huang, G.T.-J. Histologic Characterization of Regenerated Tissues in Canal Space after the Revitalization/Revascularization Procedure of Immature Dog Teeth with Apical Periodontitis. *J. Endod.* **2010**, *36*, 56–63. [CrossRef]

51. Zhang, D.-D.; Chen, X.; Bao, Z.-F.; Chen, M.; Ding, Z.-J.; Zhong, M. Histologic Comparison between Platelet-rich Plasma and Blood Clot in Regenerative Endodontic Treatment: An Animal Study. *J. Endod.* **2014**, *40*, 1388–1393. [CrossRef]

52. Palma, P.J.; Ramos, J.C.; Martins, J.B.; Diogenes, A.; Figueiredo, M.H.; Ferreira, P.; Viegas, C.; Santos, J.M. Histologic Evaluation of Regenerative Endodontic Procedures with the Use of Chitosan Scaffolds in Immature Dog Teeth with Apical Periodontitis. *J. Endod.* **2017**, *43*, 1279–1287. [CrossRef] [PubMed]

53. Saoud, T.M.A.; Zaazou, A.; Nabil, A.; Moussa, S.; Aly, H.M.; Okazaki, K.; Rosenberg, P.A.; Lin, L.M. Histological observations of pulpal replacement tissue in immature dog teeth after revascularization of infected pulps. *Dent. Traumatol.* **2015**, *31*, 243–249. [CrossRef] [PubMed]

54. Zhu, X.; Wang, Y.; Liu, Y.; Huang, G.T.-J.; Zhang, C. Immunohistochemical and Histochemical Analysis of Newly Formed Tissues in Root Canal Space Transplanted with Dental Pulp Stem Cells Plus Platelet-rich Plasma. *J. Endod.* **2014**, *40*, 1573–1578. [CrossRef] [PubMed]

55. Moradi, S.; Talati, A.; Forghani, M.; Jafarian, A.H.; Naseri, M.; Shojaeian, S. Immunohistological Evaluation of Revascularized Immature Permanent Necrotic Teeth Treated by Platelet-Rich Plasma: An Animal Investigation. *Cell J.* **2016**, *18*, 389–396.

56. Fahmy, S.H.; Hassanien, E.E.S.; Nagy, M.M.; El Batouty, K.M.; Mekhemar, M.; Fawzy El Sayed, K.; Hassanein, E.H.; Wiltfang, J.; Dörfer, C. Investigation of the regenerative potential of necrotic mature teeth following different revascularisation protocols. *Aust. Endod. J.* **2017**, *43*, 73–82. [CrossRef]

57. Thibodeau, B.; Teixeira, F.; Yamauchi, M.; Caplan, D.J.; Trope, M. Pulp Revascularization of Immature Dog Teeth With Apical Periodontitis. *J. Endod.* **2007**, *33*, 680–689. [CrossRef]

58. Rodríguez-Benítez, S.; Stambolsky, C.; Gutiérrez-Pérez, J.L.; Torres-Lagares, D.; Segura-Egea, J.J. Pulp Revascularization of Immature Dog Teeth with Apical Periodontitis Using Triantibiotic Paste and Platelet-rich Plasma: A Radiographic Study. *J. Endod.* **2015**, *41*, 1299–1304. [CrossRef]

59. Zhou, R.; Wang, Y.; Chen, Y.; Chen, S.; Lyu, H.; Cai, Z.; Huang, X. Radiographic, Histologic, and Biomechanical Evaluation of Combined Application of Platelet-rich Fibrin with Blood Clot in Regenerative Endodontics. *J. Endod.* **2017**, *43*, 2034–2040. [CrossRef]

60. El Ashry, E.A.; Farsi, N.M.; Abuzeid, S.T.; El Ashiry, M.M.; Bahammam, H.A. Dental Pulp Revascularization of Necrotic Permanent Teeth with Immature Apices. *J. Clin. Pediatr. Dent.* **2016**, *40*, 361–366. [CrossRef]

61. Da Silva, L.A.B.; Nelson-Filho, P.; da Silva, R.A.B.; Flores, D.S.H.; Heilborn, C.; Johnson, J.D.; Cohenca, N. Revascularization and periapical repair after endodontic treatment using apical negative pressure irrigation versus conventional irrigation plus triantibiotic intracanal dressing in dogs' teeth with apical periodontitis. *Oral Surg. Oral Med. Oral Pathol. Oral Radiol. Endodontology* **2010**, *109*, 779–787. [CrossRef]

62. Pagliarin, C.M.L.; de Londero, C.L.D.; Felippe, M.C.S.; Felippe, W.T.; Danesi, C.C.; Barletta, F.B. Tissue characterization following revascularization of immature dog teeth using different disinfection pastes. *Braz. Oral Res.* **2016**, *30*, S1806-83242016000100273. [CrossRef] [PubMed]

63. Zhu, X.; Zhang, C.; Huang, G.T.-J.; Cheung, G.S.P.; Dissanayaka, W.L.; Zhu, W. Transplantation of Dental Pulp Stem Cells and Platelet-rich Plasma for Pulp Regeneration. *J. Endod.* **2012**, *38*, 1604–1609. [CrossRef] [PubMed]

64. Altaii, M.; Cathro, P.; Broberg, M.; Richards, L. Endodontic regeneration and tooth revitalization in immature infected sheep teeth. *Int. Endod. J.* **2017**, *50*, 480–491. [CrossRef]

65. Ramachandran, N.; Singh, S.; Podar, R.; Kulkarni, G.; Shetty, R.; Chandrasekhar, P. A comparison of two pulp revascularization techniques using platelet-rich plasma and whole blood clot. *J. Conserv. Dent.* **2020**, *23*, 637–643. [CrossRef]

66. El Ashry, S.H.; Abu-Seida, A.M.; Bayoumi, A.A.; Hashem, A.A. Regenerative potential of immature permanent non-vital teeth following different dentin surface treatments. *Exp. Toxicol. Pathol.* **2016**, *68*, 181–190. [CrossRef] [PubMed]

67. McTigue, D.J.; Subramanian, K.; Kumar, A. Case Series: Management of Immature Permanent Teeth With Pulpal Necrosis: A Case Series. *Pediatric Dent.* **2013**, *35*, 55–60.

68. Meschi, N.; Castro, A.B.; Vandamme, K.; Quirynen, M.; Lambrechts, P. The impact of autologous platelet concentrates on endodontic healing: A systematic review. *Platelets* **2016**, *27*, 613–633. [CrossRef]

69. Bezgin, T.; Yilmaz, A.D.; Celik, B.N.; Kolsuz, M.E.; Sonmez, H. Efficacy of Platelet-rich Plasma as a Scaffold in Regenerative Endodontic Treatment. *J. Endod.* **2015**, *41*, 36–44. [CrossRef]

70. Nagy, M.M.; Tawfik, H.E.; Hashem, A.A.R.; Abu-Seida, A.M. Regenerative Potential of Immature Permanent Teeth with Necrotic Pulps after Different Regenerative Protocols. *J. Endod.* **2014**, *40*, 192–198. [CrossRef]

71. Alagl, A.; Bedi, S.; Hassan, K.; AlHumaid, J. Use of platelet-rich plasma for regeneration in non-vital immature permanent teeth: Clinical and cone-beam computed tomography evaluation. *J. Int. Med. Res.* **2017**, *45*, 583–593. [CrossRef]

72. Meschi, N.; EzEldeen, M.; Torres Garcia, A.E.; Jacobs, R.; Lambrechts, P. A Retrospective Case Series in Regenerative Endodontics: Trend Analysis Based on Clinical Evaluation and 2- and 3-dimensional Radiology. *J. Endod.* **2018**, *44*, 1517–1525. [CrossRef] [PubMed]

73. Elfrink, M.E.C.; Heijdra, J.S.C.; Krikken, J.B.; Kouwenberg-Bruring, W.H.; Kouwenberg, H.; Weerheijm, K.L.; Veerkamp, J.S.J. Regenerative endodontic therapy: A follow-up of 47 anterior traumatised teeth. *Eur. Arch. Paediatr. Dent.* **2021**, *22*, 469–477. [CrossRef] [PubMed]

74. Pereira, A.C.; de Oliveira, M.L.; Cerqueira-neto, A.C.C.L.; Gomes, B.P.F.A.; Ferraz, C.C.R.; de Almeida, J.F.A.; Marciano, M.A.; de-Jesus-Soares, A. Treatment outcomes of pulp revascularization in traumatized immature teeth using calcium hydroxide and 2% chlorhexidine gel as intracanal medication. *J. Appl. Oral Sci.* **2020**, *28*, e20200217. [CrossRef] [PubMed]

75. Chrepa, V.; Joon, R.; Austah, O.; Diogenes, A.; Hargreaves, K.M.; Ezeldeen, M.; Ruparel, N.B. Clinical Outcomes of Immature Teeth Treated with Regenerative Endodontic Procedures—A San Antonio Study. *J. Endod.* **2020**, *46*, 1074–1084. [CrossRef]
76. Mittmann, C.W.; Kostka, E.; Ballout, H.; Preus, M.; Preissner, R.; Karaman, M.; Preissner, S. Outcome of revascularization therapy in traumatized immature incisors. *BMC Oral Health* **2020**, *20*, 207. [CrossRef]
77. Linsuwanont, P.; Sinpitaksakul, P.; Lertsakchai, T. Evaluation of root maturation after revitalization in immature permanent teeth with nonvital pulps by cone beam computed tomography and conventional radiographs. *Int. Endod. J.* **2017**, *50*, 836–846. [CrossRef]
78. Estefan, B.S.; Batouty, K.M.E.; Nagy, M.M.; Diogenes, A. Influence of Age and Apical Diameter on the Success of Endodontic Regeneration Procedures. *J. Endod.* **2016**, *42*, 1620–1625. [CrossRef]
79. Peng, C.; Yang, Y.; Zhao, Y.; Liu, H.; Xu, Z.; Zhao, D.; Qin, M. Long-term treatment outcomes in immature permanent teeth by revascularisation using MTA and GIC as canal-sealing materials: A retrospective study. *Int. J. Paediatr. Dent.* **2017**, *27*, 454–462. [CrossRef]
80. Chen, S.-J.; Chen, L.-P. Radiographic outcome of necrotic immature teeth treated with two endodontic techniques: A retrospective analysis. *Biomed. J.* **2016**, *39*, 366–371. [CrossRef]
81. Jeeruphan, T.; Jantarat, J.; Yanpiset, K.; Suwannapan, L.; Khewsawai, P.; Hargreaves, K.M. Mahidol Study 1: Comparison of Radiographic and Survival Outcomes of Immature Teeth Treated with Either Regenerative Endodontic or Apexification Methods: A Retrospective Study. *J. Endod.* **2012**, *38*, 1330–1336. [CrossRef]
82. Bukhari, S.; Kohli, M.R.; Setzer, F.; Karabucak, B. Outcome of Revascularization Procedure: A Retrospective Case Series. *J. Endod.* **2016**, *42*, 1752–1759. [CrossRef] [PubMed]
83. Chan, E.K.M.; Desmeules, M.; Cielecki, M.; Dabbagh, B.; Ferraz dos Santos, B. Longitudinal Cohort Study of Regenerative Endodontic Treatment for Immature Necrotic Permanent Teeth. *J. Endod.* **2017**, *43*, 395–400. [CrossRef] [PubMed]
84. Song, M.; Cao, Y.; Shin, S.-J.; Shon, W.-J.; Chugal, N.; Kim, R.H.; Kim, E.; Kang, M.K. Revascularization-associated Intracanal Calcification: Assessment of Prevalence and Contributing Factors. *J. Endod.* **2017**, *43*, 2025–2033. [CrossRef] [PubMed]
85. Aly, M.M.; Taha, S.E.E.-D.; El Sayed, M.A.; Youssef, R.; Omar, H.M. Clinical and radiographic evaluation of Biodentine and Mineral Trioxide Aggregate in revascularization of non-vital immature permanent anterior teeth (randomized clinical study). *Int. J. Paediatr. Dent.* **2019**, *29*, 464–473. [CrossRef] [PubMed]
86. Arslan, H.; Ahmed, H.M.A.; Şahin, Y.; Doğanay Yıldız, E.; Gündoğdu, E.C.; Güven, Y.; Khalilov, R. Regenerative Endodontic Procedures in Necrotic Mature Teeth with Periapical Radiolucencies: A Preliminary Randomized Clinical Study. *J. Endod.* **2019**, *45*, 863–872. [CrossRef] [PubMed]
87. Ragab, R.A.; Lattif, A.E.A.E.; Dokky, N.A.E.W.E. Comparative Study between Revitalization of Necrotic Immature Permanent Anterior Teeth with and without Platelet Rich Fibrin: A Randomized Controlled Trial. *J. Clin. Pediatr. Dent..* **2019**, *43*, 78–85. [CrossRef] [PubMed]
88. Jiang, X.; Liu, H.; Peng, C. Clinical and Radiographic Assessment of the Efficacy of a Collagen Membrane in Regenerative Endodontics: A Randomized, Controlled Clinical Trial. *J. Endod.* **2017**, *43*, 1465–1471. [CrossRef] [PubMed]
89. Rizk, H.M.; AL-Deen, M.S.S.; Emam, A.A. Regenerative Endodontic Treatment of Bilateral Necrotic Immature Permanent Maxillary Central Incisors with Platelet-rich Plasma versus Blood Clot: A Split Mouth Double-blinded Randomized Controlled Trial. *Int. J. Clin. Pediatr. Dent.* **2019**, *12*, 332–339. [CrossRef]
90. Botero, T.M.; Tang, X.; Gardner, R.; Hu, J.C.C.; Boynton, J.R.; Holland, G.R. Clinical Evidence for Regenerative Endodontic Procedures: Immediate versus Delayed Induction? *J. Endod.* **2017**, *43*, S75–S81. [CrossRef]
91. Shivashankar, V.Y.; Johns, D.A.; Maroli, R.K.; Sekar, M.; Chandrasekaran, R.; Karthikeyan, S.; Renganathan, S.K. Comparison of the Effect of PRP, PRF and Induced Bleeding in the Revascularization of Teeth with Necrotic Pulp and Open Apex: A Triple Blind Randomized Clinical Trial. *J. Clin. Diagn. Res.* **2017**, *11*, ZC34–ZC39. [CrossRef]
92. Chen, M.Y.-H.; Chen, K.-L.; Chen, C.-A.; Tayebaty, F.; Rosenberg, P.A.; Lin, L.M. Responses of immature permanent teeth with infected necrotic pulp tissue and apical periodontitis/abscess to revascularization procedures. *Int. Endod. J.* **2012**, *45*, 294–305. [CrossRef] [PubMed]
93. Jiang, X.; Liu, H. An uncommon type of segmental root development after revitalization. *Int. Endod. J.* **2020**, *53*, 1728–1741. [CrossRef] [PubMed]
94. Sabeti, M.; Golchert, K.; Torabinejad, M. Regeneration of Pulp-Dentin Complex in a Tooth with Symptomatic Irreversible Pulpitis and Open Apex Using Regenerative Endodontic Procedures. *J. Endod.* **2021**, *47*, 247–252. [CrossRef] [PubMed]
95. Gaviño Orduña, J.F.; Caviedes-Bucheli, J.; Manzanares Céspedes, M.C.; Berástegui Jimeno, E.; Martín Biedma, B.; Segura-Egea, J.J.; López-López, J. Use of Platelet-rich Plasma in Endodontic Procedures in Adults: Regeneration or Repair? A Report of 3 Cases with 5 Years of Follow-up. *J. Endod.* **2017**, *43*, 1294–1301. [CrossRef] [PubMed]
96. Terauchi, Y.; Bakland, L.K.; Bogen, G. Combined Root Canal Therapies in Multirooted Teeth with Pulpal Disease. *J. Endod.* **2021**, *47*, 44–51. [CrossRef] [PubMed]
97. Jung, I.-Y.; Lee, S.-J.; Hargreaves, K.M. Biologically Based Treatment of Immature Permanent Teeth with Pulpal Necrosis: A Case Series. *J. Endod.* **2008**, *34*, 876–887. [CrossRef]
98. Žižka, R.; Belák, Š.; Šedý, J.; Fačevicová, K.; Voborná, I.; Marinčák, D. Clinical and Radiographic Outcomes of Immature Teeth Treated with Different Treatment Protocols of Regenerative Endodontic Procedures: A Retrospective Cohort Study. *J. Clin. Med.* **2021**, *10*, 1600. [CrossRef]

99. Li, L.; Pan, Y.; Mei, L.; Li, J. Clinical and Radiographic Outcomes in Immature Permanent Necrotic Evaginated Teeth Treated with Regenerative Endodontic Procedures. *J. Endod.* **2017**, *43*, 246–251. [CrossRef]

100. Saoud, T.M.A.; Zaazou, A.; Nabil, A.; Moussa, S.; Lin, L.M.; Gibbs, J.L. Clinical and Radiographic Outcomes of Traumatized Immature Permanent Necrotic Teeth after Revascularization/Revitalization Therapy. *J. Endod.* **2014**, *40*, 1946–1952. [CrossRef]

101. Dabbagh, B.; Alvaro, E.; Vu, D.-D.; Rizkallah, J.; Schwartz, S. Clinical Complications in the Revascularization of Immature Necrotic Permanent Teeth. *Pediatr. Dent.* **2012**, *34*, 4.

102. Dudeja, P.G.; Grover, S.; Srivastava, D.; Dudeja, K.K.; Sharma, V. Pulp Revascularization- It's your Future Whether you Know it or Not? *J. Clin. Diagn. Res.* **2015**, *9*, ZR01–ZR04. [CrossRef] [PubMed]

103. Ulusoy, A.T.; Cehreli, Z.C. Regenerative Endodontic Treatment of Necrotic Primary Molars with Missing Premolars: A Case Series. *Pediatr. Dent.* **2017**, *39*, 131–134. [PubMed]

104. Cehreli, Z.C.; Isbitiren, B.; Sara, S.; Erbas, G. Regenerative Endodontic Treatment (Revascularization) of Immature Necrotic Molars Medicated with Calcium Hydroxide: A Case Series. *J. Endod.* **2011**, *37*, 1327–1330. [CrossRef]

105. Sachdeva, G.S.; Sachdeva, L.T.; Goel, M.; Bala, S. Regenerative endodontic treatment of an immature tooth with a necrotic pulp and apical periodontitis using platelet-rich plasma (PRP) and mineral trioxide aggregate (MTA): A case report. *Int. Endod. J.* **2015**, *48*, 902–910. [CrossRef] [PubMed]

106. Lin, L.M.; Shimizu, E.; Gibbs, J.L.; Loghin, S.; Ricucci, D. Histologic and Histobacteriologic Observations of Failed Revascularization/Revitalization Therapy: A Case Report. *J. Endod.* **2014**, *40*, 291–295. [CrossRef] [PubMed]

107. Becerra, P.; Ricucci, D.; Loghin, S.; Gibbs, J.L.; Lin, L.M. Histologic Study of a Human Immature Permanent Premolar with Chronic Apical Abscess after Revascularization/Revitalization. *J. Endod.* **2014**, *40*, 133–139. [CrossRef] [PubMed]

108. Chen, X.; Bao, Z.-F.; Liu, Y.; Liu, M.; Jin, X.-Q.; Xu, X.-B. Regenerative Endodontic Treatment of an Immature Permanent Tooth at an Early Stage of Root Development: A Case Report. *J. Endod.* **2013**, *39*, 719–722. [CrossRef]

109. Chang, S.-W.; Oh, T.-S.; Lee, W.; Shun-Pan Cheung, G.; Kim, H.-C. Long-term observation of the mineral trioxide aggregate extrusion into the periapical lesion: A case series. *Int. J. Oral Sci.* **2013**, *5*, 54–57. [CrossRef]

110. Lenzi, R.; Trope, M. Revitalization Procedures in Two Traumatized Incisors with Different Biological Outcomes. *J. Endod.* **2012**, *38*, 411–414. [CrossRef]

111. Shin, S.Y.; Albert, J.S.; Mortman, R.E. One step pulp revascularization treatment of an immature permanent tooth with chronic apical abscess: A case report. *Int. Endod. J.* **2009**, *42*, 1118–1126. [CrossRef]

112. Shiehzadeh, V.; Aghmasheh, F.; Shiehzadeh, F.; Joulae, M.; Kosarieh, E.; Shiehzadeh, F. Healing of large periapical lesions following delivery of dental stem cells with an injectable scaffold: New method and three case reports. *Indian J. Dent. Res.* **2014**, *25*, 248. [CrossRef] [PubMed]

113. Plascencia, H.; Cruz, Á.; Díaz, M.; Jiménez, A.L.; Solís, R.; Bernal, C. Root Canal Filling after Revascularization/Revitalization. *J. Clin. Pediatr. Dent..* **2016**, *40*, 445–449. [CrossRef] [PubMed]

114. El-Kateb, N.M.; El-Backly, R.N.; Amin, W.M.; Abdalla, A.M. Quantitative Assessment of Intracanal Regenerated Tissues after Regenerative Endodontic Procedures in Mature Teeth Using Magnetic Resonance Imaging: A Randomized Controlled Clinical Trial. *J. Endod.* **2020**, *46*, 563–574. [CrossRef] [PubMed]

115. Brizuela, C.; Meza, G.; Urrejola, D.; Quezada, M.A.; Concha, G.; Ramírez, V.; Angelopoulos, I.; Cadiz, M.I.; Tapia-Limonchi, R.; Khoury, M. Cell-Based Regenerative Endodontics for Treatment of Periapical Lesions: A Randomized, Controlled Phase I/II Clinical Trial. *J. Dent. Res.* **2020**, *99*, 523–529. [CrossRef] [PubMed]

116. Ibrahim, L.; Tawfik, M.; Abu Naeem, F. Evaluation of The Periapical Healing Following Pulp Revascularization Using Injectable PRF VS nonsurgical Root Canal Treatment in Mature Permanent Teeth with periapical periodontitis. A Clinical Study. *Egypt. Dent. J.* **2021**, *67*, 2663–2672. [CrossRef]

117. Yoshpe, M.; Einy, S.; Ruparel, N.; Lin, S.; Kaufman, A.Y. Regenerative Endodontics: A Potential Solution for External Root Resorption (Case Series). *J. Endod.* **2020**, *46*, 192–199. [CrossRef]

118. Bakhtiar, H.; Esmaeili, S.; Fakhr Tabatabayi, S.; Ellini, M.R.; Nekoofar, M.H.; Dummer, P.M.H. Second-generation Platelet Concentrate (Platelet-rich Fibrin) as a Scaffold in Regenerative Endodontics: A Case Series. *J. Endod.* **2017**, *43*, 401–408. [CrossRef] [PubMed]

119. Mehrvarzfar, P.; Abbott, P.V.; Akhavan, H.; Savadkouhi, S.T. Modified Revascularization in Human Teeth Using an Intracanal Formation of Treated Dentin Matrix: A Report of Two Cases. *J. Int. Soc. Prev. Community Dent.* **2017**, *7*, 218–221. [CrossRef]

120. Cymerman, J.J.; Nosrat, A. Regenerative Endodontic Treatment as a Biologically Based Approach for Non-Surgical Retreatment of Immature Teeth. *J. Endod.* **2020**, *46*, 44–50. [CrossRef]

121. Nosrat, A.; Kolahdouzan, A.; Khatibi, A.H.; Verma, P.; Jamshidi, D.; Nevins, A.J.; Torabinejad, M. Clinical, Radiographic, and Histologic Outcome of Regenerative Endodontic Treatment in Human Teeth Using a Novel Collagen-hydroxyapatite Scaffold. *J. Endod.* **2019**, *45*, 136–143. [CrossRef]

122. Narang, I.; Mittal, N.; Mishra, N. A comparative evaluation of the blood clot, platelet-rich plasma, and platelet-rich fibrin in regeneration of necrotic immature permanent teeth: A clinical study. *Contemp. Clin. Dent.* **2015**, *6*, 63–68. [CrossRef] [PubMed]

123. Nakashima, M.; Iohara, K.; Murakami, M.; Nakamura, H.; Sato, Y.; Ariji, Y.; Matsushita, K. Pulp regeneration by transplantation of dental pulp stem cells in pulpitis: A pilot clinical study. *Stem Cell Res. Ther.* **2017**, *8*, 61. [CrossRef] [PubMed]

124. Meschi, N.; EzEldeen, M.; Garcia, A.E.T.; Lahoud, P.; Van Gorp, G.; Coucke, W.; Jacobs, R.; Vandamme, K.; Teughels, W.; Lambrechts, P. Regenerative Endodontic Procedure of Immature Permanent Teeth with Leukocyte and Platelet-rich Fibrin: A Multicenter Controlled Clinical Trial. *J. Endod.* **2021**, *47*, 1729–1750. [CrossRef] [PubMed]

125. Meschi, N.; Hilkens, P.; Van Gorp, G.; Strijbos, O.; Mavridou, A.; Cadenas de Llano Perula, M.; Lambrichts, I.; Verbeken, E.; Lambrechts, P. Regenerative Endodontic Procedures Posttrauma: Immunohistologic Analysis of a Retrospective Series of Failed Cases. *J. Endod.* **2019**, *45*, 427–434. [CrossRef] [PubMed]

126. Rizk, H.M.; Salah Al-Deen, M.S.M.; Emam, A.A. Comparative evaluation of Platelet Rich Plasma (PRP) versus Platelet Rich Fibrin (PRF) scaffolds in regenerative endodontic treatment of immature necrotic permanent maxillary central incisors: A double blinded randomized controlled trial. *Saudi Dent. J.* **2020**, *32*, 224–231. [CrossRef] [PubMed]

127. Shimizu, E.; Ricucci, D.; Albert, J.; Alobaid, A.S.; Gibbs, J.L.; Huang, G.T.-J.; Lin, L.M. Clinical, Radiographic, and Histological Observation of a Human Immature Permanent Tooth with Chronic Apical Abscess after Revitalization Treatment. *J. Endod.* **2013**, *39*, 1078–1083. [CrossRef]

128. De-Jesus-Soares, A.; Prado, M.C.; Nardello, L.C.L.; Pereira, A.C.; Cerqueira-Neto, A.C.C.L.; Nagata, J.Y.; Martinez, E.F.; Frozoni, M.; Gomes, B.P.F.A.; Pinheiro, E.T. Clinical and Molecular Microbiological Evaluation of Regenerative Endodontic Procedures in Immature Permanent Teeth. *J. Endod.* **2020**, *46*, 1448–1454. [CrossRef]

129. Nagata, J.Y.; Soares, A.J.; Souza-Filho, F.J.; Zaia, A.A.; Ferraz, C.C.R.; Almeida, J.F.A.; Gomes, B.P.F.A. Microbial Evaluation of Traumatized Teeth Treated with Triple Antibiotic Paste or Calcium Hydroxide with 2% Chlorhexidine Gel in Pulp Revascularization. *J. Endod.* **2014**, *40*, 778–783. [CrossRef]

130. Nazzal, H.; Kenny, K.; Altimimi, A.; Kang, J.; Duggal, M.S. A prospective clinical study of regenerative endodontic treatment of traumatized immature teeth with necrotic pulps using bi-antibiotic paste. *Int. Endod. J.* **2018**, *51*, e204–e215. [CrossRef]

131. Bose, R.; Nummikoski, P.; Hargreaves, K. A Retrospective Evaluation of Radiographic Outcomes in Immature Teeth With Necrotic Root Canal Systems Treated With Regenerative Endodontic Procedures. *J. Endod.* **2009**, *35*, 1343–1349. [CrossRef]

132. Cerqueira-Neto, A.C.C.L.; Prado, M.C.; Pereira, A.C.; Oliveira, M.L.; Vargas-Neto, J.; Gomes, B.P.F.A.; Ferraz, C.C.R.; Almeida, J.F.A.; de-Jesus-Soares, A. Clinical and Radiographic Outcomes of Regenerative Endodontic Procedures in Traumatized Immature Permanent Teeth: Interappointment Dressing or Single-Visit? *J. Endod.* **2021**, *47*, 1598–1608. [CrossRef] [PubMed]

133. Sutam, N.; Jantarat, J.; Ongchavalit, L.; Sutimuntanakul, S.; Hargreaves, K.M. A Comparison of 3 Quantitative Radiographic Measurement Methods for Root Development Measurement in Regenerative Endodontic Procedures. *J. Endod.* **2018**, *44*, 1665–1670. [CrossRef] [PubMed]

134. Jha, P.; Virdi, M.S.; Nain, S. A Regenerative Approach for Root Canal Treatment of Mature Permanent Teeth: Comparative Evaluation with 18 Months Follow-up. *Int. J. Clin. Pediatr. Dent.* **2019**, *12*, 182–188. [CrossRef] [PubMed]

135. Lovelace, T.W.; Henry, M.A.; Hargreaves, K.M.; Diogenes, A. Evaluation of the Delivery of Mesenchymal Stem Cells into the Root Canal Space of Necrotic Immature Teeth after Clinical Regenerative Endodontic Procedure. *J. Endod.* **2011**, *37*, 133–138. [CrossRef]

136. Meschi, N.; Hilkens, P.; Lambrichts, I.; Van den Eynde, K.; Mavridou, A.; Strijbos, O.; De Ketelaere, M.; Van Gorp, G.; Lambrechts, P. Regenerative endodontic procedure of an infected immature permanent human tooth: An immunohistological study. *Clin. Oral Investig.* **2016**, *20*, 807–814. [CrossRef]

137. Shah, N.; Logani, A. SealBio: A novel, non-obturation endodontic treatment based on concept of regeneration. *J. Conserv. Dent.* **2012**, *15*, 328–332. [CrossRef]

138. Adnan, S.; Ullah, R. Top-cited Articles in Regenerative Endodontics: A Bibliometric Analysis. *J. Endod.* **2018**, *44*, 1650–1664. [CrossRef]

139. Kahler, B.; Rossi-Fedele, G.; Chugal, N.; Lin, L.M. An Evidence-based Review of the Efficacy of Treatment Approaches for Immature Permanent Teeth with Pulp Necrosis. *J. Endod.* **2017**, *43*, 1052–1057. [CrossRef]

140. Wigler, R.; Kaufman, A.Y.; Lin, S.; Steinbock, N.; Hazan-Molina, H.; Torneck, C.D. Revascularization: A Treatment for Permanent Teeth with Necrotic Pulp and Incomplete Root Development. *J. Endod.* **2013**, *39*, 319–326. [CrossRef]

141. Iohara, K.; Nakashima, M.; Ito, M.; Ishikawa, M.; Nakasima, A.; Akamine, A. Dentin regeneration by dental pulp stem cell therapy with recombinant human bone morphogenetic protein 2. *J. Dent. Res.* **2004**, *83*, 590–595. [CrossRef]

142. Tziafas, D. Dynamics for Pulp-Dentin Tissue Engineering in Operative Dentistry. In *Regenerative Dentistry*; Marei, M.K., Ed.; Synthesis Lectures on Tissue Engineering; Springer International Publishing: Cham, Switzerland, 2010; pp. 111–158. ISBN 978-3-031-02581-5. [CrossRef]

143. Yoshiba, K.; Yoshiba, N.; Nakamura, H.; Iwaku, M.; Ozawa, H. Immunolocalization of fibronectin during reparative dentinogenesis in human teeth after pulp capping with calcium hydroxide. *J. Dent. Res.* **1996**, *75*, 1590–1597. [CrossRef] [PubMed]

144. Huang, G.T. Pulp and dentin tissue engineering and regeneration: Current progress. *Regen. Med.* **2009**, *4*, 697–707. [CrossRef] [PubMed]

145. Laureys, W.G.M.; Cuvelier, C.A.; Dermaut, L.R.; De Pauw, G.A.M. The Critical Apical Diameter to Obtain Regeneration of the Pulp Tissue after Tooth Transplantation, Replantation, or Regenerative Endodontic Treatment. *J. Endod.* **2013**, *39*, 759–763. [CrossRef] [PubMed]

146. Kim, J.Y.; Xin, X.; Moioli, E.K.; Chung, J.; Lee, C.H.; Chen, M.; Fu, S.Y.; Koch, P.D.; Mao, J.J. Regeneration of Dental-Pulp-like Tissue by Chemotaxis-Induced Cell Homing. *Tissue Eng. Part A* **2010**, *16*, 3023–3031. [CrossRef] [PubMed]

147. Mangione, F.; Salmon, B.; EzEldeen, M.; Jacobs, R.; Chaussain, C.; Vital, S. Characteristics of Large Animal Models for Current Cell-Based Oral Tissue Regeneration. *Tissue Eng. Part B Rev.* **2022**, *28*, 489–505. [CrossRef] [PubMed]

148. Gronthos, S.; Mankani, M.; Brahim, J.; Robey, P.G.; Shi, S. Postnatal human dental pulp stem cells (DPSCs) in vitro and in vivo. *Proc. Natl. Acad. Sci. USA* **2000**, *97*, 13625–13630. [CrossRef]
149. Yang, J.; Yuan, G.; Chen, Z. Pulp Regeneration: Current Approaches and Future Challenges. *Front. Physiol.* **2016**, *7*, 58. [CrossRef]
150. Cooke, J.P. Inflammation and Its Role in Regeneration and Repair. *Circ. Res.* **2019**, *124*, 1166–1168. [CrossRef]

International Journal of
Molecular Sciences

Review

Modulators of Wnt Signaling Pathway Implied in Dentin Pulp Complex Engineering: A Literature Review

Marion Florimond [1,2,†], Sandra Minic [1,†], Paul Sharpe [3], Catherine Chaussain [1,4], Emmanuelle Renard [5,6] and Tchilalo Boukpessi [1,7,*]

[1] Laboratory of Orofacial Pathologies, Imaging and Biotherapies, School of Dentistry, Laboratoire d'Excellence INFLAMEX, Université Paris Cité, URP 2496, 1 Rue Maurice Arnoux, 92120 Montrouge, France
[2] Dental Department, Charles Foix Hospital, AP-HP, 94200 Ivry sur Seine, France
[3] Centre for Craniofacial and Regenerative Biology, Faculty of Dentistry, Oral & Craniofacial Sciences, King's College London, London SE1 9RT, UK
[4] Dental Department, and Reference Center for Rare Diseases of Calcium and Phosphorus Metabolism, Bretonneau Hospital, AP-HP, 75018 Paris, France
[5] Inserm, UMR 1229, RMeS, Regenerative Medicine and Skeleton, Nantes Université, ONIRIS, 44000 Nantes, France
[6] CHU de Nantes, Service d'Odontologie Restauratrice et Chirurgicale, 44000 Nantes, France
[7] Dental Department, Pitié Salpétrière Hospital, DMU CHIR, AP-HP, 75013 Paris, France
* Correspondence: tchilalo.boukpessi@u-paris.fr
† These authors contributed equally to this work.

Citation: Florimond, M.; Minic, S.; Sharpe, P.; Chaussain, C.; Renard, E.; Boukpessi, T. Modulators of Wnt Signaling Pathway Implied in Dentin Pulp Complex Engineering: A Literature Review. *Int. J. Mol. Sci.* **2022**, *23*, 10582. https://doi.org/10.3390/ijms231810582

Academic Editors: Matthias Widbiller, Marco Tatullo and Kerstin M. Galler

Received: 31 July 2022
Accepted: 8 September 2022
Published: 13 September 2022

Publisher's Note: MDPI stays neutral with regard to jurisdictional claims in published maps and institutional affiliations.

Abstract: The main goal of vital pulp therapy (VPT) is to preserve the vitality of the pulp tissue, even when it is exposed due to bacterial invasion, iatrogenic mechanical preparation, or trauma. The type of new dentin formed as a result of VPT can differ in its cellular origin, its microstructure, and its barrier function. It is generally agreed that the new dentin produced by odontoblasts (reactionary dentin) has a tubular structure, while the dentin produced by pulp cells (reparative dentin) does not or has less. Thus, even VPT aims to maintain the vitality of the pulp. It does not regenerate the dentin pulp complex integrity. Therefore, many studies have sought to identify new therapeutic strategies to successfully regenerate the dentin pulp complex. Among them is a Wnt protein-based strategy based on the fact that Wnt proteins seem to be powerful stem cell factors that allow control of the self-renewal and proliferation of multiple adult stem cell populations, suitable for homeostasis maintenance, tissue healing, and regeneration promotion. Thus, this review outlines the different agents targeting the Wnt signaling that could be applied in a tooth environment, and could be a potential therapy for dentin pulp complex and bone regeneration.

Keywords: Wnt signal; dentin pulp complex regeneration engineering; small molecules

1. Introduction

The main goal of vital pulp therapy (VPT) is to preserve the vitality of the pulp tissue, even when it is exposed due to bacterial invasion, iatrogenic mechanical preparation, or trauma [1]. VPT procedures consist of direct pulp capping, and partial or full pulpotomy with bioactive capping materials. Calcium hydroxide (CH) has been extensively used for direct pulp capping and has long been considered the "gold standard" [2]. CH can release hydroxyl and calcium ions that create an alkaline bactericidal environment around the pulp tissues, prompting the formation of necrotic tissue beneath the exposed pulp, and this tissular reaction may lead to increases in cell differentiation, collagen secretion, and dentin formation [3,4].

However, the poor quality of the resulting dentin bridge and its lack of sealing within the dentin walls explain why some authors prefer the use of Tricalcium silicate-based (TCS-based) cements [4,5], such as ProRoot White MTA (Dentsply, Tulsa Dental, Tulsa, OK, USA) or Biodentine (Septodont, Saint-Maur-des-Fossés, France), which both have high clinical success rates in dentistry. The widespread clinical indications of TCS-based cements are mainly based on their ability to form CH as a by-product of hydration. MTA has been considered as potential gold standard for vital pulp therapy, because it integrated better with the pulp tissues than CH [5] and showed greater success than other capping agents when used in different conditions in clinical trials [6,7]. The type of new dentin formed can differ in different manners (cellular composition and function). Even though there is no consensus regarding the origin of this new tissue [7], it is generally admitted that reactionary dentin produced by odontoblasts has a tubular structure, while the reparative dentin does not or has much less [8].

VPT aims to maintain and, if possible, regenerate the vitality of dental pulp. This last objective is not achieved using the conventional VPT procedures. Therefore, many studies have sought to identify new therapeutic strategies to successfully regenerate the dentin pulp complex. Two different ways are currently being tested to achieve this enormous objective. The first involves the introduction of stem or progenitor cells into a site of damage, and the second aims to activate endogenous stem cells to promote tissue regeneration. This last strategy, in which endogenous stem cells are activated, avoids the risk associated with the implementation of different cells in nature or reprogrammed cells into the human body. For this purpose, a Wnt protein-based strategy, based on the fact that Wnt proteins are powerful stem cell factors [9], allows control of the self-renewal and proliferation of multiple adult stem cell populations [10].

The secreted family Wnt proteins participate in the regulation of cell differentiation, proliferation, and apoptosis, and through these mechanisms, play a key role in tissue generation, regeneration, and self-renewal [11]. Wnts induce intracellular signaling by binding to the extracellular domain of receptors encoded by Frizzled (Fz). These proteins interact also with the low-density lipoprotein receptor-related protein (LRP) 5 and LRP 6 transmembrane proteins that act as coreceptors for Wnts. They also linked to neurotrophic tyrosine kinase, receptor-related 2 (NTRK2). By means of an intracellular signal transduction pathway (activation of Dishevelled (DVL)), a protein of the destruction complex prevents activation of the destruction complex, constituted of Axin, adenomatosis polyposis coli (APC), glycogen synthase kinase 3 (GSK3), and other factors. At the same time, the cytoplasmic domain of LRP5/6 becomes phosphorylated and binds axin. This leads to disassembly of the APC-axin-β catenin complex and the release of β-catenin. Then, β-catenin can accumulate in the cytoplasm and eventually, β-catenin translocates to the nucleus, where it acts as a transcriptional activator of transcription factors in the T-cell-specific factor/lymphoid enhancing factor Tcf/Lef family and increases the transcription of Wnt target genes encoding axin, Smad6, cyclin D1 and Cx43 [12] (Figure 1). There are numerous Wnt ligands, receptors, co-factors, antagonists, and intracellular mediators.

As the regenerative capacity of multiple mammalian tissues has been shown depending on Wnt/β-catenin signaling and its activation. In the context of pulp dentin, complex regeneration the role of canonical pathway seems obvious. Numerous studies have shown this role in reactionary dentinogenesis [13]. In reparative dentinogenesis, the repair process is accompanied by increased Axin2 expression, which results in differentiation of Axin2 expressing cells from resident dental pulp stem cells into odontoblasts-like cells [14].

However, the specific role of Wnt/β-catenin signaling on odontoblast-like differentiation of hDPSCs is not completely known. Indeed, some authors such as Scheller et al. demonstrated that canonical Wnt signaling inhibited odontogenic differentiation of hDPSCs [15]. In addition, Zhang et al. reported that Wnt10a, a Wnt agonist, could negatively regulate the differentiation of DPSCs into odontoblasts by down-regulating odontoblast specific genes [16]. However, some other researchers have reported that β-catenin accumulation by various agonists promoted odontoblastic differentiation in hDPSCs [17–20]. Thus, β-catenin could play an essential role in tertiary dentinogenesis [17,21]. β-catenin could act as an activator of the transcription factor runt-related transcription factor 2 (Runx2) to enhance the odontoblastic differentiation of dental pulp stem cells [19] and stem cells from the apical papilla (SCAP) [22].

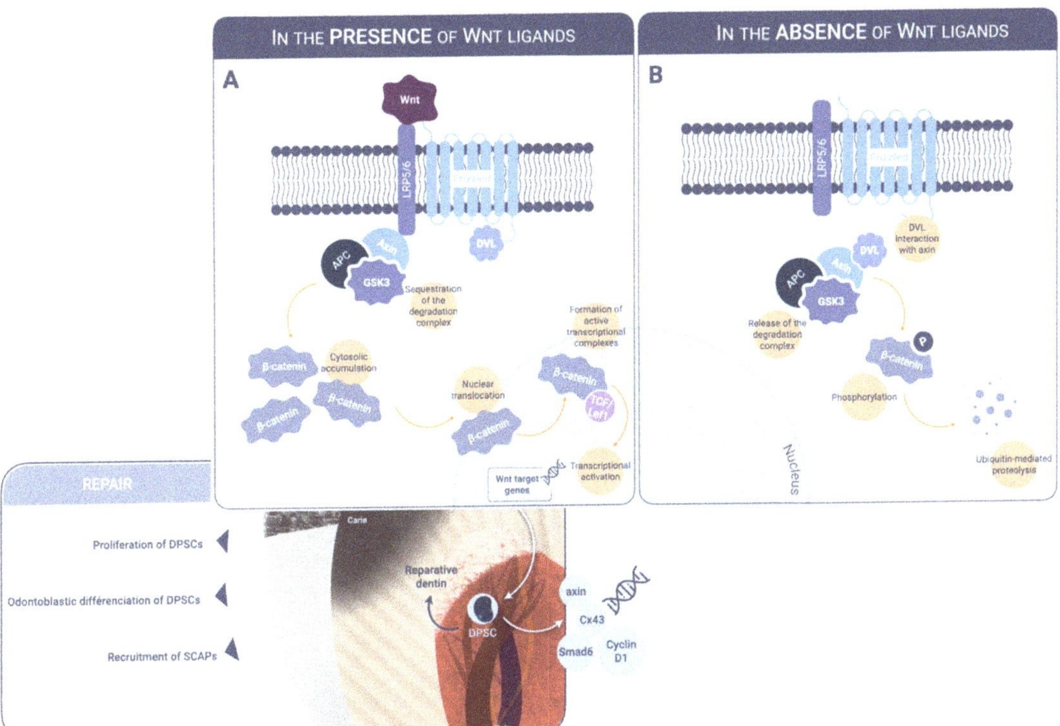

Figure 1. Schematic illustration of the canonical Wnt pathway. (**A**). In the presence of Wnt ligands interacting with LRP5/6 and frizzled, the β-catenin degradation complex is sequestrated. Cytosolic accumulation of β-catenin leads to nuclear translocation and binding to transcription factors in the Lef/Tcf family. The resulting active transcriptional complex controls the expression of target genes involved in tissue generation, regeneration and, self-renewal. (**B**). In the absence of Wnt ligands, the interaction between DVL and axin leads to the phosphorylation of cytosolic β-catenin by a protein complex involving APC, axin, and GSK3. β-catenin is then degraded by ubiquitin-mediated proteolysis. LRP: lipoprotein receptor-related protein. DVL: dishevelled-APC: adenomatosis polyposis coli. GSK: glycogen synthase kinase. TCF/Lef1: T-cell factor/lymphoid enhancer factor. DPSCs: dental pulp stem cells. SCAPs: stem cells from apical papilla.

In this review, we aimed to recapitulate new biologically based strategies to enhance this natural repair response by regulating Wnt signaling via modulatory molecules. These molecules could be a potential therapy target for dentin pulp complex.

2. Modulators of Wnt Beta-Catenin Signaling Acting on Dental Pulp Cells

2.1. Inorganic Calcium-Containing Materials

The treatment of dental caries that results in pulp exposure is currently managed by replacing lost dentine with inorganic calcium-containing materials such as CH, MTA or Biodentine that remain in the crown. Since this dentine is formed directly from new odontoblast-like cells that differentiate from resident stem cells in the pulp [14], it is imaginable that overstimulation of stem cell activity might result in increased odontoblast differentiation resulting in more efficient regenerative dentine formation. The most studied material (ProRoot MTA), in direct contact with DPSCs/DPCs, has shown significant positive results in in vitro assays assessing the involvement of the MAPK subfamilies JNK and P38, the ERK subfamily, the nuclear factor kappa B (NF-κB), and Wnt/β-catenin pathways.

2.2. Small Molecule GSK3 Inhibitors

In the context of response to damage, the pulp dentin complex can induce a Axin2 increase repair process in odontoblast-like cells that subsequently form reparative dentin [14]. Glycogen synthase kinase 3 (GSK3) is a core intracellular component of the Wnt/β-catenin signaling pathway that phosphorylates Axin and β-catenin [23–25]. When there are Wnt ligands, GSK3 activity is disabled, and β-catenin can enter the nucleus to interact with Lef/Tcf transcription and express the target genes such as Axin2. When Wnt ligands are absent, β-catenin and Axin2 are phosphorylated, causing their ubiquitination and then their degradation. GSK3 inhibitors (which are also Wnt pathway activators) could have various forms. They have shown to have natural or synthetic sources and display different mechanisms of action. A range of small molecule antagonists of GSK3 have been developed as drugs to activate the Wnt pathway in responsive cells [23,26–28]. Several GSK3 inhibitors have been shown to promote dentin repair in mice and rats with experimental pulp exposure [29,30].

2.2.1. Tideglusib

Tideglusib is the most studied GSK3 drug that has to date been shown to be safe in patients [31]. Delivery of GSK3 inhibitor drugs (20 μM CHIR99021 and 1 μM Tideglusib on biodegradable collagen sponges) directly into experimentally exposed pulp cavities in mice results in upregulation of Wnt-activity in pulp stem cells [29] and induction of high quality of reparative dentinogenesis [30]. This reparative dentine was biochemically indistinguishable from native dentine when analyzed by Raman spectroscopy. Although the extent of damage in rats is not comparable to that in large lesions in humans, the successful scaleup of reparative dentine formation in vivo seems promising, highlighting the potential of this approach. However, Tideglusib has low aqueous solubility, and in clinical trials, is delivered in a granulate form suspended in water that is less suitable in clinical practice.

2.2.2. NP928

NP928 is a new GSK3 inhibitor small-molecule drug that has increased aqueous solubility compared to other thiadiazolidinone (TDZD) drugs and can activate Wnt/β- catenin pathway similarly to tideglusib. Therefore, NP928 is a modified version of Tideglusib that removes the naphthyl moiety and increases solubility. The reparative potency and clinical usability of NP928 was evaluated in microdose concentrations of loaded MA-HA hydrogels in a pulp damage model in wild-type mice, and showed more reparative dentine was in tested groups than in controls [32].

The use of hydrogel looks superior to the sponge delivery, and the overall simpler user experience for the clinician. The use of hydrogel looks superior to the sponge delivery, and the overall simpler user experience for the clinician. These results allowed a drug called ReDent® to be transferred to clinic, on the basis of it being ready for its first human clinical trials.

A tooth cavity is a good therapeutic site for the use of a small molecule, used at very tiny concentrations. They should have a short half-life and limited range of action. They do not activate cells in the roots, for example.

2.2.3. Tivantinib

Tivantinib is a small molecule that inhibits c-Met receptor tyrosine kinase. It is a non-ATP competitive inhibitor. GSK3α and GSK3β, two structurally isoforms of GSK3, have been identified as new targets for this molecule [33]. These isoforms are negatively and positively regulated by serine or tyrosine phosphorylation, respectively. Recently, GSK3α and β were inhibited by Tivantinib in lung cancer cells [34]. Therefore, this molecule was tested in phase III trial of the treatment of hepatocellular carcinoma. Tivantinib's biocompatibility and low cytotoxicity have also been demonstrated on progenitors' murine cells [35]. It is quite interesting for the best of our knowledge and dental practice to notice that a c-Met inhibitor tyrosine kinase used in a carcinoma treatment presented a weak toxicity for dental pulp cells and the ability to activate the Wnt/β-catenin pathway at very low concentrations in vitro [35]. Nevertheless, further studies are needed to analyze and evaluate in vivo effects of delivering tivantinib on pulp injuries via collagen sponges or other vectors, such as hydrogels.

2.2.4. Lithium Chloride

Lithium compounds are inhibitors of GSK3β, used among other things to treat the bipolar patients and inhibit cancer cell metastasis [36]. Many in vitro studies reported that these agents can potentiate bone regeneration process and upregulate osteoblast differentiation and mineralization [37,38]. Ishimoto's team identified the effect of 10 mM Lithium chloride as an activator of reparative tubular dentin formation. The application of Lithium has been realized locally in rats after a pulpotomy procedure. Thus, they have shown stimulation of the Wnt/b-catenin pathway (through inhibition of the b-catenin destruction complex) when pulp cells were treated with lithium ions in vitro [39]. Interestingly, recently LiCl-100 mM was shown to activate this signaling pathway in vitro [40]. In rats, capping of pulps with surface pre-reacted glass combined with LiCl at concentrations of 10 mM or 100 mM was associated with the formation of complete reparative dentin structures that were continuous with the primary dentin without any defects, similar to that produced by MTA [40]. A research team has incorporated lithium-containing bioactive glass in a commercial GIC, so that lithium released from the GIC could naturally penetrate dentin and stimulate odontoblast activity. They succeeded to stimulate dentin formation and improve repair in a murine molar defect model [41]. Likewise, in a context injury in restorative treatment with resin polymers, the study of Bakopoulou et al. in 2015 [42] showed that human DPSCs were stimulated after lithium treatment through the accumulation of β-catenin and enhancement of its translocation in nucleus and expression of transcription factors. Exposure of lithium chloride-pre-treated cells to TEGDMA (triethylene-glycol-dimethacrylate) showed a stronger activation of the pathway. Thus, these findings stipulated that TEGDMA could continue to induce canonical Wnt pathway in DPSCs that were already "activated" by various environmental factors during pulp repair.

2.3. R-Spondin 2

Recombinant proteins such as the Wnt agonist R-spondin [43] have been used to treat oral mucositis [44]. R-spondins are secreted proteins that act as stem cell growth factors [45]. It has been shown that R-spondins clearly increase Wnt signal. R-spondin 2 (Rspo2) has been reported to show a predominant role in many differentiation processes, such as neurogenic differentiation [46], chondrogenic [47], and osteoblastic [48] differentiation. In a recent study, Gong Y. et al. succeeded in inducing odontogenic differentiation in combination with exogenously added Rspo2 in hDPSCs through an increase in levels of both mRNA and protein expression of dentin sialophosphoprotein, dentin matrix protein-1, alkaline phosphatase, bone sialoprotein, and protein expression levels of osteopontin and osteocalcin, whereas silencing Rspo2 significantly decreased the expression levels of these odontogenic markers [49]. This promotion of odontogenic differentiation is attributable to the activation of Wnt/β-catenin signaling. Thus, more investigations are needed to evaluate effects of Rspo2 in dental pulp complex repair.

2.4. Wedelolactone

Wedelolactone is a natural plant compound that has been shown to have anti-inflammatory, anticancer, and antiosteoporosis effects. The effect of wedelolactone has also been evaluated for dental treatment. For that purpose, DPSCs were treated with wedelolactone in vitro [50]. This experiment has been shown to promote odontoblast differentiation and mineralization through a direct enhancement of the nuclear accumulation of β-catenin and expression of genes involved in odontoblast differentiation. These genes included DMP-1, DSPP, and runx2. This study highlighted that wedelolactone induced the differentiation of odontoblast cells by means of semaphorin 3A/neuropilin-1 pathway-mediated β-catenin stimulation and NF-κB pathway disactivation.

2.5. Semaphorin 3A and Its Receptor Neuropilin 1

Sema3A has been shown to be an osteoprotective factor by inhibiting osteoclastic bone resorption and promoting osteoblastic bone formation through canonical Wnt/β-catenin signaling [51]. When Neuropilin-1 (NRP1) binds to Sema3A, it stimulates osteoblast differentiation through the classical Wnt/β-catenin pathway. Overexpression of NRP1 upregulated dentin matrix protein-1, dentin sialophosphoprotein, alkaline phosphatase protein level, and mineralization in DPSCs, while knockdown of NRP1 induced the opposite effects. NRP1, therefore, regulates DPSCs via the classical Wnt/β-catenin pathway [52]. Sema3A has also been showed to induce cell migration, chemotaxis, proliferation, and odontoblastic differentiation of DPSCs, and Sema3A application to dental pulp exposure sites in a rat model induced effective reparative dentin reconstruction [19].

2.6. Wnt3a Protein

In vitro, the effects of a continuous activation of Wnt/β-catenin signaling by the addition of Wnt3a on the mineralization and differentiation of pulp cells demonstrated that Wnt3a induced marked increases in the expression of Dmp1, Dspp, and Bsp, compared to controls between days 10 and 17 [53], and highlighted the role of Wnt/β-catenin signaling in the survival of resident progenitors. Indeed, a limited and early exposure to Wnt3a resulted in increased proliferation and decreased apoptosis in the undifferentiated population [54]. In vivo, the study of Hunter et al.; in 2015 [54] showed that pulp healing was positively impacted by a Wnt3a (a typical canonical Wnt ligand) amplified environment in a model of direct pulp capping. In fact, the application of Wnt3a through a lipidic vesicle, which allowed the maintenance of its activity, led to the formation of a reparative matrix resembled native dentin. In addition, this liposome delivered Wnt3a protected pulp cells from death and stimulated proliferation of undifferentiated cells in the pulp, which together significantly improved pulp healing.

2.7. Sclerostin

Sclerostin, is a secreted glycoprotein which is largely produced by osteocytes under a physiologic environment. It is an antagonist of the Wnt-BMPs signaling pathway through its binding to LRP 5/6receptor which is present on the membrane osteoblast [55]. It has been shown that when sclerostin is downregulated, an increase of osteogenesis and in bone mass are observed [56].

Secretion of sclerostin by odontoblasts has been demonstrated during tooth development [57,58]. Many studies have investigated the potential role of this molecule in the dentin pulp complex healing process [59]. In absence of sclerostin in Sost knock out mice, an increase in the pulp-healing process, following a direct pulp-capping mice model, was demonstrated [59]. In vitro, cultures of mDPCs isolated from Sost knock out germs allowed for elevated mineralization. Interestingly, the role of sclerostin in the process of human dental pulp cell (hDPCs) senescence was studied as the expression level of sclerostin varies in embryonic and adult mouse incisors and molars, and in aged individuals [58,60]. Thus, it has been shown that expression of sclerostin was increased in senescent human dental pulp and subculture-induced senescent hDPCs by immunohistochemistry and qRT-PCR analyses [61]. In addition, overexpression of sclerostin led to hDPCs senescence and inhibition of odontoblastic differentiation of hDPCs. Therefore, an anti-sclerostin treatment may be beneficial for the maintenance of the proliferation and odontoblastic differentiation potentials of hDPCs and to improve the pulp healing process in exposed pulps treatment. Indeed, Liao et al., 2019 have shown that sclerostin increased the inflammatory responses of odontoblasts under an LPS-induced environment and led to impaired dentin tissue regeneration by inhibition of odontoblastic differentiation of inflamed DPCs [62]. These findings allow new ideas toward therapeutic treatments combining anti-inflammatory effects and promotion of regeneration during dental pulp inflammation.

3. Cellular Metabolism Effect of Wnt

3.1. Epigenetic Remodeling in Human DPSCs under Wnt Ligant Exposure

The canonical Wnt signaling pathway is considered as an important regulator of stemness [11,23,63] and cell differentiation [15] in DPSCs and many other stem cell types [64]. The epigenetic regulations that Wnt pathways may exert on DPSCs have been studied. The authors have found that Wnt-3a exposure induced plural epigenetic reprogramming facets in DPSCs; especially, a global DNA hypomethylation, a global histone hyperacetylation, and an increase in both activating and repressing histone methylation marks were highlighted [65]. These findings could have seductive implications in the optimization of the clinical cell therapy.

3.2. Effect on Energetic Metabolism

Certain systemic conditions compromised the capacity of proliferation and odontogenic differentiation of human DPSCs. Diabetes is one of these pathological conditions. The reduction in human pulpal mesenchymal cells stemness in diabetic patients could affect the regenerative capacity of pulp-dentin complex and the formation of the dentin bridge. Based on these data, some authors studied the potential role of Wnt signaling in a high glucose-induced senescence model (close to diabetic conditions) [66]. Interestingly, Asghari et al. have observed the decrease in proliferation of DPSC, as well as an increased number of senescent cells and an increased p21 expression, after being exposed to different concentrations of glucose. β -catenin and Wnt1 expression in response to high glucose were significantly increased. In the same way, the authors demonstrated that in the presence of a β-catenin inhibitor PNU-74564, the amount of the senescent cells was reduced. Therefore, Wnt signaling might be the potential target for the inhibition of the senescence response in the hyperglycemic condition, suggesting the potential development of bioactive materials applied in pulp capping that would be specific for diabetic patients.

3.2.1. Famotidine

Famotidine is a competitive inhibitor of the histamine receptor, which is the dominant receptor involved in gastric acid secretion. This binding prevents the activation of adenylate cyclase normally induced by histamines. Many studies have investigated the Famotidine potential anti GSK3 effect. It seems that this inhibiting role could be attributed to the hypoglycemic aspect. Indeed, H2-receptor inhibitors could affect glucose metabolism and its role in the decrease in the glycemic response curve in vivo through binging with GSK3β [67]. Its biocompatibility and low cytotoxicity have also been demonstrated on progenitors' murine cells [35]. Its effects on DPSC could be studied in a context of glucose exposition.

3.2.2. Olanzapine

Olanzapine used to be a particular pharmacological psychiatric drug used for the treatment of schizophrenia. Its potential anti GSK3 effect has been sought by many authors. As Famotidine, it seems that this inhibiting role could be due to the hypoglycemic effect. Its biocompatibility and low cytotoxicity have also been demonstrated on progenitors' murine cells [35]. This molecule could also be studied in glucose exposition condition of DPSC.

4. Conclusions

The Wnt/b catenin pathway is a very complex intra cellular pathway.

Several studies have shown that Wnt signals are necessary for pulp dentin complex formation and repair, and in other studies, some research groups have succeeded in demonstrating that a Wnt stimulus is sufficient to induce tissue regeneration. Small-molecule drugs that stimulate Wnt/β-catenin have shown promise as a novel biological therapy for treating exposed pulpal lesions. It is necessary to consider the need of local application to avoid systemic side effects of the highly potent small molecules acting as Wnt agonists at very tiny concentrations. Interestingly, they are very cheap to make. Strong proof-of-concept data are still needed, however, along with well-crafted safety plans, to make regenerative dental medicine a reality.

After pulp exposure, molecules enhancing Wnt/β-catenin can be applied in contact with the pulp on different supports such as hydrogels or sponges (Figure 2). These materials must be sealed tightly with restorative material. The release of molecules stimulates the regeneration of the pulp-dentin complex.

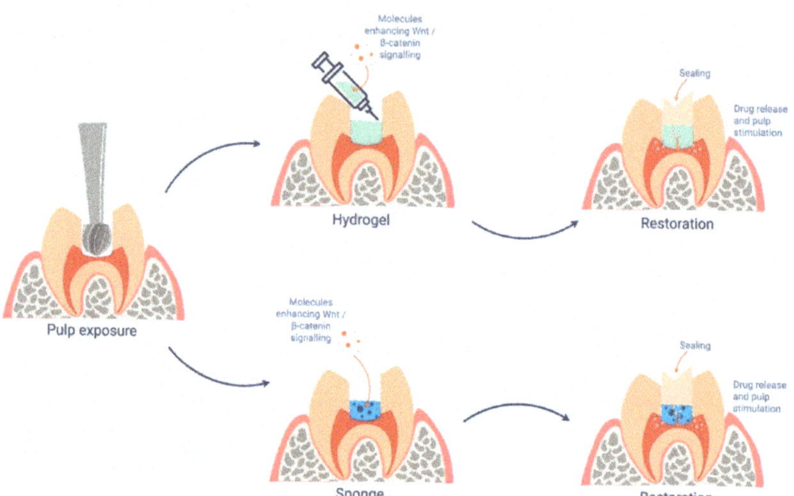

Figure 2. Schematic illustration of the clinical application of Wnt/β-catenin enhancers.

Author Contributions: All authors have contributed to the article's conceptualization, research methodology, writing—original draft preparation, review, and editing. All authors have read and agreed to the published version of the manuscript.

Funding: Minic, S.' thesis was co-funded by the French Ministry of Higher Education and Research (CIFRE fellowship No 2018/1781) and Septodont.

Institutional Review Board Statement: Not applicable.

Informed Consent Statement: Not applicable.

Data Availability Statement: Not applicable.

Conflicts of Interest: Minic, S is a student of the University of Paris whose thesis was co-funded by the French Ministry of Higher Education and Research (CIFRE fellowship No 2018/1781) and Septodont. The other authors have no potential conflicts of interest to declare in relation to the authorship and/or publication of this article.

References

1. Tziafas, D.; Economides, N. Formation of crystals on the surface of calcium hydroxide-containing materials in vitro. *J Endod.* **1999**, *25*, 539–542. [CrossRef]
2. Schroder, U. Effects of calcium hydroxide-containing pulp-capping agents on pulp cell migration, proliferation, and differentiation. *J. Dent. Res.* **1985**, *64*, 541–548. [CrossRef] [PubMed]
3. Goldberg, M.; Njeh, A.; Uzunoglu, E. Is Pulp Inflammation a Prerequisite for Pulp Healing and Regeneration? *Mediat. Inflamm.* **2015**, *2015*, 347649. [CrossRef]
4. Hilton, T.J. Keys to clinical success with pulp capping: A review of the literature. *Oper. Dent.* **2009**, *34*, 615–625. [CrossRef]
5. Tran, X.V.; Gorin, C.; Willig, C.; Baroukh, B.; Pellat, B.; Decup, F.; Opsahl Vital, S.; Chaussain, C.; Boukpessi, T. Effect of a calcium-silicate-based restorative cement on pulp repair. *J. Dent. Res.* **2012**, *91*, 1166–1171. [CrossRef]

6.   Awawdeh, L.; Al-Qudah, A.; Hamouri, H.; Chakra, R.J. Outcomes of Vital Pulp Therapy Using Mineral Trioxide Aggregate or Biodentine: A Prospective Randomized Clinical Trial. *J. Endod.* **2018**, *44*, 1603–1609. [CrossRef] [PubMed]

7.   Kuttler, Y. Classification of dentine into primary, secondary, and tertiary. *Oral Surg. Oral Med. Oral Pathol.* **1959**, *12*, 996–999. [CrossRef]

8.   Saito, K.; Nakatomi, M.; Ida-Yonemochi, H.; Ohshima, H. Osteopontin Is Essential for Type I Collagen Secretion in Reparative Dentin. *J. Dent. Res.* **2016**, *95*, 1034–1041. [CrossRef]

9.   Niehrs, C.; Acebron, S.P. Mitotic and mitogenic Wnt signalling. *EMBO J.* **2012**, *31*, 2705–2713. [CrossRef]

10.   Li, L.; Clevers, H. Coexistence of quiescent and active adult stem cells in mammals. *Science* **2010**, *327*, 542–545. [CrossRef]

11.   Clevers, H.; Loh, K.M.; Nusse, R. Stem cell signaling. An integral program for tissue renewal and regeneration: Wnt signaling and stem cell control. *Science* **2014**, *346*, 1248012. [CrossRef] [PubMed]

12.   Schulte, G. Frizzleds and WNT/beta-catenin signaling–The black box of ligand-receptor selectivity, complex stoichiometry and activation kinetics. *Eur. J. Pharmacol.* **2015**, *763*, 191–195. [CrossRef] [PubMed]

13.   Couve, E.; Osorio, R.; Schmachtenberg, O. Reactionary Dentinogenesis and Neuroimmune Response in Dental Caries. *J. Dent. Res.* **2014**, *93*, 788–793. [CrossRef] [PubMed]

14.   Babb, R.; Chandrasekaran, D.; Neves, V.C.M.; Sharpe, P.T. Axin2-expressing cells differentiate into reparative odontoblasts via autocrine Wnt/beta-catenin signaling in response to tooth damage. *Sci. Rep.* **2017**, *7*, 3102. [CrossRef]

15.   Scheller, E.L.; Chang, J.; Wang, C.Y. Wnt/beta-catenin inhibits dental pulp stem cell differentiation. *J. Dent. Res.* **2008**, *87*, 126–130. [CrossRef]

16.   Zhang, Z.; Guo, Q.; Tian, H.; Lv, P.; Zhou, C.; Gao, X. Effects of WNT10A on proliferation and differentiation of human dental pulp cells. *J. Endod.* **2014**, *40*, 1593–1599. [CrossRef]

17.   Han, N.; Zheng, Y.; Li, R.; Li, X.; Zhou, M.; Niu, Y.; Zhang, Q. beta-catenin enhances odontoblastic differentiation of dental pulp cells through activation of Runx2. *PLoS ONE* **2014**, *9*, e88890. [CrossRef]

18.   Kim, T.H.; Bae, C.H.; Lee, J.C.; Ko, S.O.; Yang, X.; Jiang, R.; Cho, E.S. beta-catenin is required in odontoblasts for tooth root formation. *J. Dent. Res.* **2013**, *92*, 215–221. [CrossRef]

19.   Yoshida, S.; Wada, N.; Hasegawa, D.; Miyaji, H.; Mitarai, H.; Tomokiyo, A.; Hamano, S.; Maeda, H. Semaphorin 3A Induces Odontoblastic Phenotype in Dental Pulp Stem Cells. *J. Dent. Res.* **2016**, *95*, 1282–1290. [CrossRef]

20.   Rahman, S.U.; Oh, J.H.; Cho, Y.D.; Chung, S.H.; Lee, G.; Baek, J.H.; Ryoo, H.M.; Woo, K.M. Fibrous Topography-Potentiated Canonical Wnt Signaling Directs the Odontoblastic Differentiation of Dental Pulp-Derived Stem Cells. *ACS Appl. Mater. Interfaces* **2018**, *10*, 17526–17541. [CrossRef]

21.   Yoshioka, S.; Takahashi, Y.; Abe, M.; Michikami, I.; Imazato, S.; Wakisaka, S.; Hayashi, M.; Ebisu, S. Activation of the Wnt/beta-catenin pathway and tissue inhibitor of metalloprotease 1 during tertiary dentinogenesis. *J. Biochem.* **2013**, *153*, 43–50. [CrossRef] [PubMed]

22.   Zhang, H.; Wang, J.; Deng, F.; Huang, E.; Yan, Z.; Wang, Z.; Deng, Y.; Zhang, Q.; Zhang, Z.; Ye, J.; et al. Canonical Wnt signaling acts synergistically on BMP9-induced osteo/odontoblastic differentiation of stem cells of dental apical papilla (SCAPs). *Biomaterials* **2015**, *39*, 145–154. [CrossRef] [PubMed]

23.   Clevers, H.; Nusse, R. Wnt/beta-catenin signaling and disease. *Cell* **2012**, *149*, 1192–1205. [CrossRef] [PubMed]

24.   Liu, C.; Li, Y.; Semenov, M.; Han, C.; Baeg, G.H.; Tan, Y.; Zhang, Z.; Lin, X.; He, X. Control of beta-catenin phosphorylation/degradation by a dual-kinase mechanism. *Cell* **2002**, *108*, 837–847. [CrossRef]

25.   Zeng, X.; Tamai, K.; Doble, B.; Li, S.; Huang, H.; Habas, R.; Okamura, H.; Woodgett, J.; He, X. A dual-kinase mechanism for Wnt co-receptor phosphorylation and activation. *Nature* **2005**, *438*, 873–877. [CrossRef]

26.   Coghlan, M.P.; Culbert, A.A.; Cross, D.A.; Corcoran, S.L.; Yates, J.W.; Pearce, N.J.; Rausch, O.L.; Murphy, G.J.; Carter, P.S.; Roxbee Cox, L.; et al. Selective small molecule inhibitors of glycogen synthase kinase-3 modulate glycogen metabolism and gene transcription. *Chem. Biol.* **2000**, *7*, 793–803. [CrossRef]

27.   Sato, N.; Meijer, L.; Skaltsounis, L.; Greengard, P.; Brivanlou, A.H. Maintenance of pluripotency in human and mouse embryonic stem cells through activation of Wnt signaling by a pharmacological GSK-3-specific inhibitor. *Nat. Med.* **2004**, *10*, 55–63. [CrossRef]

28.   Leone, A.; Volponi, A.A.; Renton, T.; Sharpe, P.T. In-Vitro regulation of odontogenic gene expression in human embryonic tooth cells and SHED cells. *Cell Tissue Res.* **2012**, *348*, 465–473. [CrossRef]

29.   Neves, V.C.; Babb, R.; Chandrasekaran, D.; Sharpe, P.T. Promotion of natural tooth repair by small molecule GSK3 antagonists. *Sci. Rep.* **2017**, *7*, 39654. [CrossRef]

30.   Zaugg, L.K.; Banu, A.; Walther, A.R.; Chandrasekaran, D.; Babb, R.C.; Salzlechner, C.; Hedegaard, M.A.B.; Gentleman, E.; Sharpe, P.T. Translation Approach for Dentine Regeneration Using GSK-3 Antagonists. *J. Dent. Res.* **2020**, *99*, 544–551. [CrossRef]

31.   Del Ser, T.; Steinwachs, K.C.; Gertz, H.J.; Andres, M.V.; Gomez-Carrillo, B.; Medina, M.; Vericat, J.A.; Redondo, P.; Fleet, D.; Leon, T. Treatment of Alzheimer's disease with the GSK-3 inhibitor tideglusib: A pilot study. *J. Alzheimers Dis.* **2013**, *33*, 205–215. [CrossRef] [PubMed]

32.   Alaohali, A.; Salzlechner, C.; Zaugg, L.K.; Suzano, F.; Martinez, A.; Gentleman, E.; Sharpe, P.T. GSK3 Inhibitor-Induced Dentinogenesis Using a Hydrogel. *J. Dent. Res.* **2022**, *101*, 46–53. [CrossRef] [PubMed]

33.   Munshi, N.; Jeay, S.; Li, Y.; Chen, C.R.; France, D.S.; Ashwell, M.A.; Hill, J.; Moussa, M.M.; Leggett, D.S.; Li, C.J. ARQ 197, a novel and selective inhibitor of the human c-Met receptor tyrosine kinase with antitumor activity. *Mol. Cancer Ther.* **2010**, *9*, 1544–1553. [CrossRef]

34. Kuenzi, B.M.; Remsing Rix, L.L.; Kinose, F.; Kroeger, J.L.; Lancet, J.E.; Padron, E.; Rix, U. Off-target based drug repurposing opportunities for tivantinib in acute myeloid leukemia. *Sci. Rep.* **2019**, *9*, 606. [CrossRef] [PubMed]
35. Birjandi, A.A.; Suzano, F.R.; Sharpe, P.T. Drug Repurposing in Dentistry; towards Application of Small Molecules in Dentin Repair. *Int. J. Mol. Sci.* **2020**, *21*, 6394. [CrossRef] [PubMed]
36. Shorter, E. The history of lithium therapy. *Bipolar Disord.* **2009**, *11*, 4–9. [CrossRef] [PubMed]
37. Li, L.; Peng, X.; Qin, Y.; Wang, R.; Tang, J.; Cui, X.; Wang, T.; Liu, W.; Pan, H.; Li, B. Acceleration of bone regeneration by activating Wnt/beta-catenin signalling pathway via lithium released from lithium chloride/calcium phosphate cement in osteoporosis. *Sci. Rep.* **2017**, *7*, 45204. [CrossRef]
38. Clement-Lacroix, P.; Ai, M.; Morvan, F.; Roman-Roman, S.; Vayssiere, B.; Belleville, C.; Estrera, K.; Warman, M.L.; Baron, R.; Rawadi, G. Lrp5-independent activation of Wnt signaling by lithium chloride increases bone formation and bone mass in mice. *Proc. Natl. Acad. Sci. USA* **2005**, *102*, 17406–17411. [CrossRef] [PubMed]
39. Ishimoto, K.; Hayano, S.; Yanagita, T.; Kurosaka, H.; Kawanabe, N.; Itoh, S.; Ono, M.; Kuboki, T.; Kamioka, H.; Yamashiro, T. Topical application of lithium chloride on the pulp induces dentin regeneration. *PLoS ONE* **2015**, *10*, e0121938. [CrossRef]
40. Ali, M.; Okamoto, M.; Komichi, S.; Watanabe, M.; Huang, H.; Takahashi, Y.; Hayashi, M. Lithium-containing surface pre-reacted glass fillers enhance hDPSC functions and induce reparative dentin formation in a rat pulp capping model through activation of Wnt/beta-catenin signaling. *Acta Biomater.* **2019**, *96*, 594–604. [CrossRef]
41. Alaohali, A.; Brauer, D.S.; Gentleman, E.; Sharpe, P.T. A modified glass ionomer cement to mediate dentine repair. *Dent. Mater.* **2021**, *37*, 1307–1315. [CrossRef]
42. Bakopoulou, A.; Leyhausen, G.; Volk, J.; Papachristou, E.; Koidis, P.; Geurtsen, W. Wnt/beta-catenin signaling regulates Dental Pulp Stem Cells' responses to pulp injury by resinous monomers. *Dent. Mater.* **2015**, *31*, 542–555. [CrossRef] [PubMed]
43. Kazanskaya, O.; Glinka, A.; del Barco Barrantes, I.; Stannek, P.; Niehrs, C.; Wu, W. R-Spondin2 is a secreted activator of Wnt/beta-catenin signaling and is required for Xenopus myogenesis. *Dev. Cell* **2004**, *7*, 525–534. [CrossRef] [PubMed]
44. Zhao, J.; Kim, K.A.; De Vera, J.; Palencia, S.; Wagle, M.; Abo, A. R-Spondin1 protects mice from chemotherapy or radiation-induced oral mucositis through the canonical Wnt/beta-catenin pathway. *Proc. Natl. Acad. Sci. USA* **2009**, *106*, 2331–2336. [CrossRef]
45. Sato, T.; van Es, J.H.; Snippert, H.J.; Stange, D.E.; Vries, R.G.; van den Born, M.; Barker, N.; Shroyer, N.F.; van de Wetering, M.; Clevers, H. Paneth cells constitute the niche for Lgr5 stem cells in intestinal crypts. *Nature* **2011**, *469*, 415–418. [CrossRef] [PubMed]
46. Takata, N.; Abbey, D.; Fiore, L.; Acosta, S.; Feng, R.; Gil, H.J.; Lavado, A.; Geng, X.; Interiano, A.; Neale, G.; et al. An Eye Organoid Approach Identifies Six3 Suppression of R-spondin 2 as a Critical Step in Mouse Neuroretina Differentiation. *Cell Rep.* **2017**, *21*, 1534–1549. [CrossRef] [PubMed]
47. Takegami, Y.; Ohkawara, B.; Ito, M.; Masuda, A.; Nakashima, H.; Ishiguro, N.; Ohno, K. R-spondin 2 facilitates differentiation of proliferating chondrocytes into hypertrophic chondrocytes by enhancing Wnt/beta-catenin signaling in endochondral ossification. *Biochem. Biophys. Res. Commun.* **2016**, *473*, 255–264. [CrossRef]
48. Arima, M.; Hasegawa, D.; Yoshida, S.; Mitarai, H.; Tomokiyo, A.; Hamano, S.; Sugii, H.; Wada, N.; Maeda, H. R-spondin 2 promotes osteoblastic differentiation of immature human periodontal ligament cells through the Wnt/beta-catenin signaling pathway. *J. Periodontal. Res.* **2019**, *54*, 143–153. [CrossRef]
49. Gong, Y.; Yuan, S.; Sun, J.; Wang, Y.; Liu, S.; Guo, R.; Dong, W.; Li, R. R-Spondin 2 Induces Odontogenic Differentiation of Dental Pulp Stem/Progenitor Cells via Regulation of Wnt/beta-Catenin Signaling. *Front. Physiol.* **2020**, *11*, 918. [CrossRef]
50. Wang, C.; Song, Y.; Gu, Z.; Lian, M.; Huang, D.; Lu, X.; Feng, X.; Lu, Q. Wedelolactone Enhances Odontoblast Differentiation by Promoting Wnt/beta-Catenin Signaling Pathway and Suppressing NF-kappaB Signaling Pathway. *Cell. Reprogram.* **2018**, *20*, 236–244. [CrossRef]
51. Hayashi, M.; Nakashima, T.; Taniguchi, M.; Kodama, T.; Kumanogoh, A.; Takayanagi, H. Osteoprotection by semaphorin 3A. *Nature* **2012**, *485*, 69–74. [CrossRef] [PubMed]
52. Song, Y.; Liu, X.; Feng, X.; Gu, Z.; Gu, Y.; Lian, M.; Xiao, J.; Cao, P.; Zheng, K.; Gu, X.; et al. NRP1 Accelerates Odontoblast Differentiation of Dental Pulp Stem Cells Through Classical Wnt/beta-Catenin Signaling. *Cell. Reprogram.* **2017**, *19*, 324–330. [CrossRef] [PubMed]
53. Vijaykumar, A.; Root, S.H.; Mina, M. Wnt/beta-Catenin Signaling Promotes the Formation of Preodontoblasts In Vitro. *J. Dent. Res.* **2021**, *100*, 387–396. [CrossRef]
54. Hunter, D.J.; Bardet, C.; Mouraret, S.; Liu, B.; Singh, G.; Sadoine, J.; Dhamdhere, G.; Smith, A.; Tran, X.V.; Joy, A.; et al. Wnt Acts as a Prosurvival Signal to Enhance Dentin Regeneration. *J. Bone Miner Res.* **2015**, *30*, 1150–1159. [CrossRef]
55. Li, X.; Zhang, Y.; Kang, H.; Liu, W.; Liu, P.; Zhang, J.; Harris, S.E.; Wu, D. Sclerostin binds to LRP5/6 and antagonizes canonical Wnt signaling. *J. Biol. Chem.* **2005**, *280*, 19883–19887. [CrossRef]
56. Li, X.; Ominsky, M.S.; Niu, Q.T.; Sun, N.; Daugherty, B.; D'Agostin, D.; Kurahara, C.; Gao, Y.; Cao, J.; Gong, J.; et al. Targeted deletion of the sclerostin gene in mice results in increased bone formation and bone strength. *J. Bone Miner Res.* **2008**, *23*, 860–869. [CrossRef]
57. Amri, N.; Djole, S.X.; Petit, S.; Babajko, S.; Coudert, A.E.; Castaneda, B.; Simon, S.; Berdal, A. Distorted Patterns of Dentinogenesis and Eruption in Msx2 Null Mutants: Involvement of Sost/Sclerostin. *Am. J. Pathol.* **2016**, *186*, 2577–2587. [CrossRef]
58. Naka, T.; Yokose, S. Spatiotemporal expression of sclerostin in odontoblasts during embryonic mouse tooth morphogenesis. *J. Endod.* **2011**, *37*, 340–345. [CrossRef]

59. Collignon, A.M.; Amri, N.; Lesieur, J.; Sadoine, J.; Ribes, S.; Menashi, S.; Simon, S.; Berdal, A.; Rochefort, G.Y.; Chaussain, C.; et al. Sclerostin Deficiency Promotes Reparative Dentinogenesis. *J. Dent. Res.* **2017**, *96*, 815–821. [CrossRef]
60. Satoh, A.; Brace, C.S.; Rensing, N.; Cliften, P.; Wozniak, D.F.; Herzog, E.D.; Yamada, K.A.; Imai, S. Sirt1 extends life span and delays aging in mice through the regulation of Nk2 homeobox 1 in the DMH and LH. *Cell Metab.* **2013**, *18*, 416–430. [CrossRef]
61. Ou, Y.; Zhou, Y.; Liang, S.; Wang, Y. Sclerostin promotes human dental pulp cells senescence. *PeerJ* **2018**, *6*, e5808. [CrossRef]
62. Liao, C.; Wang, Y.; Ou, Y.; Wu, Y.; Zhou, Y.; Liang, S. Effects of sclerostin on lipopolysaccharide-induced inflammatory phenotype in human odontoblasts and dental pulp cells. *Int. J. Biochem. Cell Biol.* **2019**, *117*, 105628. [CrossRef]
63. Uribe-Etxebarria, V.; Luzuriaga, J.; Garcia-Gallastegui, P.; Agliano, A.; Unda, F.; Ibarretxe, G. Notch/Wnt cross-signalling regulates stemness of dental pulp stem cells through expression of neural crest and core pluripotency factors. *Eur. Cell. Mater.* **2017**, *34*, 249–270. [CrossRef]
64. Huang, G.T.; Gronthos, S.; Shi, S. Mesenchymal stem cells derived from dental tissues vs. those from other sources: Their biology and role in regenerative medicine. *J. Dent. Res.* **2009**, *88*, 792–806. [CrossRef]
65. Uribe-Etxebarria, V.; Garcia-Gallastegui, P.; Perez-Garrastachu, M.; Casado-Andres, M.; Irastorza, I.; Unda, F.; Ibarretxe, G.; Subiran, N. Wnt-3a Induces Epigenetic Remodeling in Human Dental Pulp Stem Cells. *Cells* **2020**, *9*, 652. [CrossRef]
66. Asghari, M.; Nasoohi, N.; Hodjat, M. High glucose promotes the aging of human dental pulp cells through Wnt/beta-catenin signaling. *Dent. Med. Probl.* **2021**, *58*, 39–46. [CrossRef]
67. Medak, K.D.; Townsend, L.K.; Hahn, M.K.; Wright, D.C. Female mice are protected against acute olanzapine-induced hyperglycemia. *Psychoneuroendocrinology* **2019**, *110*, 104413. [CrossRef]

 International Journal of
Molecular Sciences

Article

Molecular Biological Comparison of Dental Pulp- and Apical Papilla-Derived Stem Cells

Martyna Smeda [1], Kerstin M. Galler [2], Melanie Woelflick [1], Andreas Rosendahl [1], Christoph Moehle [3], Beate Lenhardt [1], Wolfgang Buchalla [1] and Matthias Widbiller [1,*]

[1] Department of Conservative Dentistry and Periodontology, University Hospital Regensburg, 93053 Regensburg, Germany; martyna.smeda@ukr.de (M.S.); melanie.woelflick@ukr.de (M.W.); andreas.rosendahl@ukr.de (A.R.); beate.lenhardt@web.de (B.L.); wolfgang.buchalla@ukr.de (W.B.)

[2] Department of Operative Dentistry and Periodontology, Friedrich-Alexander-University Erlangen-Nürnberg, 91054 Erlangen, Germany; kerstin.galler@uk-erlangen.de

[3] Center of Excellence for Fluorescent Bioanalytics (KFB), University of Regensburg, 93053 Regensburg, Germany; christoph.moehle@exfor.uni-regensburg.de

* Correspondence: matthias.widbiller@ukr.de

Abstract: Both the dental pulp and the apical papilla represent a promising source of mesenchymal stem cells for regenerative endodontic protocols. The aim of this study was to outline molecular biological conformities and differences between dental pulp stem cells (DPSC) and stem cells from the apical papilla (SCAP). Thus, cells were isolated from the pulp and the apical papilla of an extracted molar and analyzed for mesenchymal stem cell markers as well as multi-lineage differentiation. During induced osteogenic differentiation, viability, proliferation, and wound healing assays were performed, and secreted signaling molecules were quantified by enzyme-linked immunosorbent assays (ELISA). Transcriptome-wide gene expression was profiled by microarrays and validated by quantitative reverse transcription PCR (qRT-PCR). Gene regulation was evaluated in the context of culture parameters and functionality. Both cell types expressed mesenchymal stem cell markers and were able to enter various lineages. DPSC and SCAP showed no significant differences in cell viability, proliferation, or migration; however, variations were observed in the profile of secreted molecules. Transcriptome analysis revealed the most significant gene regulation during the differentiation period, and 13 biomarkers were identified whose regulation was essential for both cell types. DPSC and SCAP share many features and their differentiation follows similar patterns. From a molecular biological perspective, both seem to be equally suitable for dental pulp tissue engineering.

Keywords: dental pulp stem cells; stem cells of the apical papilla; mesenchymal stem cells; regenerative endodontics; transcriptome

Citation: Smeda, M.; Galler, K.M.; Woelflick, M.; Rosendahl, A.; Moehle, C.; Lenhardt, B.; Buchalla, W.; Widbiller, M. Molecular Biological Comparison of Dental Pulp- and Apical Papilla-Derived Stem Cells. *Int. J. Mol. Sci.* **2022**, 23, 2615. https://doi.org/10.3390/ijms23052615

Academic Editor: Takayoshi Yamaza

Received: 25 January 2022
Accepted: 25 February 2022
Published: 27 February 2022

Publisher's Note: MDPI stays neutral with regard to jurisdictional claims in published maps and institutional affiliations.

1. Introduction

Regenerative endodontic procedures aim to replace an irreversibly inflamed or necrotic dental pulp. In order to generate new pulp-like tissue, researchers have successfully made use of stem cells, which is one of the three pillars in tissue engineering next to scaffold materials and signaling molecules [1–4]. Currently, two tissue engineering concepts for pulp regeneration can be differentiated, the first being based on cell transplantation and the second on cell-homing [5]. For the transplantation approach, stem cells and growth factors are inserted into a suitable scaffold and injected directly into the root canal. This requires storage and laboratory processing of stem cells beforehand, which is afflicted with high costs. However, a primarily cell-free approach based on cell-homing seems to be more practical for use in dental offices. In this case, no cells have to be transplanted but local stem cells are attracted from periapical tissues by recombinant signaling molecules or endogenous, dentin-derived growth factors and migrate into the root canal. Moreover, cell-homing can not only be used to restore the whole pulp but also parts of the tissue that are lost due to local inflammatory or necrotic processes [3,6,7].

Among various types of stem cells associated with dental tissues [8], especially dental pulp stem cells (DPSC) [9] and stem cells from the apical papilla (SCAP) [10] appear to be suitable sources for cell-homing as they are located in the root canal or in the apical papilla at the root tip and can thus give rise to new tissue. Interestingly, they share a common developmental origin as derivates of the pluripotent cranial neural crest cells that migrate to the first branchial arch and form the dental mesenchyme or ectomesenchyme [11–13]. During tooth development, interactions between the ectomesenchyme and the primitive oral epithelium result in the formation of a tooth bud. Subsequently, ectomesenchymal cells start to condense beneath and around the bud which leads to the formation of the dental papilla and the dental follicle. As the enamel organ continues to grow, forming first a cap and later a bell shape, the epithelial cervical loops enclose the cells of the dental papilla, initiating their transformation into the dental pulp. As soon as crown development is near completion, root formation starts with the apical proliferation of the cervical loops which now form a two-layered structure called Hertwig's epithelial root sheath (HERS). HERS determines the shape of the later tooth root(s) and harbors mesenchymal cells, but has only limited growth potential. During root development, the dental papilla gradually transforms into radicular pulp tissue, whereas the follicle turns into periodontium [13–15]. Thus, the remaining dental papilla, which is termed apical papilla, can be found at the root end of immature teeth until root formation is completed. Histologically, it appears as a densified connected tissue separated from the pulpal tissue by a cell-rich zone.

Since both tissues can provide cells for pulp regeneration, the question arises whether the originating stem cells are equally suitable for this purpose. In addition to general qualities such as migration and proliferation, the ability to differentiate into a mineralizing odontoblast-like phenotype plays a particularly important role. Currently, there is still much to find out about genetic regulation during the differentiation of DPSC and SCAP, as only a modest number of studies directly compare the two stem cell types [10,16,17].

Therefore, the aim of this study was to outline parallels as well as differences between DPSC and SCAP isolated from the same donor regarding stemness, proliferation and viability, migration, and production of signaling molecules. The main focus was placed on the gene expression profiling of both cell types to identify genome-wide regulatory genes during induced differentiation.

2. Results

2.1. Stem Cell Characterization

Overall, forward- and side-scatter signals revealed that both cell types were similar in size and granularity from a cytomorphological perspective (Figure 1a,b). A high proportion of DPSC and SCAP expressed mesenchymal stem cell markers, however, the markers of different origin (CD34, CD45, CD11b, CD19, and HLA-DR) were consistently undetected (Figure 1c–e). Cell culture experiments showed that both SCAP (Figure 1f–h) and DPSC (Figure 1i–k) were able to enter osteogenic, chondrogenic, and adipogenic lineage. Generally, Alizarin Red S stained large, widespread areas of mineralization whereas Oil Red O staining showed primarily scattered deposits of neutral lipids inside the cultured cells. The Alcian Blue 8GX dye revealed clusters of glycosaminoglycans as part of the recently formed cartilaginous matrix.

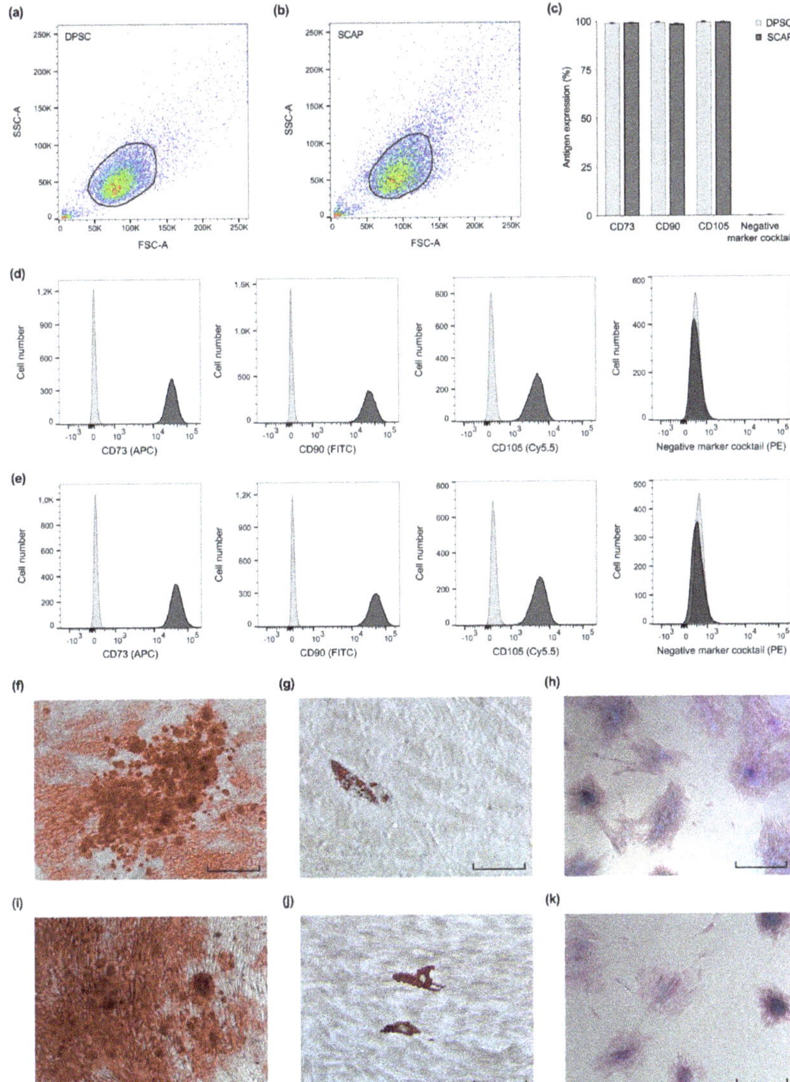

Figure 1. Stem cell characterization. Flow cytometric analysis (**a–e**) revealed similarities in size and granularity for dental pulp stem cells (DPSC) (**a**) and stem cells from the apical papilla (SCAP) (**b**). (**c**) DPSC and SCAP both expressed characteristic mesenchymal stem cell markers (CD73, CD90, and CD105), however, the markers CD34, CD45, CD11b, CD19, and HLA-DR were not detected. Exemplary overlay histograms of DPSC (**d**) and SCAP (**e**) show the control populations (light grey) and the specifically stained cells (dark grey). DPSC (**f–h**) and SCAP (**i–k**) both successfully entered the osteogenic (**f,i**), adipogenic (**g,j**), and chondrogenic lineage (**h,k**). Scale bars: 80 µm.

2.2. Cell Viability and Proliferation

Cell viability and proliferation assays showed similar patterns for SCAP and DPSC (Figure 2a,b). Viability and cell number increased until day 7 for both cell types and fetal bovine serum (FBS) concentrations. From day 5 on, DPSC and SCAP cultivated with 10% FBS showed a significantly higher viability and cell number compared to the ones cultivated with 1% FBS ($p \leq 0.003$). However, cell number and viability of DPSC and SCAP showed no significant differences at the same culture conditions ($p \geq 0.1657$).

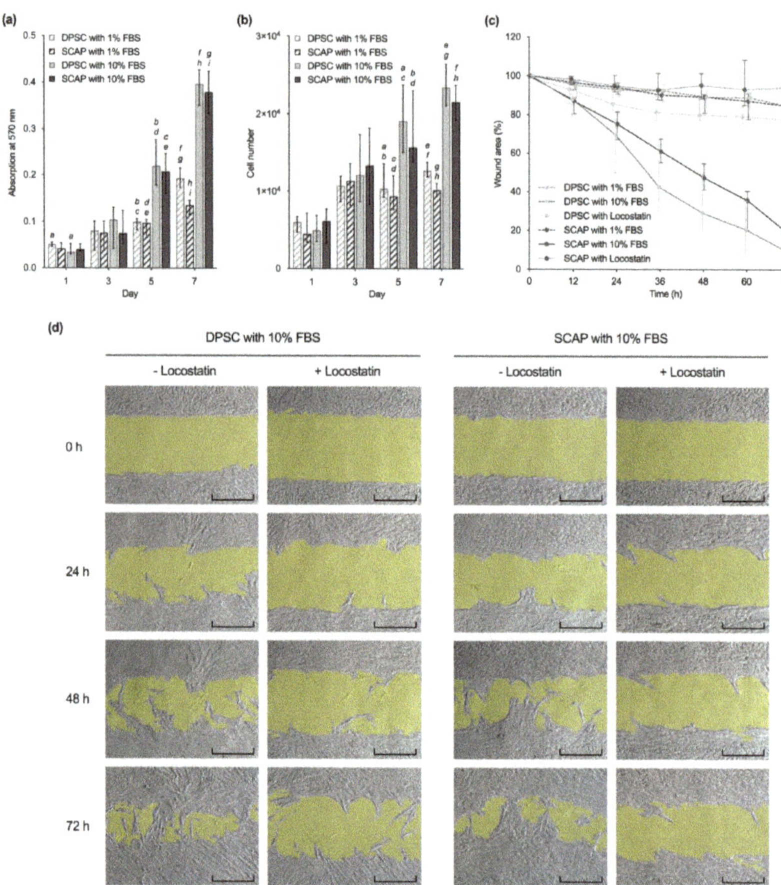

Figure 2. Comparison of (**a**) viability, (**b**) proliferation, and (**c**,**d**) migration of dental pulp stem cells (DPSC) and stem cells from the apical papilla (SCAP). (**a**,**b**) Viability and proliferation of DPSC and SCAP at different serum concentrations. Median values and 25–75% percentiles are based on three independent experiments performed in triplicates ($n = 9$). For each time point, the same lower-case letters indicate bars with statistically significant difference ($p \leq 0.05$). (**c**) Cell migration of both cell types was determined over 72 h in triplicates and repeated two times ($n = 12$). Medians and 25–75% percentiles were normalized to the initial wound area, which was set to 100%. (**d**) Microscopic images of gap closure (gap area highlighted in yellow) by DPSC and SCAP cultured in alpha minimum essential medium (αMEM) with 10% fetal bovine serum (FBS) with and without Locostatin over 72 h. Scale bars: 400 µm.

2.3. Cell Migration

In the course of 72 h, wound healing was observed in all groups without the addition of inhibitors, whereby a clear influence of the serum concentration was observed (Figure 2c). At all times, cells cultivated with 10% FBS showed significantly higher migration rates than the ones cultivated with 1% FBS ($p \leq 0.0212$). However, there were no statistically significant differences between DPSC with 10% FBS and SCAP with 10% FBS during 72 h ($p > 0.9999$). The addition of the migration inhibitor Locostatin to 10% FBS decreased migration rates of both DPSC and SCAP after 24 h with statistical significance ($p \leq 0.0002$).

The microscopic images show continuous sheath migration over 72 h for SCAP as well as DPSC cultures (Figure 2d). Locostatin largely suppressed this without evidence of cytotoxic effects.

2.4. Release of Signaling Molecules

In general, DPSC released more osteoprotegerin (OPG), tissue inhibitor of metallo-proteinase (TIMP), vascular endothelial growth factor (VEGF), and transforming growth factor beta 1 (TGF-β1) than SCAP (Figure 3). While the amount of interleukin 6 (IL-6) was similarly reduced in DPSC and SCAP during osteogenic differentiation ($p \leq 0.0056$), significantly more interleukin 8 (IL-8) was secreted in the SCAP cultured with StemPro® compared to medium with 10% FBS ($p < 0.0001$). Regarding the culture conditions, less IL-6 was released from both cell types during induced differentiation ($p \leq 0.0056$). At the same time, more IL-8 was secreted in osteogenic cultures with statistical significance for SCAP ($p < 0.0001$). While DPSC showed an increasing VEGF release during osteogenic differentiation, SCAP showed no such tendencies and VEGF secretion was even significantly lower compared to DPSC in osteogenic culture ($p \leq 0.0082$). Similarly, the release of TIMP was higher for DPSC compared to SCAP in the respective culture conditions ($p \leq 0.0331$). The release of TGF-β1 was significantly higher during induced differentiation of DPSC compared to SCAP for the intermediate and late phase ($p \leq 0.0038$).

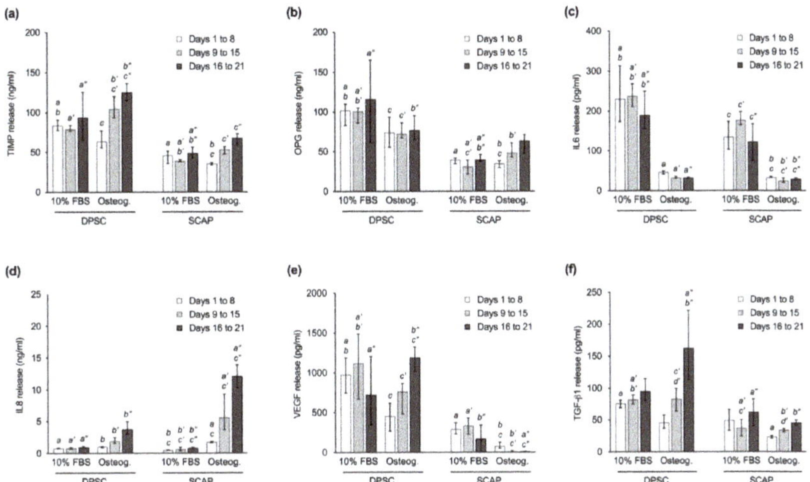

Figure 3. Release of signaling molecules by dental pulp stem cells (DPSC) and stem cells from the apical papilla (SCAP) in the initial phase (days 1 to 8), intermediate phase (days 9 to 15), and late phase (days 16 to 21) of culture. Graphs show the secretion of (**a**) tissue inhibitor of metalloproteinase (TIMP), (**b**) osteoprotegerin (OPG), (**c**) interleukin 6 (IL-6), (**d**) interleukin 8 (IL-8), (**e**) vascular endothelial growth factor (VEGF), and (**f**) transforming growth factor beta 1 (TGF-β1). Median values and 25–75% percentiles are based on three independent experiments performed in triplicates ($n = 9$). For each culture phase, equal lower-case letters indicate pairs that were found significantly different ($p \leq 0.05$). One or two apostrophes were added for the intermediate or late phase for a better overview.

With regard to the time course, cells cultured with StemPro® showed a continuous increase in the release of TIMP, IL-8, and TGF-β1 from day 1 to 21, which was statistically significant for all three signaling molecules when comparing the initial and the late phase ($p < 0.0001$). In addition, during osteogenic differentiation, an increasing production of VEGF and OPG was observed in DPSC and SCAP, respectively, with statistical significance between the initial and the late phase ($p \leq 0.0001$). No relevant changes of expression levels were observed for IL-6 for both cell types.

2.5. Gene Expression Profiling

The transcriptome analysis showed that, with regard to the culture parameters, the culture time in particular has an impact on gene regulation, whereas the culture medium as well as the cell type were only secondary variables. The comparison of the repeats demonstrates the high reproducibility of the experiments (Figure 4a). The observation of up- and downregulation of genes over a culture period of 14 days revealed for both cell types that fewer genes were regulated during osteogenic differentiation compared to the control with 10% FBS (Figure 4b). The majority of genes that were differentially expressed during induced osteogenesis were upregulated on day 14, only a few were downregulated.

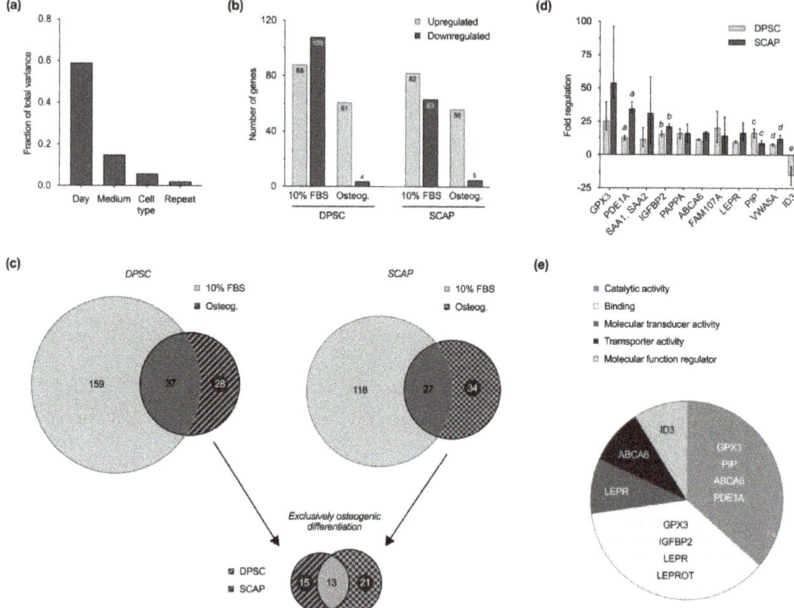

Figure 4. Transcriptome analysis. (**a**) Fraction of total variance highlights the impact of culture time on gene regulation among other culture parameters. (**b**) Number of up- and downregulated genes within the investigated groups on day 14 compared to day 1. Similarities between dental pulp stem cells (DPSC) and stem cells from the apical papilla (SCAP) are visible in standard cell culture and osteogenic differentiation. Fewer genes were regulated during differentiation and most of them were upregulated on day 14 of cell culture. (**c**) VENN diagrams for each cell type reveal subsets of genes that are specifically regulated in standard culture or osteogenic differentiation. Comparison of exclusively regulated genes during osteogenic differentiation of DPSC (28) and SCAP (34) enables the identification of 13 key genes that play a key role in osteogenesis of both cell types. (**d**) Microarray validation by quantitative reverse transcription PCR (qRT-PCR). Regulation of selected genes in DPSC and SCAP during induced differentiation at day 14 as median values and 25–75% percentiles. Regulations of all genes were statistically significant compared to baseline ($p \leq 0.0236$), whereas equal lower-case letters for each gene indicate pairs (DPSC and SCAP) with significant differences

($p \leq 0.05$). (**e**) PANTHER classification of 13 key genes according to molecular function. Genes with unknown or unapproved function are not shown (*SAA2/SAA2-SAA4/SAA4, SAA1, GPM6B, FAM107A, PAPPA, VWA5A*).

Of all the regulated genes, those that are differentially regulated exclusively in the course of osteogenic differentiation were identified for both cell types; these were 28 in DPSC and 34 in SCAP (Supplementary File 1). Of particular interest is the intersection of 13 regulated genes that were reproducibly regulated in the course of induced osteogenesis in both DPSC and SCAP (Figure 4c). Thus, 13 genes were identified that were exclusively regulated in both cell types during osteogenic differentiation, 12 of which were over-expressed and 1 of which was diminished (Table 1). The results of the microarray analysis were validated and confirmed by quantitative reverse transcription PCR (qRT-PCR) of selected genes. Figure 4d shows the regulation of gene expression in DPSC and SCAP during induced differentiation at day 14 compared to day 1.

Table 1. Genes that were exclusively regulated in both cell types during induced osteogenic differentiation.

Gene	Description	Fold Change	*p*-Value	FDR
GPX3	glutathione peroxidase 3	48.34	3.33×10^{-8}	8.93×10^{-5}
PIP	prolactin-induced protein	29.21	1.02×10^{-5}	1.40×10^{-3}
IGFBP2	insulin like growth factor binding protein 2	23.67	1.97×10^{-7}	2.00×10^{-4}
SAA2; SAA2-SAA4; SAA4	serum amyloid A2; SAA2-SAA4 readthrough; serum amyloid A4, constitutive	21.85	2.06×10^{-7}	2.00×10^{-4}
SAA1	serum amyloid A1	20.62	1.66×10^{-6}	5.00×10^{-4}
GPM6B	glycoprotein M6B	19.01	3.08×10^{-6}	7.00×10^{-5}
FAM107A	family with sequence similarity 107, member A	15.8	2.88×10^{-6}	7.00×10^{-4}
LEPR; LEPROT	leptin receptor; leptin receptor overlapping transcript	14.87	3.35×10^{-5}	2.60×10^{-3}
PAPPA	pregnancy-associated plasma protein A, pappalysin 1	14.22	5.79×10^{-6}	1.00×10^{-3}
ABCA6	ATP binding cassette subfamily A member 6	13.67	3.75×10^{-5}	2.70×10^{-3}
VWA5A	von Willebrand factor A domain containing 5A	12.06	9.66×10^{-6}	1.40×10^{-3}
PDE1A	phosphodiesterase 1A, calmodulin-dependent	10.26	6.00×10^{-7}	3.00×10^{-4}
ID3	inhibitor of DNA binding 3, dominant negative helix-loop-helix protein	−46.66	1.54×10^{-6}	5.00×10^{-4}

With the help of the PANTHER classification system, the respective genes were attributed to their respective molecular functions (Figure 4e). Several genes were assorted to the categories of catalytic activity (*GPX3, PIP, ABCA6, PDE1A*), binding (*GPX3, IGFBP2, LEPR/LEPROT*), molecular transducer activity (*LEPR*), transporter activity (*ABCA6*), and molecular function regulator (*ID3*); however, the molecular functions of some remain unknown (*SAA2/SAA2-SAA4/SAA4, SAA1, GPM6B, FAM107A, PAPPA, VWA5A*).

3. Discussion

Based on these results, both DPSC and SCAP appear to be suitable cells for regenerative endodontic approaches. It is important to consider, however, that their availability also depends on the stage of inflammation or necrosis in the dental pulp. When inflammation spreads, e.g., during carious decay, it is initially located adjacent to the area of bacterial invasion, whereas the rest of the pulp is initially not affected [7,18]. In this scenario, the irreversibly damaged tissue can be removed selectively, leaving behind a healthy pulp rich in mesenchymal stem cells and capable to regenerate the lost tissue [3,5]. Once the whole pulp is irreversibly inflamed or necrotic, this source disappears, leaving the apical papilla at the root tip of juvenile patients [19]. Residing stem cells in the apical papilla reportedly survive pulpal necrosis [20] and thus remain available for endodontic regeneration. They can either be actively brought into the canal via induced bleeding as part of revitalization procedures or migrate in the course of cell-homing approaches to form pulp-like tissue.

3.1. Stem Cell Characterization

Obviously, both mesenchymal stem cell types examined in this study originate from two evolutionary and anatomically closely linked tissues and reside in niches accessible for cell homing strategies for pulpal regeneration. While SCAP can be found in the apical papilla at the tips of immature roots, DPSC are located perivascularly inside the dental pulp and express vasculature-specific antigens such as α-smooth muscle actin, CD146, and the pericyte marker 3G5 [10,21]. Though they reflect different developmental stages of the former dental papilla, DPSC and SCAP have common features, e.g., the expression of the mesenchymal progenitor marker STRO-1 [10].

In accordance with previous studies, our analysis revealed both cell types to be positive for the mesenchymal stem cell markers CD73, CD90, and CD105, as well as a multi-lineage differentiation potential [10,22–24], both of which characterize mesenchymal stem cells according to Dominici et al. [25]. As reported previously, a pronounced osteogenic but rather weak adipogenic differentiation potential was observed for both DPSC and SCAP after Alizarin Red S and Oil Red O staining [17,26,27]. Thus, especially when comparing the two tissues of one donor, both the apical papilla and the dental pulp seem to represent a comparably suitable reservoir for multipotent stem cells. In regard to the pulp tissue's functionality, it is particularly their capability to transform into a mineralizing phenotype that is welcome in regenerative endodontic approaches.

3.2. Cell Viability, Proliferation, and Migration

Likewise, cell viability and proliferation as well as the ability for migration are important cell properties for endodontic regeneration. The goal is to achieve cell migration from their niches into a three-dimensional scaffold and transformation into a pulpal tissue [5].

In this regard, no significant differences were observed between DPSC and SCAP under the same culture conditions. These findings coincide with results from a recent study by Park et al. [17], which also reported a similar proliferation and colony-forming potential for SCAP and DPSC. Moreover, higher serum concentrations in cell culture media (10% FBS) had an impact on viability, proliferation, and migration of both cell types, as was already seen for SCAP [28]. The addition of Locostatin, a migration inhibitor targeting Raf kinase inhibitor protein (RKIP), significantly decreased the migration of DPSC and SCAP in a similar manner. In this context, it was shown that Locostatin inhibits not only the migration of terminally differentiated cells, but also of dental stem cells [29].

Nevertheless, controversial statements are made about the performance of DPSC and SCAP in the literature as well. Previous studies reported higher viability, proliferation, and migration rates for SCAP compared to DPSC [10,16,30]. Sonoyama et al. [10] defined SCAP as early progenitors from a developing tissue that are more suitable for the use in regenerative procedures due to higher expression levels of survivin, an inhibitor of apoptosis, and a higher telomerase activity, both relevant in terms of cell proliferation. However, these studies did not generally use cells from the same donor in the experiments and also differed in the culture conditions, e.g., serum content or passage. It also has to be mentioned that cell cultures, especially the ones from developing tissues such as SCAP, might contain several types of undifferentiated cells, resulting in different cell proliferation and differentiation potential [31].

The findings of this study suggest that DPSC and SCAP have similar viability and capacity for cell proliferation and migration, therefore they both fulfil the requirements for tissue engineering techniques. Nevertheless, it should also be mentioned here that although the results are based on cells derived from one donor, it cannot be ruled out that donor-dependent differences may also occur.

3.3. Release of Signaling Molecules

In addition to the culture behavior of the DPSC and SCAP, their signaling characteristics play an important role in the course of tissue formation. A variety of signaling molecules are known to be secreted by mesenchymal cells with different functions in terms of migration, differentiation, and inflammation, however, stem cell cultures from the dental pulp or apical papilla from a single donor have rarely been investigated in comparison [32]. Thus, a selection of relevant proteins was specifically quantified during induced osteogenic differentiation in order to reveal the signaling potential of DPSC and SCAP in terms of pulp tissue formation (TIMP, OPG, IL-6, IL-8, VEGF, and TGF-β1). In general, relevant differences regarding the cell type, the time of culture, and/or the culture conditions were observed for all signal molecules investigated, which shall be discussed in more detail subsequently.

TIMP, a multifunctional cytokine, was secreted by both DPSC and SCAP during standard cell culture with slightly higher levels in DPSC cultures. Despite the type of medium having no considerable impact, a time-dependent increase of TIMP was observed during osteogenic differentiation. Assuming that TIMP not only influences processes such as cell growth, apoptosis and angiogenesis, but also controls the activity of matrix metalloproteinases and therefore plays a key role in the remodeling of extracellular matrix, it can be suggested that DPSC have a slightly higher differentiation potential than SCAP [33]. As SCAP, which originate from a developing tissue, probably provide a less differentiated phenotype according to Sonoyama et al. [10], they probably release less TIMP at the beginning, which increases during osteogenic differentiation.

A similar observation was made for OPG, showing lower secretion levels in SCAP cultures and an increase during induced osteogenic differentiation. According to literature, OPG is supposed to be a pro-osteogenic factor that has the ability to prime undifferentiated mesenchymal stem cells towards mineralization [34]. Furthermore, it counteracts osteoclastogenesis and is released constitutively to limit the differentiation of osteoclasts and thus controls bone remodeling processes [35].

TGF-β1, a promotor of odontoblast formation and key protein in dentin mineralization [36], was found in a greater extent in DPSC cultures. Furthermore, increasing secretion was evident in the course of differentiation of both DPSC and SCAP. Interestingly, previous studies also reported higher concentrations of neurotrophins and growth factors (NT-3, BMP-4, and TGF-β3) in DPSC [32], leading to the suggestion that they give more pronounced paracrine signals related to odontoblast differentiation compared to SCAP. Since TGFs and BMPs play an important role in dentin secretion and are also embedded within it [37], it is conceivable that they are prevalent in terminally differentiated DPSC.

Considering that TGF-β1 is also a known promotor of angiogenesis by stimulating the production of VEGF mRNA [38,39], it is not surprising that higher levels of VEGF were detected in DPSC cultures. However, it was unexpected that SCAP, unlike DPSC, secreted moderate amounts of VEGF in the standard medium but almost none in the course of induced differentiation. A previous study on osteogenesis of mesenchymal stem cells reported that VEGF does not only stimulate mineralization but is also secreted in a differentiation dependent manner [40]. A possible conclusion would be that SCAP are isolated in a less differentiated state and accordingly lag behind in culture.

IL-6, a pleiotropic cytokine, was secreted in high levels by both DPSC and SCAP under standard culture conditions. It is not only involved in regeneration, inflammation, and the activation of immune cells, but also an important factor to maintain homeostasis [41]. Studies showed that regenerative and anti-inflammatory properties are mediated by a classic signaling pathway where the IL-6 receptor is membrane-bound on target cells. In contrast, binding to the soluble IL-6 receptor leads to the activation of pro-inflammatory activities through the trans-signaling pathway [42,43]. Both pathways seem to be used by DPSC and SCAP, showing the versatility needed for tissue engineering and the maintenance of a healthy dental pulp [44,45].

In contrast, IL-8 was released only in low levels by DPSC as well as SCAP under standard conditions; however, a considerable increase was observed during osteogenic differentiation, especially in SCAP. Interestingly, a recent study also measured high levels of proinflammatory cytokines such as IL-8 released by SCAP exposed to different anaerobic oral bacteria [46]. In general, IL-8 promotes osteoclastogenesis and is associated with bone resorption [47]. Moreover, it is supposed to have a chemotactic effect on mesenchymal stem cells [48]. These findings align with the results of this study, showing IL-8 as a key factor during osteogenic differentiation.

Overall, DPSC and SCAP appear to have many similarities in terms of migration, proliferation, and differentiation. In view of observations from animal studies or case reports, this leads to the assumption that both cell types have a similar potential to migrate into the root canal and form mineralizing tissue during regenerative procedures. Interestingly, the profile of secreted signaling molecules suggests that SCAP are in a less differentiated state than DPSC and thus may be more versatile.

3.4. Gene Expression Profiling

This study was one of the first to comprehensively compare gene expression of DPSC and SCAP isolated from the same donor and thus eliminating donor and culture-specific variables such as age, developmental state, isolation technique, or cell passage. Deeper insights in regulatory mechanisms were gained by comparison of stem cells cultivated under standard culture osteogenic conditions, where cells undergoing induced osteogenic differentiation represent the processes that are expected in the context of regenerative endodontic applications. Mesenchymal progenitor cells must form a mineralizing phenotype in the course of cell homing, i.e., differentiate osteogenically or odontogenically, as it is often termed in the endodontic context. The comprehensive transcriptome analysis allowed to define genome-wide regulatory genes that play a key role during the differentiation processes of odontogenic stem cells. Strict analysis settings were established to focus on relevant genes that play a central role in the course of differentiation, regardless of cell type. The aim was to gain more insight into the regulation of cell differentiation in order to possibly control or optimize this process clinically. Comparisons concerning differential gene expression were made between day 1 and day 14 as the culture parameter "time" appeared to be the most influential one. In both DPSC and SCAP, fewer genes were significantly regulated during the differentiation process than during standard cell culture. Thus, unspecifically regulated genes of the control cultures were deliberately excluded in the course of further analysis to identify exactly those genes that were exclusively up- or downregulated during induced osteogenic differentiation. Particularly, the final comparison of both cell types allowed to narrow this search down to only 13 genes that play a key role in induced differentiation of both DPSC and SCAP. Finally, all identified genes were categorized according to their molecular function by the PANTHER system: catalytic activity (*GPX3*, *PIP*, *ABCA6*, *PDE1A*), binding (*GPX3*, *IGFBP2*, *LEPR/LEPROT*), molecular transducer activity (*LEPR*), transporter activity (*ABCA6*), molecular function regulator (*ID3*) and unknown molecular function (*SAA2/SAA2-SAA4/SAA4*, *SAA1*, *GPM6B*, *FAM107A*, *PAPPA*, *VWA5A*).

The upregulated genes that are associated with catalytic activity usually code for macromolecules with enzyme function that can catalyze biochemical reactions. Accordingly, glutathione peroxidase 3 (*GPX3*) was the most highly expressed gene detected with microarray as well as PCR analysis. In analogy to previous reports on osteogenically differentiated human mesenchymal stem cells, it was upregulated in both DPSC and SCAP on day 14 [49]. Glutathione peroxidase 3 protects cells from oxidative damage by reduction of hydrogen peroxide and is suspected to be a key factor during osteogenic differentiation of mesenchymal stem cells [50].

Furthermore, the prolactin-induced protein (*PIP*) is a secreted glycoprotein with endonuclease activity that is involved in proteolysis and immunological processes. Though little is known about its exact physiological function, studies observed that prolactin-induced protein is able to bind to CD4 receptors of T lymphocytes or macrophages and the Fc fragment of immunoglobulin G, and therefore supposedly has immunomodulatory capabilities [51,52]. Its high expression on day 14 of osteogenic differentiation is in line with the findings of Li et al. [53], where *PIP* upregulations were observed during the osteogenic induction of periodontal ligament stem cells (PDLSC). They concluded that the upregulation seems to be caused by dexamethasone, a glucocorticoid which is also part the StemPro® Osteogenesis Differentiation Kit. Dexamethasone supposedly stimulates *PIP* expression by a glucocorticoid receptor-dependent transcriptional activation. Moreover, they observed that a knockdown of *PIP* and its fibronectin-degrading properties even enhanced mineralization of PDLSC [53].

The ATP-binding cassette sub-family A member 6 belongs to the ABC transporter family and *ABCA6* was significantly upregulated on day 14. So far, its function is not described sufficiently; however, it is supposed to be involved in the lipid transport and homeostasis of macrophages due to its cholesterol-responsive regulation [54]. Another member of the ABC transporter family, ATP-binding cassette sub-family A member 1, has been described as a mediator of cortisol and dexamethasone transport [55]. This physiological ability is also conceivable for ATP-binding cassette sub-family A member 6 considering the presence of dexamethasone in the differentiation medium.

PDE1A, short for calcium/calmodulin-dependent 3′,5′-cyclic nucleotide phosphodiesterase 1A, was also upregulated [56]. According to literature, it is responsible for signal transduction and ion and calmodulin binding. Calmodulin, a multifunctional calcium-binding messenger protein, takes part in the differentiation of osteoblasts by regulating bone morphogenetic protein-2 (BMP-2) signaling, BMP-2 being a known osteogenic differentiation factor [57]. In a previous study, *PDE1A* was upregulated in human PDLSCs cultivated in osteogenic differentiation medium during matrix maturation, therefore concluding that genes related to calcium binding might be vital for the differentiation of stem cells into osteoblasts [58].

If binding functions are attributed to proteins by PANTHER, these can interact specifically with other molecules or selectively occupy binding sites. This is the case, for example, with the insulin-like growth factor-binding protein 2, which was highly expressed in this study. A publication on mesenchymal stromal cells revealed that the expression of insulin-like growth factor 2 (*IGF2*), insulin-like growth factor-binding protein 2 (*IGFBP2*), and integrin alpha5 (*ITGA5*) was upregulated during induced osteogenic differentiation [59]. The hypothesis of Hamidouche et al. [59] was that dexamethasone from the differentiation medium induced *ITGA5* expression, leading to an upregulated production of insulin-like growth factor 2 and insulin-like growth factor-binding protein 2 and finally triggering osteoblast gene expression, increasing mRNA levels of *RUNX2*, *ALP*, and *COL1A1*. Because this coincides with the findings of this study where a higher expression of the gene *IGFBP2* was seen on day 14, it is possible that a similar crosstalk between the mentioned molecules can be observed in dental stem cell cultures.

The leptin receptor (*LEPR*) and the leptin receptor overlapping transcript (*LEPROT*) were also upregulated on day 14. While information on leptin receptors in the context of osteogenic differentiation of dental stem cells is rare, it was shown that mesenchymal stromal cells that express leptin receptors not only give rise to most of the new formed bone in adult bone marrow but are also the ones responsible for regeneration after a trauma [60]. Therefore, it allows the assumption that there is a link between the expression of leptin receptors in both DPSC and SCAP and the formation of mineralized matrix during induced osteogenic differentiation.

Interestingly, *ID3*, short for inhibitor of DNA-binding 3, was the only key gene found downregulated on day 14 of the microarray analysis. These findings align with the study of Peng et al. [61] which reported that at an early stage, bone morphogenetic proteins induce an overexpression of ID helix-loop-helix proteins needed for the proliferation of osteoblast progenitor cells. This is followed by an obligatory downregulation of ID proteins, a process that is crucial for the terminal differentiation of cells committed to the osteoblast lineage.

The serum amyloid A (SAA) proteins are a family of lipophilic molecules that are relevant during acute phase response, e.g., during infectious attack, and also responsible for the transport of high-density lipoproteins and cholesterol. Furthermore, they are suspected to influence tissue remodeling through their interaction with metalloproteinases. After synthesis in the liver, they circulate in the blood serum [62]. In this experiment, *SAA1*, *SAA2*, and *SAA4* were upregulated on day 14 during osteogenic differentiation. Ebert et al. [63] reported a similar observation, stating that SAA proteins hold the potential to induce mineralization in mesenchymal stem cells via Toll-like receptor 4 activation. They also promote the expression of proinflammatory cytokines such as IL-6, IL-8, interleukin 1 beta (IL-1β), C-X-C motif chemokine ligand 1 (CXCL1), and C-X-C motif chemokine ligand 2 (CXCL2).

GPM6B, a gene encoding for the membrane glycoprotein M6-b, showed high expression levels on day 14 of induced osteogenic differentiation. Its involvement in osteoblast differentiation and bone formation can be explained by its influence on the activity of alkaline phosphatase, whereby a reduced activity leads to a weaker mineralization of the extracellular matrix. The cytoskeleton organization, respectively the distribution of actin filaments and focal adhesions, seem to be influenced by *GPM6B* expression as well [64].

The genes *FAM107A* (actin-associated protein FAM107A), *PAPPA* (pappalysin-1), and *VWA5A* (von Willebrand factor A domain-containing protein 5A) that were also significantly upregulated on day 14 have not been described in the context of osteogenic differentiation yet. Further research is required to define their exact function during the mineralization process. So far, actin-associated protein FAM107A with its nuclear localization and coiled-coil domain is suspected to be a gene transcription and cell cycle regulator and a tumor suppressor gene [65]. Pappalysin-1 is known as a metalloproteinase, mainly investigated as a marker of acute coronary syndromes [66] or pathological birth disorders [67]. Interestingly, a reported function of pappalysin-1 is also the cleavage of the complex between insulin-like growth factor and insulin like growth factor binding protein [68]. *VWA5A*, also known as breast cancer suppressor candidate-1 (BCSC-1), is investigated as a tumor suppressor gene [69].

Overall, the gene expression profiling revealed extended insight into the regulation during induced osteogenic differentiation. The analysis of the transcriptome of DPSC and SCAP cultures showed many parallels, once again highlighting their shared evolutionary origin, leading to comparable gene expression patterns. *GPX3, PIP, IGFBP2, SAA2/SAA2-SAA4/SAA4, SAA1, GPM6B, FAM107A, LEPR/LEPROT, PAPPA, ABCA6, VWA5A, PDE1A*, and *ID3* appear to be crucial genes during mineralization processes. While some of them have already been described in the context of osteogenic differentiation for other stem cell types, further research is needed to explore the exact molecular functions of some of the genes.

4. Materials and Methods

4.1. Cell Isolation

Cells were isolated from both pulp tissue and the apical papilla of an extracted third molar of an 18-year-old patient with informed consent and according to a previously described protocol approved by an appropriate review board at the University of Regensburg [70]. The tooth was removed due to lack of space, was not impacted, and did not show carious decay or other pathological alterations. The primary cells were cultured in αMEM supplemented with 10% FBS, 50 μg/mL L-ascorbic acid 2-phosphate, 100 U/mL penicillin, and 100 μg/mL streptomycin at 37 °C with 5% CO_2. All cell culture reagents were purchased from Gibco™ (Thermo Fisher Scientific, Waltham, MA, USA). The cells were used in the experiments with the same passages and at most in passage 3.

4.2. Stem Cell Characterization

Cells obtained from the pulp and apical papilla were analyzed for the expression of characteristic mesenchymal stem cell markers (CD73, CD90, and CD105) as suggested by the International Society for Cellular Therapy [25]. At the same time, cells with the expression of markers of different origin (CD34, CD45, CD11b, CD19, and HLA-DR) were excluded. Therefore, flow cytometric analysis was performed using a Human MSC Analysis Kit (BD Stemflow™, BD Bioscience, San Jose, CA, USA). Fluorescence was determined by FACSCanto™ (BD Bioscience, San Jose, CA, USA) on basis of at least 2×10^4 events for each sample. Data from three independent experiments were collected ($n = 3$) and analyzed by FlowJo™ (BD Bioscience, San Jose, CA, USA).

Furthermore, the cells' potential for multi-lineage differentiation was investigated. Pulp- and papilla-derived cells were incubated with adipogenic, chondrogenic, and osteogenic culture media (StemPro® Adipogenesis, Osteogenesis and Chondrogenesis Differentiation Kit, Invitrogen Corporation, Carlsbad, CA, USA). Chondrogenic differentiation was evaluated by staining with Alcian Blue 8GX (Sigma-Aldrich, St. Louis, MO, USA) after 10 days, adipogenic differentiation with Oil Red O (Sigma-Aldrich, St. Louis, MO, USA), and osteogenic differentiation with Alizarin Red S (Carl Roth, Karlsruhe, Germany) after 21 days.

4.3. Cell Viability and Proliferation

DPSC and SCAP were cultured either with αMEM and 1% FBS or αMEM and 10% FBS in 96-well plates (4000 cells/well) and both cell viability and cell proliferation were determined after 1, 3, 5, and 7 days.

For the MTT assay, cells were then incubated with 100 μL/well of a 0.5 mg/mL MTT solution (Thiazolyl Blue Tetrazolium Bromide, Sigma-Aldrich, St. Louis, MO, USA) for 60 min at 37 °C and 5% CO_2. Subsequently, the dye was dissolved in 200 μL/well of dimethyl sulfoxide (DMSO, Merck Millipore, Billerica) and optical density was measured on a microplate reader at $\lambda = 570$ nm (Infinite® 200, Tecan, Männedorf, Switzerland). Cell number of DPSC and SCAP was determined by CyQUANT™ Cell Proliferation Assay (Life Technologies, Carlsbad, CA, USA) as described previously [37].

Median and 25–75% percentiles were calculated on basis of three independent experiments performed in triplicate ($n = 9$).

4.4. Cell Migration

To describe migration activity, a wound healing assay was performed. First, 70 μL of cell suspension (7×10^5 cells/mL) was applied into each chamber of a 2-well silicone insert (Culture-Insert 2 Well in μ-Dish 35 mm, Ibidi, Gräfelfing, Germany) and incubated at 37 °C and 5% CO_2 for 24 h. After removal of the silicone insert with sterile tweezers, the cell layer was covered with 2 mL of medium and cultivated for 72 h.

The following groups were established for each cell type: (1) αMEM with 1% FBS, (2) αMEM with 10% FBS and (3) αMEM with 10% FBS and 20 μM Locostatin (Santa Cruz Biotechnology, CA, USA). Then, 8 μL of migration inhibitor Locostatin was added twice a day.

The gap closure by migrating cell sheets was monitored by imaging of the same site of the standardized gaps in 12 h-intervals with a Zeiss Axio Lab.A1 microscope at 10× magnification (Zeiss, Jena, Germany). Images were edited with Fiji (National Institutes of Health, Bethesda, MD, USA) and the gap area was quantified. The wound healing assay was performed in triplicate and repeated twice ($n = 12$). Median values and 25–75% percentiles were normalized to the initial wound area, which was set to 100%.

4.5. Release of Signaling Molecules

Levels of biologically active proteins secreted by DPSC and SCAP during standard cell culture (αMEM and 10% FBS) and induced osteogenic differentiation (StemPro® Osteogenesis Differentiation Kit, Invitrogen Corporation, Carlsbad, CA, USA) were quantified. Both cell types were cultured in 12-well plates (19,000 cells/well) for 3 weeks. Media were replaced regularly and stored frozen at −20 °C. The collected supernatants were pooled in the initial phase (days 1 to 8), the intermediate phase (days 9 to 15) and the late phase (days

16 to 21) of culture. Finally, the concentrations of TIMP, OPG, IL-6, IL-8, VEGF and TGF-β1 were determined by enzyme-linked immunosorbent assays (Quantikine® ELISA Kit, R&D Systems, Minneapolis, MN, USA). Medians with 25–75% percentiles were calculated from three experiments performed in triplicates ($n = 9$).

4.6. Gene Expression Profiling

Transcriptome-wide gene expression of DPSC and SCAP during induced osteogenic differentiation was profiled by Clariom™ S Human Arrays (Applied Biosystems™ by Thermo Fisher Scientific Waltham, MA, USA), which covers over 20,000 well-annotated genes.

Therefore, DPSC and SCAP were cultured in 12-well plates (19,000 cells/well) with either αMEM and 10% FBS or StemPro® Osteogenesis Differentiation Kit (Invitrogen Corporation, Carlsbad, CA, USA). RNA was isolated (RNeasy® Mini Kit, Qiagen, Hilden, Germany) on days 1 and 14 and quantified spectrophotometrically (NanoDrop™ 2000, Thermo Fisher Scientific, Waltham, MA, USA). Further sample processing was carried out by the Genomics Core Unit based at the University of Regensburg according to the Affymetrix GeneChip WT PLUS Reagent Kit instructions (Affymetrix, Santa Clara, CA, USA). To confirm the obtained results and increase data quality, the microarray experiment was performed in replicates and reproduced independently ($n = 2$). Raw data of all experiments can be accessed in the Supplementary File 2.

The Transcriptome Analysis Console (TAC) 4.0 Software (Applied Biosystems™ by Thermo Fisher Scientific Waltham, MA, USA) was used for comprehensive analysis of the microarray data and genes were ranked by the empirical Bayes method. Results of all replicates and repeats were analyzed collectively with the TAC software and the *p*-value and false discovery rate (FDR) were calculated according to the given algorithms. After data import, gene regulation (fold-change > 10; *p*-value ≤ 0.01) was evaluated in the context of various parameters such as timepoint, type of medium, and cell type. Group selections were made to identify similarities and differences. Finally, genes were submitted to PANTHER (version 16.0, released 2020/12/01), an organized database located in the Gene Ontology Consortium [71].

To validate the genetic profiling by the microarray, qRT-PCR was performed for selected genes using the TaqMan Fast Advanced Master Mix (Applied Biosystems, Thermo Fisher Scientific, Waltham, USA) and probes for following genes: *GPX3* (Hs00173566_m1), *PDE1A* (Hs00897273_m1), *SAA1*, *SAA2* (Hs00761940_s1), *IGFBP2* (Hs01040718_m1), *PAPPA* (Hs01029908_m1), *ABCA6* (Hs00979431_mH), *FAM107A* (Hs01100593_g1), *LEPR* (Hs00900252_g1), *PIP* (Hs00160082_m1), *VWA5A* (Hs00938346_g1), *ID3* (Hs00171409_m1), and the housekeeping genes *RPS18* (Hs99999901_s1), *ACTB* (Hs01060665_g1), *GAPDH* (Hs02786624_g1). Finally, results were normalized to the arithmetic mean of all housekeeping genes and related to day 1 by the comparative CT method ($2^{-\Delta\Delta CT}$) [72] to compute medians with 25–75% percentiles ($n = 4$).

4.7. Statistical Analysis

Data were not normally distributed and, therefore, analyzed nonparametrically at a significance level of $\alpha = 0.05$. For situations with two unpaired groups, Mann–Whitney U-tests were performed and *p*-values were adjusted for multiple comparisons by the Holm-Šídák method ($\alpha = 0.05$). In case of three or more unpaired groups, the Kruskal–Wallis test followed by Dunn's multiple comparison test was applied. Statistically significant differences between the groups were indicated by equal lower-case letters in the respective figures. All statistical analyses were computed with GraphPad Prism 9 (GraphPad Software, La Jolla, CA, USA) and detailed record of all comparisons can be found in in Supplementary File 3.

5. Conclusions

Excluding parameters such as donor, age, passage, or culture conditions, this study shows that DPSC and SCAP not only share a common evolutionary origin, but also behave similarly in terms of viability, proliferation, migration, and regulation of gene expression. Based on the results of cytokine secretion, it can be assumed that SCAP can be isolated in a less differentiated state than DPSC. One possible explanation is their origin from the apical papilla, a tissue found only in immature teeth. However, both dental stem cell types seem equally suitable as reservoirs for regenerative procedures based on cell homing, which relies on migration into the root canal and differentiation into a mineralizing phenotype. Future studies on the identified regulatory genes may help to understand the exact molecular processes during osteogenic differentiation of dental stem cells.

Supplementary Materials: The following supporting information can be downloaded at: https://www.mdpi.com/article/10.3390/ijms23052615/s1.

Author Contributions: Conceptualization, M.W. (Matthias Widbiller); formal analysis, M.S., M.W. (Melanie Woelflick), A.R. and M.W. (Matthias Widbiller); investigation, M.S., M.W. (Melanie Woelflick), A.R. and B.L.; methodology, M.W. (Melanie Woelflick), A.R., C.M. and M.W. (Matthias Widbiller); project administration, M.W. (Matthias Widbiller); resources, K.M.G. and W.B.; supervision, M.W. (Matthias Widbiller); writing—original draft, M.S. and M.W. (Matthias Widbiller); writing—review and editing, K.M.G. and W.B. All authors have read and agreed to the published version of the manuscript.

Funding: This research received no external funding.

Institutional Review Board Statement: Not applicable.

Informed Consent Statement: Not applicable.

Data Availability Statement: The data presented in this study are available on request from the corresponding author.

Conflicts of Interest: The authors declare no conflict of interest.

References

1. Nakashima, M.; Akamine, A. The application of tissue engineering to regeneration of pulp and dentin in endodontics. *J. Endod.* **2005**, *31*, 711–718. [CrossRef]
2. Galler, K.M.; D'Souza, R.N. Tissue engineering approaches for regenerative dentistry. *Regen. Med.* **2011**, *6*, 111–124. [CrossRef] [PubMed]
3. Galler, K.M.; Widbiller, M. Perspectives for Cell-homing Approaches to Engineer Dental Pulp. *J. Endod.* **2017**, *43*, S40–S45. [CrossRef]
4. Langer, R.; Vacanti, J.P. Tissue engineering. *Science* **1993**, *260*, 920–926. [CrossRef] [PubMed]
5. Widbiller, M.; Schmalz, G. Endodontic regeneration: Hard shell, soft core. *Odontology* **2020**, *109*, 303–312. [CrossRef] [PubMed]
6. Widbiller, M.; Eidt, A.; Hiller, K.-A.; Buchalla, W.; Schmalz, G.; Galler, K.M. Ultrasonic activation of irrigants increases growth factor release from human dentine. *Clin. Oral Investig.* **2017**, *21*, 879–888. [CrossRef] [PubMed]
7. Schmalz, G.; Widbiller, M.; Galler, K.M. Clinical Perspectives of Pulp Regeneration. *J. Endod.* **2020**, *46*, S161–S174. [CrossRef] [PubMed]
8. Miura, M.; Gronthos, S.; Zhao, M.; Lu, B.; Fisher, L.W.; Robey, P.G.; Shi, S. SHED: Stem cells from human exfoliated deciduous teeth. *Proc. Natl. Acad. Sci. USA* **2003**, *100*, 5807–5812. [CrossRef] [PubMed]
9. Gronthos, S.; Mankani, M.; Brahim, J.; Robey, P.G.; Shi, S. Postnatal human dental pulp stem cells (DPSCs) in vitro and in vivo. *Proc. Natl. Acad. Sci. USA* **2000**, *97*, 13625–13630. [CrossRef]
10. Sonoyama, W.; Liu, Y.; Fang, D.; Yamaza, T.; Seo, B.-M.; Zhang, C.; Liu, H.; Gronthos, S.; Wang, C.-Y.; Wang, S.; et al. Mesenchymal stem cell-mediated functional tooth regeneration in swine. *PLoS ONE* **2006**, *1*, e79. [CrossRef] [PubMed]
11. Janebodin, K.; Horst, O.V.; Ieronimakis, N.; Balasundaram, G.; Reesukumal, K.; Pratumvinit, B.; Reyes, M. Isolation and characterization of neural crest-derived stem cells from dental pulp of neonatal mice. *PLoS ONE* **2011**, *6*, e27526. [CrossRef] [PubMed]
12. Cordero, D.R.; Brugmann, S.; Chu, Y.; Bajpai, R.; Jame, M.; Helms, J.A. Cranial neural crest cells on the move: Their roles in craniofacial development. *Am. J. Med. Genet. A* **2011**, *155*, 270–279. [CrossRef] [PubMed]
13. Chai, Y.; Jiang, X.; Ito, Y.; Bringas, P., Jr.; Han, J.; Rowitch, D.H.; Soriano, P.; McMahon, A.P.; Sucov, H.M. Fate of the mammalian cranial neural crest during tooth and mandibular morphogenesis. *Development* **2000**, *127*, 1671–1679. [CrossRef] [PubMed]
14. Ten Cate, A.R. The histochemistry of human tooth development. *Proc. Nutr. Soc.* **1959**, *18*, 65–70. [CrossRef] [PubMed]
15. Rothová, M.; Peterková, R.; Tucker, A.S. Fate map of the dental mesenchyme: Dynamic development of the dental papilla and follicle. *Dev. Biol.* **2012**, *366*, 244–254. [CrossRef]

16. Yao, J.; Chen, N.; Wang, X.; Zhang, L.; Huo, J.; Chi, Y.; Li, Z.; Han, Z. Human Supernumerary Teeth-Derived Apical Papillary Stem Cells Possess Preferable Characteristics and Efficacy on Hepatic Fibrosis in Mice. *Stem Cells Int.* **2020**, *2020*, 6489396. [CrossRef] [PubMed]
17. Park, M.-K.; Kim, S.; Jeon, M.; Jung, U.-W.; Lee, J.-H.; Choi, H.-J.; Choi, J.-E.; Song, J.S. Evaluation of the Apical Complex and the Coronal Pulp as a Stem Cell Source for Dentin-pulp Regeneration. *J. Endod.* **2020**, *46*, 224–231.e3. [CrossRef] [PubMed]
18. Galler, K.M.; Weber, M.; Korkmaz, Y.; Widbiller, M.; Feuerer, M. Inflammatory Response Mechanisms of the Dentine-Pulp Complex and the Periapical Tissues. *Int. J. Mol. Sci.* **2021**, *22*, 1480. [CrossRef]
19. Huang, G.T.-J.; Sonoyama, W.; Liu, Y.; Liu, H.; Wang, S.; Shi, S. The hidden treasure in apical papilla: The potential role in pulp/dentin regeneration and bioroot engineering. *J. Endod.* **2008**, *34*, 645–651. [CrossRef]
20. Chrepa, V.; Pitcher, B.; Henry, M.A.; Diogenes, A. Survival of the Apical Papilla and Its Resident Stem Cells in a Case of Advanced Pulpal Necrosis and Apical Periodontitis. *J. Endod.* **2017**, *43*, 561–567. [CrossRef]
21. Shi, S.; Gronthos, S. Perivascular niche of postnatal mesenchymal stem cells in human bone marrow and dental pulp. *J. Bone Miner. Res.* **2003**, *18*, 696–704. [CrossRef] [PubMed]
22. Hadaegh, Y.; Niknam, M.; Attar, A.; Maharlooei, M.K.; Tavangar, M.S.; Aarabi, A.M.; Monabati, A. Characterization of stem cells from the pulp of unerupted third molar tooth. *Indian J. Dent. Res.* **2014**, *25*, 14–21. [CrossRef] [PubMed]
23. Gronthos, S.; Brahim, J.; Li, W.; Fisher, L.W.; Cherman, N.; Boyde, A.; DenBesten, P.; Robey, P.G.; Shi, S. Stem cell properties of human dental pulp stem cells. *J. Dent. Res.* **2002**, *81*, 531–535. [CrossRef]
24. Genova, T.; Cavagnetto, D.; Tasinato, F.; Petrillo, S.; Ruffinatti, F.A.; Mela, L.; Carossa, M.; Munaron, L.; Roato, I.; Mussano, F. Isolation and Characterization of Buccal Fat Pad and Dental Pulp MSCs from the Same Donor. *Biomedicines* **2021**, *9*, 265. [CrossRef]
25. Dominici, M.; Le Blanc, K.; Mueller, I.; Slaper-Cortenbach, I.; Marini, F.; Krause, D.; Deans, R.; Keating, A.; Prockop, D.; Horwitz, E. Minimal criteria for defining multipotent mesenchymal stromal cells. The International Society for Cellular Therapy position statement. *Cytotherapy* **2006**, *8*, 315–317. [CrossRef] [PubMed]
26. Huang, G.T.-J.; Yamaza, T.; Shea, L.D.; Djouad, F.; Kuhn, N.Z.; Tuan, R.S.; Shi, S. Stem/progenitor cell-mediated de novo regeneration of dental pulp with newly deposited continuous layer of dentin in an in vivo model. *Tissue Eng. Part A* **2010**, *16*, 605–615. [CrossRef]
27. Nada, O.A.; El Backly, R.M. Stem Cells From the Apical Papilla (SCAP) as a Tool for Endogenous Tissue Regeneration. *Front. Bioeng. Biotechnol.* **2018**, *6*, 103. [CrossRef]
28. Schneider, R.; Holland, G.R.; Chiego, D., Jr.; Hu, J.C.C.; Nör, J.E.; Botero, T.M. White mineral trioxide aggregate induces migration and proliferation of stem cells from the apical papilla. *J. Endod.* **2014**, *40*, 931–936. [CrossRef]
29. Janjusevic, M.; Greco, S.; Islam, M.S.; Castellucci, C.; Ciavattini, A.; Toti, P.; Petraglia, F.; Ciarmela, P. Locostatin, a disrupter of Raf kinase inhibitor protein, inhibits extracellular matrix production, proliferation, and migration in human uterine leiomyoma and myometrial cells. *Fertil. Steril.* **2016**, *106*, 1530–1538.e1. [CrossRef]
30. Cantore, S.; Ballini, A.; de Vito, D.; Martelli, F.S.; Georgakopoulos, I.; Almasri, M.; Dibello, V.; Altini, V.; Farronato, G.; Dipalma, G.; et al. Characterization of human apical papilla-derived stem cells. *J. Biol. Regul. Homeost. Agents* **2017**, *31*, 901–910.
31. Prateeptongkum, E.; Klingelhöffer, C.; Morsczeck, C. The influence of the donor on dental apical papilla stem cell properties. *Tissue Cell* **2015**, *47*, 382–388. [CrossRef] [PubMed]
32. Joo, K.H.; Song, J.S.; Kim, S.; Lee, H.-S.; Jeon, M.; Kim, S.-O.; Lee, J.-H. Cytokine Expression of Stem Cells Originating from the Apical Complex and Coronal Pulp of Immature Teeth. *J. Endod.* **2018**, *44*, 87–92.e1. [CrossRef] [PubMed]
33. Ries, C. Cytokine functions of TIMP-1. *Cell. Mol. Life Sci.* **2014**, *71*, 659–672. [CrossRef] [PubMed]
34. Palumbo, S.; Li, W.-J. Osteoprotegerin enhances osteogenesis of human mesenchymal stem cells. *Tissue Eng. Part A* **2013**, *19*, 2176–2187. [CrossRef]
35. Kanji, S.; Sarkar, R.; Pramanik, A.; Kshirsagar, S.; Greene, C.J.; Das, H. Dental pulp-derived stem cells inhibit osteoclast differentiation by secreting osteoprotegerin and deactivating AKT signalling in myeloid cells. *J. Cell. Mol. Med.* **2021**, *25*, 2390–2403. [CrossRef] [PubMed]
36. Sloan, A.J.; Smith, A.J. Stimulation of the dentine–pulp complex of rat incisor teeth by transforming growth factor-β isoforms 1–3 in vitro. *Arch. Oral Biol.* **1999**, *44*, 149–156. [CrossRef]
37. Widbiller, M.; Eidt, A.; Lindner, S.R.; Hiller, K.-A.; Schweikl, H.; Buchalla, W.; Galler, K.M. Dentine matrix proteins: Isolation and effects on human pulp cells. *Int. Endod. J.* **2018**, *51* (Suppl. 4), e278–e290. [CrossRef] [PubMed]
38. Ucuzian, A.A.; Gassman, A.A.; East, A.T.; Greisler, H.P. Molecular mediators of angiogenesis. *J. Burn Care Res.* **2010**, *31*, 158–175. [CrossRef]
39. Ferrara, N.; Davis-Smyth, T. The biology of vascular endothelial growth factor. *Endocr. Rev.* **1997**, *18*, 4–25. [CrossRef]
40. Mayer, H.; Bertram, H.; Lindenmaier, W.; Korff, T.; Weber, H.; Weich, H. Vascular endothelial growth factor (VEGF-A) expression in human mesenchymal stem cells: Autocrine and paracrine role on osteoblastic and endothelial differentiation. *J. Cell. Biochem.* **2005**, *95*, 827–839. [CrossRef]
41. Gallorini, M.; Widbiller, M.; Bolay, C.; Carradori, S.; Buchalla, W.; Cataldi, A.; Schweikl, H. Relevance of Cellular Redox Homeostasis for Vital Functions of Human Dental Pulp Cells. *Antioxidants* **2022**, *11*, 23. [CrossRef] [PubMed]
42. Hunter, C.A.; Jones, S.A. IL-6 as a keystone cytokine in health and disease. *Nat. Immunol.* **2015**, *16*, 448–457. [CrossRef] [PubMed]

43. Scheller, J.; Chalaris, A.; Schmidt-Arras, D.; Rose-John, S. The pro- and anti-inflammatory properties of the cytokine interleukin-6. *Biochim. Biophys. Acta* **2011**, *1813*, 878–888. [CrossRef] [PubMed]

44. Sancilio, S.; Marsich, E.; Schweikl, H.; Cataldi, A.; Gallorini, M. Redox Control of IL-6-Mediated Dental Pulp Stem-Cell Differentiation on Alginate/Hydroxyapatite Biocomposites for Bone Ingrowth. *Nanomaterials* **2019**, *9*, 1656. [CrossRef]

45. Park, Y.-T.; Lee, S.-M.; Kou, X.; Karabucak, B. The Role of Interleukin 6 in Osteogenic and Neurogenic Differentiation Potentials of Dental Pulp Stem Cells. *J. Endod.* **2019**, *45*, 1342–1348. [CrossRef]

46. Rakhimova, O.; Schmidt, A.; Landström, M.; Johansson, A.; Kelk, P.; Romani Vestman, N. Cytokine Secretion, Viability, and Real-Time Proliferation of Apical-Papilla Stem Cells Upon Exposure to Oral Bacteria. *Front. Cell. Infect. Microbiol.* **2020**, *10*, 620801. [CrossRef]

47. Bendre, M.S.; Margulies, A.G.; Walser, B.; Akel, N.S.; Bhattacharrya, S.; Skinner, R.A.; Swain, F.; Ramani, V.; Mohammad, K.S.; Wessner, L.L.; et al. Tumor-derived interleukin-8 stimulates osteolysis independent of the receptor activator of nuclear factor-kappaB ligand pathway. *Cancer Res.* **2005**, *65*, 11001–11009. [CrossRef]

48. Yang, A.; Lu, Y.; Xing, J.; Li, Z.; Yin, X.; Dou, C.; Dong, S.; Luo, F.; Xie, Z.; Hou, T.; et al. IL-8 Enhances Therapeutic Effects of BMSCs on Bone Regeneration via CXCR2-Mediated PI3k/Akt Signaling Pathway. *Cell. Physiol. Biochem.* **2018**, *48*, 361–370. [CrossRef]

49. Zhang, W.; Dong, R.; Diao, S.; Du, J.; Fan, Z.; Wang, F. Differential long noncoding RNA/mRNA expression profiling and functional network analysis during osteogenic differentiation of human bone marrow mesenchymal stem cells. *Stem Cell Res. Ther.* **2017**, *8*, 30. [CrossRef]

50. Brigelius-Flohé, R.; Maiorino, M. Glutathione peroxidases. *Biochim. Biophys. Acta* **2013**, *1830*, 3289–3303. [CrossRef]

51. Urbaniak, A.; Jablonska, K.; Podhorska-Okolow, M.; Ugorski, M.; Dziegiel, P. Prolactin-induced protein (PIP)-characterization and role in breast cancer progression. *Am. J. Cancer Res.* **2018**, *8*, 2150–2164. [PubMed]

52. Chiu, W.W.-C.; Chamley, L.W. Human seminal plasma prolactin-inducible protein is an immunoglobulin G-binding protein. *J. Reprod. Immunol.* **2003**, *60*, 97–111. [CrossRef]

53. Li, X.; Zhang, Y.; Jia, L.; Xing, Y.; Zhao, B.; Sui, L.; Liu, D.; Xu, X. Downregulation of Prolactin-Induced Protein Promotes Osteogenic Differentiation of Periodontal Ligament Stem Cells. *Med. Sci. Monit.* **2021**, *27*, e930610. [CrossRef] [PubMed]

54. Kaminski, W.E.; Wenzel, J.J.; Piehler, A.; Langmann, T.; Schmitz, G. ABCA6, a novel a subclass ABC transporter. *Biochem. Biophys. Res. Commun.* **2001**, *285*, 1295–1301. [CrossRef]

55. Pohl, A.; Devaux, P.F.; Herrmann, A. Function of prokaryotic and eukaryotic ABC proteins in lipid transport. *Biochim. Biophys. Acta* **2005**, *1733*, 29–52. [CrossRef] [PubMed]

56. Michibata, H.; Yanaka, N.; Kanoh, Y.; Okumura, K.; Omori, K. Human Ca2+/calmodulin-dependent phosphodiesterase PDE1A: Novel splice variants, their specific expression, genomic organization, and chromosomal localization. *Biochim. Biophys. Acta* **2001**, *1517*, 278–287. [CrossRef]

57. Zayzafoon, M. Calcium/calmodulin signaling controls osteoblast growth and differentiation. *J. Cell. Biochem.* **2006**, *97*, 56–70. [CrossRef]

58. Choi, H.-D.; Noh, W.-C.; Park, J.-W.; Lee, J.-M.; Suh, J.-Y. Analysis of gene expression during mineralization of cultured human periodontal ligament cells. *J. Periodontal Implant Sci.* **2011**, *41*, 30–43. [CrossRef]

59. Hamidouche, Z.; Fromigué, O.; Ringe, J.; Häupl, T.; Marie, P.J. Crosstalks between integrin alpha 5 and IGF2/IGFBP2 signalling trigger human bone marrow-derived mesenchymal stromal osteogenic differentiation. *BMC Cell Biol.* **2010**, *11*, 44. [CrossRef]

60. Zhou, B.O.; Yue, R.; Murphy, M.M.; Peyer, J.G.; Morrison, S.J. Leptin-receptor-expressing mesenchymal stromal cells represent the main source of bone formed by adult bone marrow. *Cell Stem Cell* **2014**, *15*, 154–168. [CrossRef]

61. Peng, Y.; Kang, Q.; Luo, Q.; Jiang, W.; Si, W.; Liu, B.A.; Luu, H.H.; Park, J.K.; Li, X.; Luo, J.; et al. Inhibitor of DNA binding/differentiation helix-loop-helix proteins mediate bone morphogenetic protein-induced osteoblast differentiation of mesenchymal stem cells. *J. Biol. Chem.* **2004**, *279*, 32941–32949. [CrossRef] [PubMed]

62. Sack, G.H., Jr. Serum Amyloid A (SAA) Proteins. In *Vertebrate and Invertebrate Respiratory Proteins, Lipoproteins and Other Body Fluid Proteins*; Subcellular Biochemistry; Springer: Berlin/Heidelberg, Germany, 2020; Volume 94, pp. 421–436. [CrossRef]

63. Ebert, R.; Benisch, P.; Krug, M.; Zeck, S.; Meißner-Weigl, J.; Steinert, A.; Rauner, M.; Hofbauer, L.; Jakob, F. Acute phase serum amyloid A induces proinflammatory cytokines and mineralization via toll-like receptor 4 in mesenchymal stem cells. *Stem Cell Res.* **2015**, *15*, 231–239. [CrossRef] [PubMed]

64. Drabek, K.; van de Peppel, J.; Eijken, M.; van Leeuwen, J.P. GPM6B regulates osteoblast function and induction of mineralization by controlling cytoskeleton and matrix vesicle release. *J. Bone Miner. Res.* **2011**, *26*, 2045–2051. [CrossRef] [PubMed]

65. Pastuszak-Lewandoska, D.; Czarnecka, K.H.; Migdalska-Sęk, M.; Nawrot, E.; Domańska, D.; Kiszałkiewicz, J.; Kordiak, J.; Antczak, A.; Górski, P.; Brzeziańska-Lasota, E. Decreased FAM107A Expression in Patients with Non-small Cell Lung Cancer. *Adv. Exp. Med. Biol.* **2015**, *852*, 39–48. [CrossRef]

66. Cirillo, P.; Cimmino, G.; Conte, S.; Pellegrino, G.; Morello, A.; Golino, P.; Trimarco, B. Relationship between Pregnancy-associated Plasma Protein-A and tissue factor levels in the coronary circulation of patients with acute coronary syndrome. *Int. J. Cardiol.* **2018**, *258*, 14–16. [CrossRef]

67. Sifakis, S.; Androutsopoulos, V.P.; Pontikaki, A.; Velegrakis, A.; Papaioannou, G.I.; Koukoura, O.; Spandidos, D.A.; Papantoniou, N. Placental expression of PAPPA, PAPPA-2 and PLAC-1 in pregnacies is associated with FGR. *Mol. Med. Rep.* **2018**, *17*, 6435–6440. [CrossRef]

68. Monget, P.; Mazerbourg, S.; Delpuech, T.; Maurel, M.-C.; Manière, S.; Zapf, J.; Lalmanach, G.; Oxvig, C.; Overgaard, M.T. Pregnancy-associated plasma protein-A is involved in insulin-like growth factor binding protein-2 (IGFBP-2) proteolytic degradation in bovine and porcine preovulatory follicles: Identification of cleavage site and characterization of IGFBP-2 degradation. *Biol. Reprod.* **2003**, *68*, 77–86. [CrossRef]

69. Martin, E.S.; Cesari, R.; Pentimalli, F.; Yoder, K.; Fishel, R.; Himelstein, A.L.; Martin, S.E.; Godwin, A.K.; Negrini, M.; Croce, C.M. The BCSC-1 locus at chromosome 11q23-q24 is a candidate tumor suppressor gene. *Proc. Natl. Acad. Sci. USA* **2003**, *100*, 11517–11522. [CrossRef]

70. Galler, K.M.; Schweikl, H.; Thonemann, B.; D'Souza, R.N.; Schmalz, G. Human pulp-derived cells immortalized with Simian Virus 40 T-antigen. *Eur. J. Oral Sci.* **2006**, *114*, 138–146. [CrossRef]

71. Mi, H.; Ebert, D.; Muruganujan, A.; Mills, C.; Albou, L.-P.; Mushayamaha, T.; Thomas, P.D. PANTHER version 16: A revised family classification, tree-based classification tool, enhancer regions and extensive API. *Nucleic Acids Res.* **2021**, *49*, D394–D403. [CrossRef]

72. Schmittgen, T.D.; Livak, K.J. Analyzing real-time PCR data by the comparative C(T) method. *Nat. Protoc.* **2008**, *3*, 1101–1108. [CrossRef] [PubMed]

International Journal of
Molecular Sciences

Article

Human Amnion Epithelial Cells: A Potential Cell Source for Pulp Regeneration?

Cristina Bucchi [1,*], Ella Ohlsson [2], Josep Maria de Anta [3], Melanie Woelflick [2], Kerstin Galler [4], María Cristina Manzanares-Cespedes [3] and Matthias Widbiller [2]

1   Research Centre for Dental Sciences (CICO), Department of Integral Adult Dentistry, Faculty of Dentistry, Universidad de La Frontera, Temuco 4811230, Chile
2   Department of Conservative Dentistry and Periodontology, University Hospital Regensburg, 93053 Regensburg, Germany; ella.ohlsson@ukr.de (E.O.); melanie.woelflick@ukr.de (M.W.); matthias.widbiller@ukr.de (M.W.)
3   Human Anatomy and Embryology Unit, Department of Pathology and Experimental Therapeutics, Faculty of Medicine and Health Sciences, Campus de Bellvitge, Universitat de Barcelona, 08907 L'Hospitalet de Llobregat, Spain; janta@ub.edu (J.M.d.A.); mcmanzanares@ub.edu (M.C.M.-C.)
4   Department of Conservative Dentistry and Periodontology, Friedrich-Alexander-University Erlangen-Nürnberg, 91054 Erlangen, Germany; kerstin.galler@uk-erlangen.de
*   Correspondence: cristina.bucchi@ufrontera.cl

Abstract: The aim of this study was to analyze the suitability of pluripotent stem cells derived from the amnion (hAECs) as a potential cell source for revitalization in vitro. hAECs were isolated from human placentas, and dental pulp stem cells (hDPSCs) and dentin matrix proteins (eDMPs) were obtained from human teeth. Both hAECs and hDPSCs were cultured with 10% FBS, eDMPs and an osteogenic differentiation medium (StemPro). Viability was assessed by MTT and cell adherence to dentin was evaluated by scanning electron microscopy. Furthermore, the expression of mineralization-, odontogenic differentiation- and epithelial–mesenchymal transition-associated genes was analyzed by quantitative real-time PCR, and mineralization was evaluated through Alizarin Red staining. The viability of hAECs was significantly lower compared with hDPSCs in all groups and at all time points. Both hAECs and hDPSCs adhered to dentin and were homogeneously distributed. The regulation of odontoblast differentiation- and mineralization-associated genes showed the lack of transition of hAECs into an odontoblastic phenotype; however, genes associated with epithelial–mesenchymal transition were significantly upregulated in hAECs. hAECs showed small amounts of calcium deposition after osteogenic differentiation with StemPro. Pluripotent hAECs adhere on dentin and possess the capacity to mineralize. However, they presented an unfavorable proliferation behavior and failed to undergo odontoblastic transition.

Keywords: human amnion epithelial cells; dental pulp stem cells; dentin matrix proteins; odontoblastic differentiation; revitalization

Citation: Bucchi, C.; Ohlsson, E.; Anta, J.M.d.; Woelflick, M.; Galler, K.; Manzanares-Cespedes, M.C.; Widbiller, M. Human Amnion Epithelial Cells: A Potential Cell Source for Pulp Regeneration? *Int. J. Mol. Sci.* **2022**, *23*, 2830. https://doi.org/10.3390/ijms23052830

Academic Editor: Nicola Maffulli

Received: 29 January 2022
Accepted: 1 March 2022
Published: 4 March 2022

Publisher's Note: MDPI stays neutral with regard to jurisdictional claims in published maps and institutional affiliations.

1. Introduction

Regenerative endodontics refers to biologically based treatment procedures, e.g., revitalization, for immature necrotic teeth [1]. It aims at the restoration of the pulp's physiology, including its immune, sensory and secretory functions, to improve the long-term prognosis of the tooth [2]. Over the last two decades, in vivo studies have shown satisfactory clinical outcomes with healing of periapical lesions [3] and resolution of clinical symptoms [4,5], as well as root thickening and lengthening [4] or apical closure [6]. However, the newly formed hard tissue does not resemble dentin but an ectopic tissue similar to cementum [7] or osteodentin [8], while the soft tissue lacks the pulp's characteristic organization and cells with a distinct odontoblast phenotype [9]. The absence of odontoblasts after revitalization and the formation of a tissue other than dentin might compromise both the capability of the treated teeth to react to future injuries and also their biomechanical performance [10].

Novel and more elaborate approaches are being tested to overcome the lack of regeneration after the classical approach of revitalization and to achieve more predictable histological outcomes [11]. In this context, tissue engineering relies on the delivery of stem cells and/or recombinant growth factors in a scaffold into the root canal to facilitate pulp regeneration. These methods can be subcategorized into the cell homing approach, which utilizes signaling molecules to induce the migration, proliferation and differentiation of stem cells from the periapical tissues [12,13], and the cell transplantation approach [14]. The latter relies on the delivery of stem cells able to form new pulp tissue in the root canal [15].

To be considered as regenerated dental pulp, the newly formed tissue in the root canal must be vascularized as well as innervated, contain a similar cell density and microarchitecture to natural pulp and give rise to new odontoblast cells located at the dentin–pulp interface that are able to secrete tubular dentin in the course of tooth development but also at later time points [16]. In the context of pulp regeneration, the re-establishment of an odontoblast layer seems to be crucial due to its central role in tooth physiology and pathology [17–19]. Located at the dentin–pulp interface, these cells are the first line of defense against a bacterial invasion [20], they release antimicrobial agents [21] and have an immunomodulatory potential [22], allowing the tooth to immediately respond to stimuli, e.g., by secretion of dentin [17]. They also possess sensory functions by transducing pH changes and pressure as well as other pain-related stimuli [18]. Thus, it is of great interest for dental pulp tissue engineering to identify cell sources that are capable of differentiating into odontoblasts.

Recent in vivo studies have shown that pulp regeneration is possible after stem cell transplantation [23,24]. Histology revealed newly differentiated odontoblast-like mineralizing cells in contact with dentin. This approach is based on the transplantation of previously isolated and expanded autologous dental pulp stem cells (hDPSCs) and has proven successful in clinical trials [14,25]. In vitro studies also display odontogenic differentiation and the mineralization potential of hDPSCs when cultured with dentin matrix proteins (eDMPs) [26]. Dental pulp stem cells express dentin sialoprotein and differentiate into odontoblast-like cells with cellular processes extending into the dentinal tubules when seeded into EDTA-conditioned dentin cylinders and transplanted subcutaneously into immunocompromised mice [27]. However, the cell transplantation approach using hDPSCs and other tooth-derived cell types is challenging due to the necessity of cell expansion to obtain a sufficient number of cells, the need for a donor tooth and the limited differentiation potential of the multipotent stem cells compared to pluripotent stem cells [28].

A potential cell source to overcome those obstacles might be the amnion [29], the innermost layer of the human placenta. It contains amniotic epithelial cells (hAECs), which are formed by day 8 after fertilization and therefore maintain the plasticity of pre-gastrulation cells. Thus, hAECs are able to differentiate into cells of all three embryological layers [30], whereas multipotent stem cells, such as hDPSCs, are only capable of differentiating into cell types of one germ layer [31]. Human AECs showed the expression of human embryonic and pluripotent stem cell markers [30], such as stage-specific embryonic antigen-4 (*SSEA4*), octamer-binding transcription factor 4 (*OCT4*) and nanog homebox (*NANOG*) [32]. Moreover, hAECs reportedly have antimicrobial properties [33], immunomodulating potential [34] and can induce angiogenesis [35], which makes a useful cell type for regenerative therapies [29]. Up to 300 million hAECs can be obtained from one human placenta by a simple isolation protocol [29] which may be either expanded, directly applied or cryopreserved, e.g., in cell banks [29], which would ease the provision of these pluripotent stem cells. They are already used to treat several medical conditions, e.g., liver diseases and Parkinson's disease [36,37]. Moreover, amnion epithelial cells are not tumorigenic [32,38] and do not elicit an immune response upon heterologous transplantation, since they express very low levels of leukocyte antigens [30,32].

Although some exceptional studies have tested the transplantation of amnion membrane into the root canal with satisfactory clinical outcomes [39], no study has assessed the potential of hAECs to be used for endodontic regeneration. Thus, the aim of this study was to evaluate hAECs as a potential cell source for dental pulp tissue engineering. It was hypothesized that both hAECs and hDPSCs provide similar qualities in terms of viability, dentin adherence, odontoblast-like differentiation and mineralization.

2. Results

2.1. Isolation and Characterization of hAECs

As shown by the histological analysis, the first digestion of the amnion only partially detached hAECs (Figure 1A,B); however, the second digestion released nearly all cells (Figure 1C). Interestingly, the flow cytometric analysis of hAECs in culture revealed both epithelial (CD49f, CD326) and mesenchymal (CD105, CD44) surface antigens (Figure 1D,E).

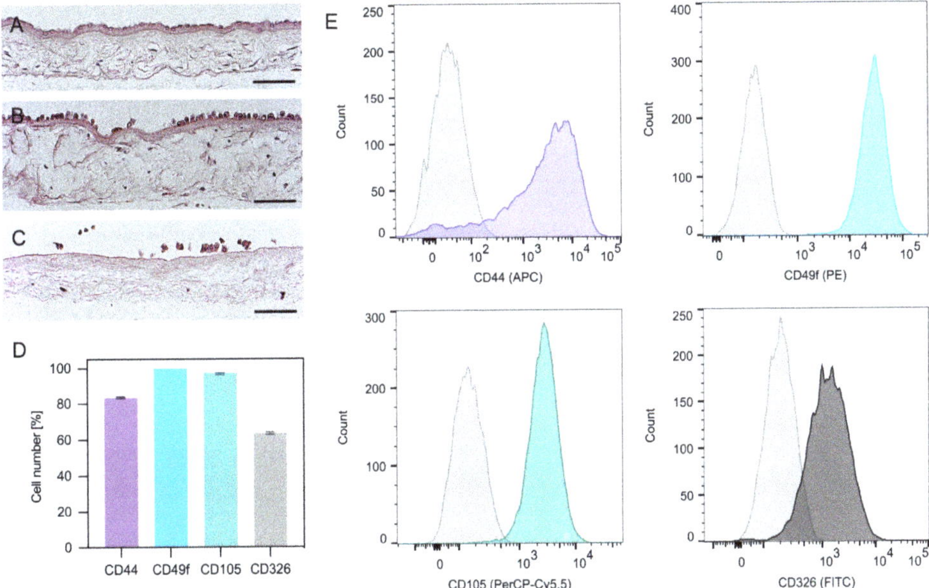

Figure 1. Amnion staining and expression profile of hAECs. Amnion before digestion (**A**) and after the first (**B**) and second digestion (**C**). The hAECs were attached to a collagen membrane forming a monolayer of columnar/cuboidal cells (hematoxylin and eosin; scale bars: 100 µm). Expression profile of hAECs determined by flow cytometry analysis (**D**). The hAECs in culture expressed both mesenchymal markers (CD44 and CD105) as well as epithelial markers (CD49f and CD326) (**D,E**).

2.2. Cell Viability

Human amnion epithelial cells showed a reduced viability compared to hDPSCs in all groups and at all time points (Figure 2A). Neither eDMPs nor StemPro had a significant impact on the viability of hAECs and hDPSCs at days 2 and 4; however, eDMP revealed a reduction at day 8 (Figure 2A).

2.3. Fluorescence Microscopy

Morphologically, the primary culture of the hAECs appeared homogenous with cobblestone-like morphology (Figure 2B–D), whereas hDPSCs were spindle-shaped and considerably smaller (Figure 2E–G). Overall, no relevant medium-dependent changes in cellular morphology were displayed by either cell type.

2.4. Cell Adhesion to Dentin

Representative scanning electron microscopic images of hDPSCs and hAECs on dentin disks are shown in Figure 3. Scanning electron microscope images revealed that hDPSCs

(Figure 3A,B) and hAECs (Figure 3C,D) were homogeneously distributed on the dentin. Moreover, both cell types showed adhesion to dentin and spread their processes over the surface in both EDTA-conditioned (Figure 3B,D) and unconditioned (Figure 3A,C) dentin. EDTA-conditioned dentin exhibited a clean dentin surface where dentin tubules were visible, while tubules were covered with a smear layer in unconditioned disks. Both hDPSCs and hAECs extended processes to form cellular contacts (Figure 3B,D). Whereas the hDPSCs adhered to dentin appeared spindle-shaped (Figure 3A,B), the hAECs retained their typical cubic morphology (Figure 3C,D).

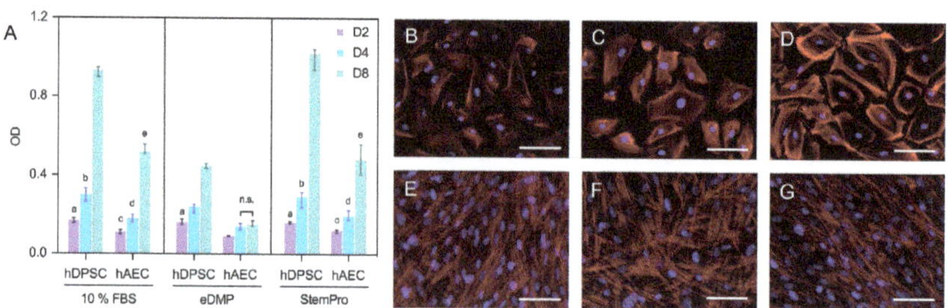

Figure 2. Viability and morphology of hAECs and hDPSCs. Cell viability of hAECs and hDPSCs cultured with eDMP and StemPro after 2, 4 and 8 days (**A**). Median values and 25–75% percentiles were calculated from three independent experiments performed in triplicate (*n* = 9). Fluorescence microscopy of hAECs and hDPSCs cultured with different media after 7 days and stained with DAPI and phalloidin (**B–G**). Cells were cultured in DMEM with 10% FBS (**B,E**), with eDMP (**C,F**) and with StemPro (**D,G**). hAECs exhibit a cobblestone-like morphology (**B–D**) while hDPSCs exhibit a mesenchymal stem cell phenotype (**E–G**). (Scale bars: 50 μm).

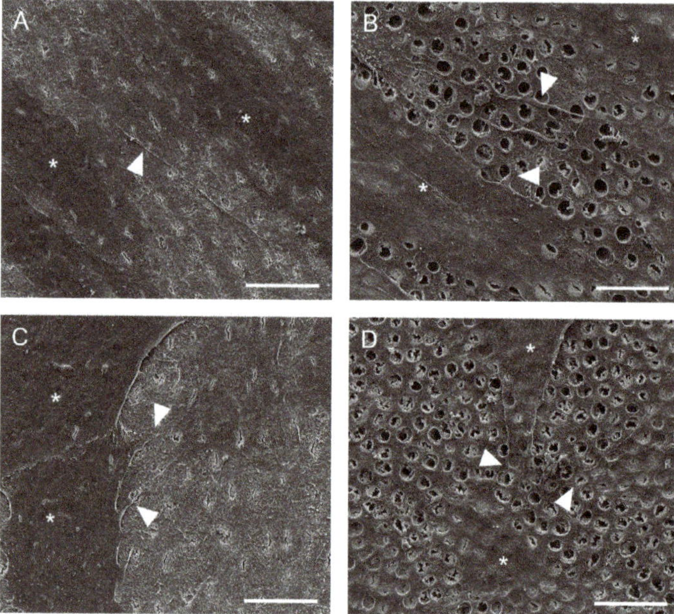

Figure 3. Adhesion of hDPSC and hAECs onto dentin surface. Representative SEM images of dentin surface with hDPSC (**A,B**) and hAECs (**C,D**) after 48 h (cells marked by asterisks). Cell adhesion and spreading on the surface of dentin was evident with (**B,D**) and without (**A,C**) EDTA conditioning. Some cytoplasmic processes (arrowheads) were evident in both cell types. (Scale bars: 20 μm).

2.5. Gene Expression

Genes associated with odontoblast differentiation and mineralization (collagen type I alpha 1 chain (*COL1A1*), bone morphogenetic protein 4 (*BMP4*), integrin binding sialoprotein *(IBSP)*, nestin *(NES)* and bone gamma-carboxyglutamate protein or osteocalcin (*BGLAP)*) were either not expressed in hAECs or the expression was significantly downregulated in comparison to the hDPSCs (Figure 4A). However, genes associated with epithelial–mesenchymal transition were upregulated in hAECs compared to hDPSCs. Specifically, the insulin like growth factor binding protein 2 (*IGFBP2*) gene was significantly upregulated in hAECs cultured with StemPro or eDMPs at days 1 and 7, and S100 calcium binding protein A4 (*S100A4*) was considerably upregulated in hAECs at all time points (Figure 4A). Glutathione peroxidase 3 (*GPX3*), a gene associated with the reduction of hydrogen peroxide, which arises from oxidative stress, was significantly upregulated in hAECs in almost all groups and at all time points.

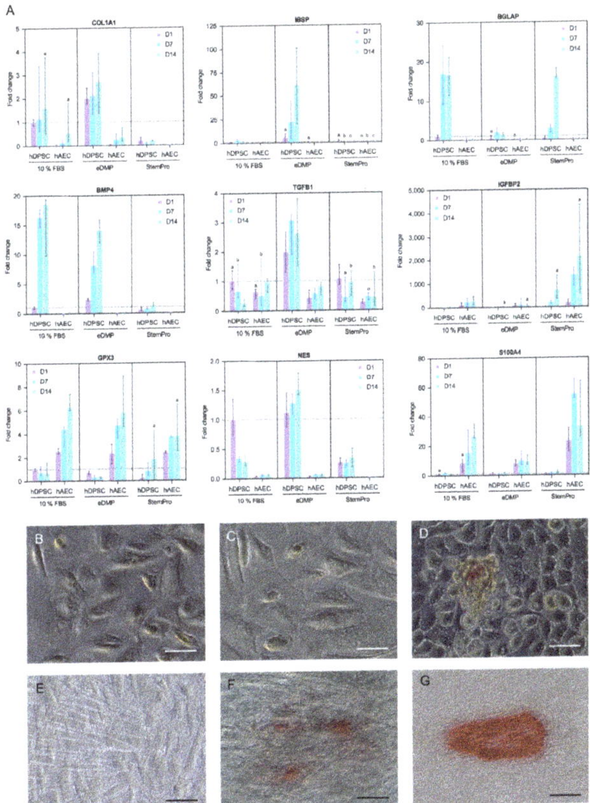

Figure 4. Expression of odontogenic and mineralization-associated genes. Effect of eDMPs and StemPro on expression of odontogenic and mineralization-associated marker genes (*COL1A1*, *BMP4*, *IBSP*, *IGFBP-2*, *NES*, *TGFB1* and *BGLAP*) in hAECs and hDPSCs at days 1, 7 and 14 (**A**). Genes indicative of epithelial–mesenchymal transition (*S100A4*) and protection against oxidative damage (*GPX3*) are also depicted (**A**). Target gene expressions are depicted relative to the untreated control (hDPSCs with 10% FBS at day 1) and median values were calculated from two independent experiments in duplicated samples (*n* = 4). Non-significant differences between hAECs and hDPSCs for each medium and follow-up point are marked with lowercase letters (a, b, c). The effect of eDMPs and StemPro on mineralization of hAECs (**B–D**) and hDPSCs (**E–G**) using Alizarin Red staining assay. Calcium deposits were evident in hAECs cultured with StemPro (**D**) and hDPCS cultured with eDMPs (**F**) and StemPro (**G**). (Scale bars: 50 μm).

2.6. Mineralization

Representative images of the mineralization capability of hAECs and hDPSCs after cultivation with eDMP and StemPro are shown in Figure 4B–G. Calcium deposits were observed in hAECs and hDPSCs cultured with StemPro at day 21 (Figure 4D,G) and in hDPSCs cultured with eDMPs (Figure 4F). Mineralization appeared in the form of small and dense nodules, which were significantly smaller in hAECs. However, no calcium deposits were observed in either cell type cultured in their respective standard media (Figure 4B,E) or in hAECs cultured with eDMPs (Figure 4C).

3. Discussion

Revitalization is a promising endodontic therapy for immature necrotic teeth with excellent clinical results [40]. However, the newly generated tissues are reported to be reparative tissues, which show microanatomical deficits, such as the lack odontoblast cells at the dentin–pulp interface [41]. In order to restore the pulp to its original form and function, numerous tissue-engineering-approaches are currently being investigated based on the concepts of cell transplantation as well as cell homing. Due to their pluripotency, hAECs differentiate into various cell types in vivo depending on their local environment. This has been demonstrated by injection of hAECs into the liver and into bone defects and heart tissue and observing the adequate differentiation into functional hepatocytes [42], osteoblasts [43] and cardiomyocytes [44], respectively. To the best of our knowledge, no study has evaluated the odontogenic differentiation of hAECs so far. The results revealed that human hAECs can adhere and spread on dentin and that they are able to mineralize; however, the transition into an odontoblast lineage was not observed.

3.1. Cell Adhesion to Dentin

In the context of pulp regeneration, cell attachment to the dentin walls of the root canal is essential [45]. Anatomical and functional restoration of the pulp–dentin complex is only possible with a stable adhesion of the transplanted or recruited stem cells to the collagenous extracellular matrix of dentin. The establishment of an odontoblast layer [27] is necessary as an immunological cellular barrier of the pulp [20] and enables a continuous mineralization in contact with the dentin walls upon the receipt of external stimuli. In the present study, the dentin disks were conditioned with EDTA prior to cell seeding. As a chelator of calcium, it removes the smear layer and provides a clean surface with exposed dentin tubules [46]. Furthermore, its demineralizing effect releases bioactive proteins and growth factors from the dentin extracellular matrix [47] which facilitate stem cell migration, mineralization and odontogenic differentiation [26]. In the clinical situation, EDTA also reverses the deleterious effects of the disinfectant sodium hypochlorite on the survival of stem cells [48], which makes it a crucial irrigation step in the current recommendations for revitalization procedures [49]. To investigate the interaction of hAECs with dentin, cells were isolated from human placentas and cultured directly on dentin disks for 48 h. Scanning electron microscope images revealed that hAECs were able to adhere and spread on dentin in a similar fashion to the dental pulp stem cells irrespective of EDTA-conditioning. As expected, the remaining smear layer did not affect the cells' survival; however, unconditioned dentin did not provide the necessary access to the tubules.

3.2. Mineralization

After proper adhesion of the cells to the dentin matrix, they are expected to differentiate into an odontogenic or osteogenic phenotype [27] and to start to secrete a mineralized matrix. While osteogenic differentiation was induced with StemPro as commercial differentiation medium, development into an odontogenic lineage was promoted by the addition of extracted human eDMPs, as has been carried out and described in previous studies [26,50]. Alizarin Red staining after 21 days revealed that human amnion epithelial cells cultured with StemPro medium were able to produce calcium deposits, confirming the basic mineralization capacity of these cells, as shown previously [43,51]; however, the calcification nodules were small and scarce compared to those of hDPSCs. Morphologically, a slight enlargement of the hAECs was observed after induced osteogenesis, which did not mirror previous investigations that reported a two-to-three-fold expansion of cell bodies following osteogenic differentiation [30]. Overall, hAECs appeared not to respond to osteogenic culture medium as well as was described in other studies [51,52]. Importantly, hAECs cultured with 500 pg/mL of eDMPs did not reveal any calcium deposits [26], in contrast to hDPSCs, which allows the conclusion that hAECs are not responsive to eDMPs in the tested concentrations either.

3.3. Cell Differentiation

StemPro osteogenic differentiation medium and eDMPs induce osteogenic and odontogenic differentiation of mesenchymal stem cells, respectively [26]. In previous studies, hAECs cultured with standard osteogenic medium [43,51], i.e., medium containing β-glycerophosphate and dexamethasone, showed an increase in alkaline phosphatase activity as well as the expression of bone-related genes [43,51]. As far as we know, the effect of eDMPs and StemPro on amniotic epithelial stem cells has not been investigated so far. In accordance with existing research, hDPSCs were able to differentiate into an odontogenic cell type [53,54]. This was shown by the upregulation of typical marker genes for both osteogenic and odontogenic lineage: *COL1A1, IBSP, BGLAP, BMP4* and transforming growth factor beta 1 (*TGF-β1*) in qRT-PCR. In this context, COL1A1 is classified as a marker for early mineralization [55–57]. IBSP is secreted during crystallization [58,59]. *BGLAP* can actively bind calcium and is therefore a late marker for mineralization [26,56,58,60,61]. *BMP4* is known to stimulate odontogenesis and bone formation [62–64]. *TGF-β1* is expressed by odontoblasts during maturation and dentinogenesis [65]. Furthermore, the upregulation of the dental specific marker *NES* confirms a differentiation of the hDPSCs along the odontogenic cell line [59,66]. To evaluate the odontoblastic transition of hAECs cultured with eDMPs, the expression of genes related to odontoblast differentiation and mineralization, including *COL1A1, BMP4, IBSP, IGFBP2, NES* and *BGLAP* (osteocalcin) [67], was analyzed and compared with hDPSCs. We expected to see similar results for hAECs in eDMP; however, the hAECs did not show any significant upregulation of mineralization markers under odontogenic culture conditions. We therefore conclude that, under the settings of this study, the hAECs were unable to differentiate into an odontogenic phenotype.

3.4. Epithelial–Mesenchymal Transition

A possible explanation for the compromised differentiation displayed by the otherwise pluripotent hAECs could be that they underwent epithelial–mesenchymal transition (EMT), which has been associated with a reduced osteogenic differentiability [68]. During this process the epithelial cells lose their characteristics, such as polarization or cell–cell connections, and change their phenotype to that of mesenchymal cells. EMT can be physiological, e.g., during embryonic development, inflammation, wound healing or fibrosis, but is also part of pathological processes, such as tumor progression or oncogenesis. Interestingly, it has even been observed to occur in cell culture [69,70]. Reportedly, freshly isolated hAECs do not express mesenchymal surface markers, such as CD105 and CD44, but display primarily epithelial markers, such as CD49f and CD326 [29,71–73]. However, cells in this study also expressed CD105 and CD44, which is in line with research concerning cultured and expanded hAECs [74–77]. This increasing change in phenotype over the cultivation period has previously been described [68] and can also be seen as an indication that the cells undergo EMT. Furthermore, S100A4, a calcium-binding protein also called fibroblast specific protein 1, has been described as a marker for this transition [78]. *IGFBP2*, which was upregulated by hAECs up to 4000-fold in comparison to the expression in DPSC, can also be classified as an EMT marker [79]. A way to induce the EMT process in vitro can be the addition of epidermal growth factor (EGF) and TGF-β, both of which were necessarily in the media [80]. Autocrine TGF-β production can also stimulate this process [81]. An additional explanation for why EMT is occurring could be unintentional selection of the mesenchymal phenotype by the culture protocol. Cells without CD44 expression are more likely to detach from the culture flask and thereby be removed during medium change [68]. However, the change from an epithelial to a mesenchymal phenotype is described as being accompanied by a change in morphology [68], which was not observed in our experiment. Further research needs to be undertaken comparing the differentiability of freshly isolated and expanded hAECs.

3.5. Impact of Culture Conditions

Glutathione peroxidase 3 (*GPX3*) aims at the reduction of hydrogen peroxide, which may arise from oxidative stress in cell metabolism. In this case, increased *GPX3* levels might accompany stressful culture conditions or trypsination of the cells, which is commonly paraphrased as "culture shock" [82,83]. Reportedly, hAECs are a highly sensitive cell type and quite challenging in in vitro culture [84]. Their viability was significantly lower compared to hDPSCs over all time points and EGF was essential to provide a more physiological environment; however, analogously to the observations by other research groups, the cells did not thrive outside their specific stem cell niche [38]. The upregulation of *GPX3* by hAECs, in combination with the reduced metabolic activity in eDMP, as indicated by the MTT assay, could indicate unfavorable culture conditions that might affect differentiation. This assumption is to be verified in further experiments by, e.g., determining intracellular reactive oxygen species (ROS) or antioxidative enzymes.

4. Materials and Methods

4.1. Isolation and Characterization of hAECs

Human placentas were obtained from caesarean deliveries of healthy donors with informed consent and the approval of the Bioethical Commission of the University of Barcelona, Spain (No.: IRB00003099). The placentas were transported to the laboratory for further processing in sterile saline at 4 °C. The hAECs were isolated in a laminar flow cabinet following a previously published protocol [30]. Briefly, the amnion was detached from the underlying chorion and washed with 200 mL of Ringer's acetate solution (pH 6.5; Baxter, Deerfield, MA, USA) for up to 10 min and 200 mL of PBS (PBS, Biochrom, Berlin, Germany) to remove all blood particles. Subsequently, 2–3 g portions of amnion were digested in Falcon tubes with 20 mL of $10\times$ TryPLE solution (Life Technologies, Gaithersburg, USA) on a shaker (35 rpm) at 37 °C for 30 min. The membrane underwent a second digestion step with fresh digestion solution. Cells from both digestions were centrifuged at $300\times g$ for 10 min and suspended in DMEM (DMEM, high glucose; Life Technologies, Gaithersburg, MD, USA) with 10% FBS, 100 U/mL penicillin and 100 µg/mL streptomycin, 2 mM L-glutamine, 1 mM sodium pyruvate and 10 ng/mL of epidermal growth factor (EGF; StemCell Technologies, Vancouver, BC, Canada). This medium containing EGF, which is an essential supplement for hAEC growth, is referred to as DMEM 10% FBS in the following text. Cells from passage 2 were used in all experiments and cultured at 37 °C and 5% CO_2. To assess the effectiveness of the isolation procedure, amniotic tissue before digestion as well as after the first and second digestion step was fixed in formalin for 2 h and processed for histology. Histological processing and HE staining was performed according to a previously published protocol [85]. Furthermore, flow cytometry was conducted to evaluate the antigen profile of the isolated cells. Immediately after isolation, hAECs were seeded in T75 flasks and cultured to 80% confluence with DMEM 10% FBS. Finally, a suspension of 2×10^5 cells in 81 µL was incubated with 2 µL of mouse anti-human CD44 (APC; 560890, BD Biosciences, San Jose, CA, USA), 5 µL of mouse anti-human CD105 (PerCP-CY 5.5; 562245, BD Biosciences, San Jose, CA, USA), 8 µL of anti-CD326 (FITC; 347197, BD Biosciences, San Jose, CA, USA) and 4 µL of rat anti-human CD49f (PE; 561894, BD Biosciences, San Jose, CA, USA). Flow cytometry was conducted with at least 10,000 events per sample (FACSCant, BD Biosciences, San Jose, CA, USA) and data was analyzed with FlowJo (version 10.8, BD Biosciences, San Jose, CA, USA). Cells from two different donors were investigated in triplicate and median values with 25–75% percentiles were computed ($n = 6$).

4.2. Isolation and Characterization of hDPSCs

Human dental pulp stem cells were isolated from human third molars and cultured as described previously [86]. Dental pulp stem cells were maintained in αMEM supplemented with 100 U/mL penicillin, 100 μg/mL streptomycin and 10% FBS. Finally, to ensure mesenchymal stem cell character, the cells were sorted for the surface antigen STRO-1using the MACS-System (magnetic-activated cell sorting; Miltenyi Biotec, Bergisch Gladbach, Germany). Furthermore, mesenchymal stem cell antigens, following Dominici et al. [87], were determined in accordance with a previous work [50].

4.3. Extraction of Dentin Matrix Proteins (eDMPs)

Human caries-free third molars were collected from donors (15–25 years old) after informed consent and with approval by an appropriate review board at the University of Regensburg (No.: 19-1327-101; Faculty of Medicine, University of Regensburg, Regensburg, Germany). eDMPs were extracted from human teeth according to a validated protocol [26]. Briefly, dentin was pulverized (Mixer Mill MM 200, RETSCH, Haan, Germany) after removal of the enamel, cementum and pulp. Dentin powder was suspended in 10% EDTA (AppliChem, Darmstadt, Germany) and the solution was purified with syringe filters (1.2, 0.45, and 0.2 μm Acrodisc Syringe Filters with Supor Membrane; Pall Corporation, Port Washington, WI, USA) and enriched by centrifugal filtration with a molecular weight cut-off of at 3000 Da (Amicon Ultra-15 3K; Merck Millipore, Billerica, MA, USA). The solvent was exchanged for phosphate-buffered saline (PBS without Ca^{2+}, Mg^{2+}; Biochrom, Berlin, 137 Germany). Finally, TGF-β1 was quantified as a representative growth factor (Quantikine 138 ELISA Kit; R&D Systems Inc., Minneapolis, MN, USA) to facilitate standardized supplementation to culture media.

4.4. Cell Viability

To quantify the cell viability, hAECs (3.2×10^3 cells/well) and hDPSCs (3.2×10^3 cells/well) were seeded in 96-well plates to reach 80% confluency. The hAECs were exposed to the following media: (i) DMEM 10% FBS; (ii) DMEM 10% FBS and 500 pg/mL eDMPs; and (iii) osteogenic differentiation medium with 10 ng/mL EGF (StemPro Osteogenesis Differentiation Kit; Thermo Fisher Scientific, Waltham, MA, USA). Likewise, hDPSCs were cultured in (i) α-MEM with 10% FBS, (ii) α-MEM with 10% FBS and 500 pg/mL eDMPs and (iii) osteogenic differentiation medium (StemPro Osteogenesis Differentiation Kit; Thermo Fisher Scientific, Waltham, MA, USA). MTT assays were performed after 2, 4 and 8 days. The cells were incubated with 100 μL/well of a 0.5 mg/mL MTT solution (Thiazolyl Blue Tetrazolium Bromide; Sigma-Aldrich, Saint Louis, MO, USA) for 60 min at 37 °C and 5% CO_2. Subsequently, the dye was dissolved in 200 μL/well of dimethyl sulfoxide (DMSO; Merck Millipore, Billerica, MA, USA) on a shaker (540 rpm) at room temperature for 10 min. Optical density readings were performed on a microplate reader at $\lambda = 540$ nm (Infinite 200; Tecan, Männedorf, Switzerland) and the results were summarized as median values with 25–75% percentiles ($n = 8$).

4.5. Fluorescence Microscopy

To evaluate morphological changes induced by the three different types of media, hAECs (7.5×10^3 cells/well) and hDPSCs (5×10^3 cells/well) were seeded on coverslips in 24-well plates and cultured as described above. After 7 days, cells were fixed with 4% formalin (10 min), permeabilized with 0.1% Triton X (5 min) and stained with phalloidin (30 min) and DAPI (1 min). Coverslips were mounted on slides (ProLong Glass Antifade Mountant, Thermo Fisher Scientific, Waltham, USA) and imaged on a ZEISS microscope (Axio Vert.A1, Carl Zeiss Microscopy, Jena, Germany) with the ZEISS Axiocam 503 color camera (Carl Zeiss Microscopy, Jena, Germany). Images with filters for blue

(Carl Zeiss Microscopy, Jena, Germany) and red fluorescence (Set 43 and Set 49, Carl Zeiss Microscopy, Jena, Germany) in place were taken independently and digitally superimposed. ZEN software was used for microscopy and imaging (version 3.1, Carl Zeiss Microscopy, Jena, Germany).

4.6. Cell Adhesion to Dentin

The adherence and phenotypic changes of hAECs and hDPSCs seeded on dentin disks were evaluated by scanning electron microscopy. Dentin disks of 0.2 mm thickness were obtained from the crown of human molars. They were optionally rinsed in 10% EDTA for 15 min and washed with distilled water afterwards. Subsequently, hAECs and hDPSCs (3.8×10^4 cells/well) were seeded on the dentin disks in 24-well plates. The hAECs were cultured in DMEM with 10% FBS and DPSCs in α-MEM with 10% FBS for 48 h (37 °C, 5% CO_2). Samples were fixed with 2.5% glutaraldehyde in 0.1 M Sørensen's phosphate buffer for 30 min and analyzed on a FEI Quanta 400 environmental scanning electron microscope (SEM) with a field emitter and an Everhart–Thornley detector at 4.0 kV and high-vacuum conditions (FEI Europe B.V., Eindhoven, The Netherlands).

4.7. Gene Expression

Cultures were established with hAECs (7.5×10^4 cells/well) and hDPSCs (4.9×10^4 cells/well) in 12-well plates, as described above. After 1, 7 and 14 days, mRNA was extracted using the RNeasy Mini Kit (Qiagen, Hilden, Germany) and quantified spectrophotometrically (NanoDrop 2000, Thermo Fisher Scientific, Waltham, USA). Then, 500 ng of nucleic acids were transcribed into cDNA (Omniscript RT Kit, Qiagen, Hilden, Germany) using oligo-dT primers (Qiagen, Hilden, Germany). To assess the effect of eDMP and StemPro on gene expression, qRT-PCR was performed using the TaqMan Fast Advanced Master 188 Mix (4444557, Applied Biosystems, Thermo Fisher Scientific, Waltham, MA, USA) and probes for the following genes: collagen type I alpha 1 (COL1A1; Hs00164004_m1), integrin binding sialoprotein (IBSP; Hs00913377_m1), bone gamma-carboxyglutamate protein (BGLAP; Hs01587814_g1), bone morphogenic protein 4 (BMP4; Hs00370078_m1), transforming growth factor beta 1 (TGFB1; Hs00998133_m1), nestin (NES; Hs04187831_g1), insulin-like growth factor binding protein 2 (IGFBP2; Hs01040719_m1), S100 calcium binding protein A4 (S100A4; Hs00243202_m1), glutathione peroxidase 3 (GPX3; Hs01078668_m1) and 40S ribosomal protein S18 (RPS18; Hs99999901_s1) as the housekeeping gene. Finally, measurements for all target genes were normalized to RPS18 and related to hDPSCs cultured in 10% by the comparative CT method ($\Delta\Delta$CT) [88]. Medians with 25–75% percentiles were calculated on the basis of two experiments with cells from different donors ($n = 4$).

4.8. Mineralization

To visualize calcium deposition, hAECs (3.7×10^4 cells/well) and hDPSCs (2.4×10^4 cells/well) were seeded in 24-well plates to reach 80% confluency. Both cell types were cultured according to the previously described groups. After 21 days, cells were fixed with formalin for 10 min and incubated with 40 mM alizarin (Alizarin Red S, Carl Roth, Karlsruhe, Germany) at pH 4.2 and room temperature for 30 min. Images were taken with an inverted microscope (Axio Vert.A1, Carl Zeiss Microscopy GmbH, Jena, Germany).

4.9. Statistical Analysis

Data were treated nonparametrically and pairwise Mann–Whitney U tests were conducted at a significance level of $\alpha = 0.05$. Statistical analysis was performed, comparing hAECs and hDPSCs for each medium for each follow-up point. All statistical analyses were computed with GraphPad Prism 9 (GraphPad Software, La Jolla, CA, USA) and non-statistical significance ($p > 0.05$) was indicated in the respective figures by lowercase letters.

5. Conclusions

Human amnion epithelial cells can adhere and spread on dentin and are able to differentiate and mineralize. Nevertheless, hAECs failed to reveal an odontoblastic transition under in vitro conditions. Even if hAECs show great promise in other regenerative applications, they do not seem to be a feasible alternative stem cell source for dental pulp tissue engineering. In addition to the difficult culture behavior, their cellular reactions are difficult to control by signaling molecules and are not as reliable as, for example, mesenchymal stem cells from pulp, which was particularly evident in EMT. Thus, the advantages of hAECs in terms of pluripotency do not come into play and their advantages in terms of high availability can presumably not be used.

Author Contributions: Conceptualization, C.B., J.M.d.A., M.C.M.-C., M.W. (Matthias Widbiller) and M.W. (Melanie Woelflick); methodology, C.B., J.M.d.A., M.W. (Matthias Widbiller) and M.W. (Melanie Woelflick); formal analysis, C.B., E.O., J.M.d.A., M.C.M.-C. and M.W. (Matthias Widbiller); investigation, C.B., E.O. and M.W. (Melanie Woelflick); resources, C.B., J.M.d.A., M.C.M.-C., M.W. (Matthias Widbiller) and K.G.; writing—original draft preparation, C.B., E.O. and M.W. (Matthias Widbiller); writing—review and editing, J.M.d.A., M.C.M.-C., M.W. (Matthias Widbiller), E.O. and K.G.; visualization, C.B. and M.W. (Matthias Widbiller); supervision, K.G., J.M.d.A. and M.W. (Matthias Widbiller); project administration, C.B.; funding acquisition, C.B., M.W. (Matthias Widbiller) and K.G. All authors have read and agreed to the published version of the manuscript.

Funding: This research was partially funded by Universidad de La Frontera (Temuco, Chile), grant number DI19-0024.

Institutional Review Board Statement: The study was conducted according to the guidelines of the Declaration of Helsinki and approved by the Bioethical Commission of the University of Barcelona, Spain (No.: IRB00003099).

Informed Consent Statement: Informed consent was obtained from all subjects involved in the study.

Data Availability Statement: Data is contained within the article.

Conflicts of Interest: The authors declare no conflict of interest.

References

1. Murray, P.; Garcia-Godoy, F.; Hargreaves, K.M. Regenerative Endodontics: A Review of Current Status and a Call for Action. *J. Endod.* **2007**, *33*, 377–390. [CrossRef] [PubMed]
2. Cvek, M. Prognosis of luxated non-vital maxillary incisors treated with calcium hydroxide and filled with gutta-percha. A retrospective clinical study. *Dent. Traumatol.* **1992**, *8*, 45–55. [CrossRef] [PubMed]
3. Tong, H.J.; Rajan, S.; Bhujel, N.; Kang, J.; Duggal, M.; Nazzal, H. Regenerative Endodontic Therapy in the Management of Nonvital Immature Permanent Teeth: A Systematic Review—Outcome Evaluation and Meta-analysis. *J. Endod.* **2017**, *43*, 1453–1464. [CrossRef] [PubMed]
4. Lin, J.; Zeng, Q.; Wei, X.; Zhao, W.; Cui, M.; Gu, J.; Lu, J.; Yang, M.; Ling, J. Regenerative Endodontics Versus Apexification in Immature Permanent Teeth with Apical Periodontitis: A Prospective Randomized Controlled Study. *J. Endod.* **2017**, *43*, 1821–1827. [CrossRef]
5. Jiang, X.; Liu, H.; Peng, C. Clinical and Radiographic Assessment of the Efficacy of a Collagen Membrane in Regenerative Endodontics: A Randomized, Controlled Clinical Trial. *J. Endod.* **2017**, *43*, 1465–1471. [CrossRef]
6. Nazzal, H.; Kenny, K.; Altimimi, A.; Kang, J.; Duggal, M.S. A prospective clinical study of regenerative endodontic treatment of traumatized immature teeth with necrotic pulps using bi-antibiotic paste. *Int. Endod. J.* **2018**, *51*, e204–e215. [CrossRef]
7. Zhu, W.; Zhu, X.; Huang, G.T.-J.; Cheung, G.S.P.; Dissanayaka, W.; Zhang, C. Regeneration of dental pulp tissue in immature teeth with apical periodontitis using platelet-rich plasma and dental pulp cells. *Int. Endod. J.* **2013**, *46*, 962–970. [CrossRef]
8. Meschi, N.; Hilkens, P.; Lambrichts, I.; Van den Eynde, K.; Mavridou, A.; Strijbos, O.; De Ketelaere, M.; Van Gorp, G.; Lambrechts, P. Regenerative endodontic procedure of an infected immature permanent human tooth: An immunohistological study. *Clin. Oral Investig.* **2015**, *20*, 807–814. [CrossRef]
9. Wang, X.; Thibodeau, B.; Trope, M.; Lin, L.M.; Huang, G.T.-J. Histologic Characterization of Regenerated Tissues in Canal Space after the Revitalization/Revascularization Procedure of Immature Dog Teeth with Apical Periodontitis. *J. Endod.* **2010**, *36*, 56–63. [CrossRef]
10. Bucchi, C.; Marcé-Nogué, J.; Galler, K.M.; Widbiller, M. Biomechanical performance of an immature maxillary central incisor after revitalization: A finite element analysis. *Int. Endod. J.* **2019**, *52*, 1508–1518. [CrossRef]

11. Orti, V.; Collart-Dutilleul, P.-Y.; Piglionico, S.; Pall, O.; Cuisinier, F.; Panayotov, I. Pulp Regeneration Concepts for Nonvital Teeth: From Tissue Engineering to Clinical Approaches. *Tissue Eng. Part B Rev.* **2018**, *24*, 419–442. [CrossRef] [PubMed]

12. Galler, K.M.; Widbiller, M. Perspectives for Cell-homing Approaches to Engineer Dental Pulp. *J. Endod.* **2017**, *43*, S40–S45. [CrossRef] [PubMed]

13. He, L.; Zhong, J.; Gong, Q.; Cheng, B.; Kim, S.G.; Ling, J.; Mao, J.J. Regenerative Endodontics by Cell Homing. *Dent. Clin.* **2017**, *61*, 143–159. [CrossRef] [PubMed]

14. Nakashima, M.; Iohara, K.; Murakami, M.; Nakamura, H.; Sato, Y.; Ariji, Y.; Matsushita, K. Pulp regeneration by transplantation of dental pulp stem cells in pulpitis: A pilot clinical study. *Stem Cell Res. Ther.* **2017**, *8*, 61. [CrossRef]

15. Ducret, M.; Fabre, H.; Celle, A.; Mallein-Gerin, F.; Perrier-Groult, E.; Alliot-Licht, B.; Farges, J.-C. Current challenges in human tooth revitalization. *Bio.-Med. Mater. Eng.* **2017**, *28*, S159–S168. [CrossRef]

16. Huang, G.T.-J. Dental pulp and dentin tissue engineering and regeneration advancement and challenge. *Front. Biosci.* **2011**, *3*, 788–800. [CrossRef]

17. Couve, E.; Osorio, R.; Schmachtenberg, O. Reactionary Dentinogenesis and Neuroimmune Response in Dental Caries. *J. Dent. Res.* **2014**, *93*, 788–793. [CrossRef]

18. Couve, E.; Osorio, R.; Schmachtenberg, O. The Amazing Odontoblast. *J. Dent. Res.* **2013**, *92*, 765–772. [CrossRef]

19. Farges, J.-C.; Keller, J.-F.; Carrouel, F.; Durand, S.H.; Romeas, A.; Bleicher, F.; Lebecque, S.; Staquet, M.-J. Odontoblasts in the dental pulp immune response. *J. Exp. Zoöl. Part B Mol. Dev. Evol.* **2009**, *312*, 425–436. [CrossRef]

20. Farges, J.-C.; Alliot-Licht, B.; Renard, E.; Ducret, M.; Gaudin, A.; Smith, A.J.; Cooper, P.R. Dental Pulp Defence and Repair Mechanisms in Dental Caries. *Mediat. Inflamm.* **2015**, *2015*, 230251. [CrossRef]

21. Farges, J.-C.; Bellanger, A.; Ducret, M.; Aubert-Foucher, E.; Richard, B.; Alliot-Licht, B.; Bleicher, F.; Carrouel, F. Human odontoblast-like cells produce nitric oxide with antibacterial activity upon *TLR2* activation. *Front. Physiol.* **2015**, *6*, 185. [CrossRef] [PubMed]

22. Farges, J.-C.; Carrouel, F.; Keller, J.-F.; Baudouin, C.; Msika, P.; Bleicher, F.; Staquet, M.-J. Cytokine production by human odontoblast-like cells upon Toll-like receptor-2 engagement. *Immunobiology* **2011**, *216*, 513–517. [CrossRef]

23. Itoh, Y.; Sasaki, J.I.; Hashimoto, M.; Katata, C.; Hayashi, M.; Imazato, S. Pulp Regeneration by 3-dimensional Dental Pulp Stem Cell Constructs. *J. Dent. Res.* **2018**, *97*, 1137–1143. [CrossRef] [PubMed]

24. Iohara, K.; Imabayashi, K.; Ishizaka, R.; Watanabe, A.; Nabekura, J.; Ito, M.; Matsushita, K.; Nakamura, H.; Nakashima, M. Complete Pulp Regeneration after Pulpectomy by Transplantation of CD105+ Stem Cells with Stromal Cell-Derived Factor-1. *Tissue Eng. Part A* **2011**, *17*, 1911–1920. [CrossRef] [PubMed]

25. Xuan, K.; Li, B.; Guo, H.; Sun, W.; Kou, X.; He, X.; Zhang, Y.; Sun, J.; Liu, A.; Liao, L.; et al. Deciduous autologous tooth stem cells regenerate dental pulp after implantation into injured teeth. *Sci. Transl. Med.* **2018**, *10*, eaaf3227. [CrossRef] [PubMed]

26. Widbiller, M.; Eidt, A.; Lindner, S.R.; Hiller, K.-A.; Schweikl, H.; Buchalla, W.; Galler, K.M. Dentine matrix proteins: Isolation and effects on human pulp cells. *Int. Endod. J.* **2018**, *51*, e278–e290. [CrossRef]

27. Galler, K.M.; D'Souza, R.; Federlin, M.; Cavender, A.C.; Hartgerink, J.; Hecker, S.; Schmalz, G. Dentin Conditioning Codetermines Cell Fate in Regenerative Endodontics. *J. Endod.* **2011**, *37*, 1536–1541. [CrossRef]

28. Miki, T. Stem cell characteristics and the therapeutic potential of amniotic epithelial cells. *Am. J. Reprod. Immunol.* **2018**, *80*, e13003. [CrossRef]

29. Gramignoli, R.; Srinivasan, R.C.; Kannisto, K.; Strom, S.C. Isolation of Human Amnion Epithelial Cells According to Current Good Manufacturing Procedures. *Curr. Protoc. Stem Cell Biol.* **2016**, *37*, 1–13. [CrossRef]

30. Ilancheran, S.; Michalska, A.; Peh, G.; Wallace, E.M.; Pera, M.; Manuelpillai, U. Stem Cells Derived from Human Fetal Membranes Display Multilineage Differentiation Potential. *Biol. Reprod.* **2007**, *77*, 577–588. [CrossRef]

31. Gronthos, S.; Mankani, M.; Brahim, J.; Robey, P.G.; Shi, S. Postnatal human dental pulp stem cells (DPSCs) in vitro and in vivo. *Proc. Natl. Acad. Sci. USA* **2000**, *97*, 13625–13630. [CrossRef] [PubMed]

32. Yang, P.-J.; Yuan, W.-X.; Liu, J.; Li, J.-Y.; Tan, B.; Qiu, C.; Zhu, X.-L.; Qiu, C.; Lai, D.-M.; Guo, L.-H.; et al. Biological characterization of human amniotic epithelial cells in a serum-free system and their safety evaluation. *Acta Pharmacol. Sin.* **2018**, *39*, 1305–1316. [CrossRef]

33. Nemr, W.; Bashandy, M.; Araby, E.; Khamiss, O. Molecular displaying of differential immunoresponse to various infections of amniotic epithelia. *Am. J. Reprod. Immunol.* **2017**, *77*, e12662. [CrossRef]

34. Motedayyen, H.; Fathi, F.; Fasihi-Ramandi, M.; Taheri, R.A. The effect of lipopolysaccharide on anti-inflammatory and pro-inflammatory cytokines production of human amniotic epithelial cells. *Reprod. Biol.* **2018**, *18*, 404–409. [CrossRef]

35. Zhu, D.; Muljadi, R.; Chan, S.T.; Vosdoganes, P.; Lo, C.; Mockler, J.C.; Wallace, E.; Lim, R. Evaluating the Impact of Human Amnion Epithelial Cells on Angiogenesis. *Stem Cells Int.* **2015**, *2016*, 4565612. [CrossRef] [PubMed]

36. Miki, T.; Lehmann, T.; Cai, H.; Stolz, D.B.; Strom, S.C. Stem Cell Characteristics of Amniotic Epithelial Cells. *Stem Cells* **2005**, *23*, 1549–1559. [CrossRef] [PubMed]

37. Lim, R.; Hodge, A.; Moore, G.; Wallace, E.M.; Sievert, W. A Pilot Study Evaluating the Safety of Intravenously Administered Human Amnion Epithelial Cells for the Treatment of Hepatic Fibrosis. *Front. Pharmacol.* **2017**, *8*, 549. [CrossRef]

38. Miki, T.; Strom, S.C. Amnion-derived pluripotent/multipotent stem cells. *Stem Cell Rev. Rep.* **2006**, *2*, 133–141. [CrossRef]

39. Bajaj, M.; Soni, A.J. Revascularization of a Nonvital, Immature Permanent Tooth Using Amniotic Membrane: A Novel Approach. *Int. J. Clin. Pediatr. Dent.* **2019**, *12*, 150–152. [CrossRef]

40. Chrepa, V.; Joon, R.; Austah, O.; Diogenes, A.; Hargreaves, K.M.; Ezeldeen, M.; Ruparel, N.B. Clinical Outcomes of Immature Teeth Treated with Regenerative Endodontic Procedures—A San Antonio Study. *J. Endod.* **2020**, *46*, 1074–1084. [CrossRef]

41. Becerra, P.; Ricucci, D.; Loghin, S.; Gibbs, J.L.; Lin, L.M. Histologic Study of a Human Immature Permanent Premolar with Chronic Apical Abscess after Revascularization/Revitalization. *J. Endod.* **2014**, *40*, 133–139. [CrossRef] [PubMed]

42. Lin, J.S.; Zhou, L.; Sagayaraj, A.; Jumat, N.H.B.; Choolani, M.; Chan, J.K.Y.; Biswas, A.; Wong, P.C.; Lim, S.G.; Dan, Y.Y. Hepatic differentiation of human amniotic epithelial cells and in vivo therapeutic effect on animal model of cirrhosis. *J. Gastroenterol. Hepatol.* **2015**, *30*, 1673–1682. [CrossRef] [PubMed]

43. Mattioli, M.; Gloria, A.; Turriani, M.; Mauro, A.; Curini, V.; Russo, V.; Tetè, S.; Marchisio, M.; Pierdomenico, L.; Berardinelli, P.; et al. Stemness characteristics and osteogenic potential of sheep amniotic epithelial cells. *Cell Biol. Int.* **2011**, *36*, 7–19. [CrossRef] [PubMed]

44. Fang, C.-H.; Jin, J.; Joe, J.-H.; Song, Y.-S.; So, B.-I.; Lim, S.M.; Cheon, G.J.; Woo, S.-K.; Ra, J.-C.; Lee, Y.-Y.; et al. In Vivo Differentiation of Human Amniotic Epithelial Cells into Cardiomyocyte-Like Cells and Cell Transplantation Effect on Myocardial Infarction in Rats: Comparison with Cord Blood and Adipose Tissue-Derived Mesenchymal Stem Cells. *Cell Transplant.* **2012**, *21*, 1687–1696. [CrossRef] [PubMed]

45. Galler, K.M.; Widbiller, M.; Buchalla, W.; Eidt, A.; Hiller, K.-A.; Hoffer, P.C.; Schmalz, G.H. EDTA conditioning of dentine promotes adhesion, migration and differentiation of dental pulp stem cells. *Int. Endod. J.* **2016**, *49*, 581–590. [CrossRef]

46. Morago, A.; Ruiz-Linares, M.; Ferrer-Luque, C.M.; Baca, P.; Archilla, A.R.; Arias-Moliz, M.T. Dentine tubule disinfection by different irrigation protocols. *Microsc. Res. Tech.* **2019**, *82*, 558–563. [CrossRef]

47. Galler, K.M.; Buchalla, W.; Hiller, K.-A.; Federlin, M.; Eidt, A.; Schiefersteiner, M.; Schmalz, G. Influence of Root Canal Disinfectants on Growth Factor Release from Dentin. *J. Endod.* **2015**, *41*, 363–368. [CrossRef]

48. Martin, D.E.; de Almeida, J.F.A.; Henry, M.A.; Khaing, Z.; Schmidt, C.E.; Teixeira, F.B.; Diogenes, A. Concentration-dependent Effect of Sodium Hypochlorite on Stem Cells of Apical Papilla Survival and Differentiation. *J. Endod.* **2014**, *40*, 51–55. [CrossRef]

49. Galler, K.M.; Krastl, G.; Simon, S.; Van Gorp, G.; Meschi, N.; Vahedi, B.; Lambrechts, P. European Society of Endodontology position statement: Revitalization procedures. *Int. Endod. J.* **2016**, *49*, 717–723. [CrossRef]

50. Widbiller, M.; Eidt, A.; Wölflick, M.; Lindner, S.R.; Schweikl, H.; Hiller, K.-A.; Buchalla, W.; Galler, K.M. Interactive effects of LPS and dentine matrix proteins on human dental pulp stem cells. *Int. Endod. J.* **2018**, *51*, 877–888. [CrossRef]

51. Si, J.; Zhang, J.; Dai, J.; Yu, D.; Yu, H.; Shi, J.; Wang, X.; Shen, S.G.F.; Guo, L. Osteogenic Differentiation of Human Amniotic Epithelial Cells and Its Application in Alveolar Defect Restoration. *Stem Cells Transl. Med.* **2014**, *3*, 1504–1513. [CrossRef]

52. Fatimah, S.S.; Ng, S.L.; Chua, K.H.; Hayati, A.R.; Tan, A.E.; Tan, G.C. Value of human amniotic epithelial cells in tissue engineering for cornea. *Hum. Cell* **2010**, *23*, 141–151. [CrossRef] [PubMed]

53. Huang, G.T.-J.; Shagramanova, K.; Chan, S.W. Formation of Odontoblast-Like Cells from Cultured Human Dental Pulp Cells on Dentin In Vitro. *J. Endod.* **2006**, *32*, 1066–1073. [CrossRef]

54. Liu, G.; Xu, G.; Gao, Z.; Liu, Z.; Xu, J.; Wang, J.; Zhang, C.; Wang, S. Demineralized Dentin Matrix Induces Odontoblastic Differentiation of Dental Pulp Stem Cells. *Cells Tissues Organs* **2016**, *201*, 65–76. [CrossRef] [PubMed]

55. Braut, A.; Kollar, E.J.; Mina, M. Analysis of the odontogenic and osteogenic potentials of dental pulp in vivo using a *Col1a1-2.3-GFP* transgene. *Int. J. Dev. Biol.* **2003**, *47*, 281–292. [PubMed]

56. Huang, W.; Yang, S.; Shao, J.; Li, Y.P. Signaling and transcriptional regulation in osteoblast commitment and differentiation. *Front. Biosci.* **2007**, *12*, 3068–3092. [CrossRef]

57. Kaneto, C.M.; Lima, P.S.P.; Zanette, D.L.; Oliveira, T.Y.K.; Pereira, F.D.A.; Lorenzi, J.C.C.; Dos Santos, J.L.; Prata, K.L.; Neto, J.M.P.; De Paula, F.J.A.; et al. Osteoblastic differentiation of bone marrow mesenchymal stromal cells in Bruck Syndrome. *BMC Med Genet.* **2016**, *17*, 38. [CrossRef]

58. Simon, S.; Smith, A.; Lumley, P.; Berdal, A.; Smith, G.; Finney, S.; Cooper, P. Molecular characterization of young and mature odontoblasts. *Bone* **2009**, *45*, 693–703. [CrossRef]

59. Smith, A.J.; Scheven, B.A.; Takahashi, Y.; Ferracane, J.L.; Shelton, R.M.; Cooper, P.R. Dentine as a bioactive extracellular matrix. *Arch. Oral Biol.* **2012**, *57*, 109–121. [CrossRef]

60. Chen, L.; Jacquet, R.; Lowder, E.; Landis, W.J. Refinement of collagen–mineral interaction: A possible role for osteocalcin in apatite crystal nucleation, growth and development. *Bone* **2015**, *71*, 7–16. [CrossRef]

61. Dacic, S.; Kalajzic, I.; Visnjic, D.; Lichtler, A.C.; Rowe, D.W. Col1a1-Driven Transgenic Markers of Osteoblast Lineage Progression. *J. Bone Miner. Res.* **2001**, *16*, 1228–1236. [CrossRef] [PubMed]

62. Nakashima, M.; Nagasawa, H.; Yamada, Y.; Reddi, A.H. Regulatory Role of Transforming Growth Factor-β, Bone Morphogenetic Protein-2, and Protein-4 on Gene Expression of Extracellular Matrix Proteins and Differentiation of Dental Pulp Cells. *Dev. Biol.* **1994**, *162*, 18–28. [CrossRef] [PubMed]

63. Asgary, S.; Nazarian, H.; Khojasteh, A.; Shokouhinejad, N. Gene Expression and Cytokine Release during Odontogenic Differentiation of Human Dental Pulp Stem Cells Induced by 2 Endodontic Biomaterials. *J. Endod.* **2014**, *40*, 387–392. [CrossRef] [PubMed]

64. MacDougall, M.J.; Javed, A. Dentin and Bone: Similar Collagenous Mineralized Tissues. In *Bone and Development*; Bronner, F., Farach-Carson, M.C., Roach, H.T., Eds.; Springer London: London, UK, 2010; pp. 183–200. [CrossRef]

65. Niwa, T.; Yamakoshi, Y.; Yamazaki, H.; Karakida, T.; Chiba, R.; Hu, J.C.C.; Nagano, T.; Yamamoto, R.; Simmer, J.P.; Margo-lis, H.C.; et al. The dynamics of TGF-β in dental pulp, odontoblasts and dentin. *Sci. Rep.* **2018**, *8*, 4450. [CrossRef] [PubMed]

66. About, I.; Laurent-Maquin, D.; Lendahl, U.; Mitsiadis, T.A. Nestin Expression in Embryonic and Adult Human Teeth under Normal and Pathological Conditions. *Am. J. Pathol.* **2000**, *157*, 287–295. [CrossRef]

67. Dds, M.W.; Bucchi, C.; Rosendahl, A.; Spanier, G.; Buchalla, W.; Galler, K.M. Isolation of primary odontoblasts: Expectations and limitations. *Aust. Endod. J.* **2019**, *45*, 378–387. [CrossRef]

68. Stadler, G.; Hennerbichler, S.; Lindenmair, A.; Peterbauer, A.; Hofer, K.; van Griensven, M.; Gabriel, C.; Redl, H.; Wolbank, S. Phenotypic shift of human amniotic epithelial cells in culture is associated with reduced osteogenic differentiation in vitro. *Cytotherapy* **2008**, *10*, 743–752. [CrossRef]

69. Okada, H.; Danoff, T.M.; Kalluri, R.; Neilson, E.G. Early role of *Fsp1* in epithelial-mesenchymal transformation. *Am. J. Physiol. Physiol.* **1997**, *273*, F563–F574. [CrossRef]

70. Hay, E.D. An Overview of Epithelio-Mesenchymal Transformation. *Cells Tissues Organs* **1995**, *154*, 8–20. [CrossRef]

71. Pratama, G.; Vaghjiani, V.; Tee, J.Y.; Liu, Y.H.; Chan, J.; Tan, C.; Murthi, P.; Gargett, C.; Manuelpillai, U. Changes in Culture Expanded Human Amniotic Epithelial Cells: Implications for Potential Therapeutic Applications. *PLoS ONE* **2011**, *6*, e26136. [CrossRef]

72. Tabatabaei, M.; Mosaffa, N.; Nikoo, S.; Bozorgmehr, M.; Ghods, R.; Kazemnejad, S.; Rezania, S.; Keshavarzi, B.; Arefi, S.; Ramezani-Tehrani, F.; et al. Isolation and partial characterization of human amniotic epithelial cells: The effect of tryp-sin. *Avicenna. J. Med. Biotechnol.* **2014**, *6*, 10–20. [PubMed]

73. Murphy, S.V.; Kidyoor, A.; Reid, T.; Atala, A.; Wallace, E.M.; Lim, R. Isolation, Cryopreservation and Culture of Human Amnion Epithelial Cells for Clinical Applications. *J. Vis. Exp.* **2014**, *94*, e52085. [CrossRef] [PubMed]

74. De Coppi, P.; Atala, A. Stem Cells from the Amnion. In *Principles of Regenerative Medicine*; Elsevier: Amsterdam, The Netherlands, 2019; pp. 133–148. [CrossRef]

75. Portmann-Lanz, C.B.; Schoeberlein, A.; Huber, A.; Sager, R.; Malek, A.; Holzgreve, W.; Surbek, D.V. Placental mesenchymal stem cells as potential autologous graft for pre- and perinatal neuroregeneration. *Am. J. Obstet. Gynecol.* **2006**, *194*, 664–673. [CrossRef] [PubMed]

76. Wolbank, S.; Peterbauer, A.; Fahrner, M.; Hennerbichler, S.; van Griensven, M.; Stadler, G.; Redl, H.; Gabriel, C. Dose-Dependent Immunomodulatory Effect of Human Stem Cells from Amniotic Membrane: A Comparison with Human Mesenchymal Stem Cells from Adipose Tissue. *Tissue Eng.* **2007**, *13*, 1173–1183. [CrossRef]

77. Toda, A.; Okabe, M.; Yoshida, T.; Nikaido, T. The Potential of Amniotic Membrane/Amnion-Derived Cells for Regeneration of Various Tissues. *J. Pharmacol. Sci.* **2007**, *105*, 215–228. [CrossRef]

78. Ambartsumian, N.; Klingelhöfer, J.; Grigorian, M. The Multifaceted S100A4 Protein in Cancer and Inflammation. In *Calcium Binding Proteins of the EF-Hand Superfamily*; Heizmann, C.W., Ed.; Springer: New York, NY, USA, 2019; Volume 1929, pp. 339–365. [CrossRef]

79. Li, T.; Forbes, M.E.; Fuller, G.N.; Li, J.; Yang, X.; Zhang, W. IGFBP2: Integrative hub of developmental and oncogenic signaling network. *Oncogene* **2020**, *39*, 2243–2257. [CrossRef]

80. Zavadil, J.; Böttinger, E.P. TGF-β and epithelial-to-mesenchymal transitions. *Oncogene* **2005**, *24*, 5764–5774. [CrossRef]

81. Alcaraz, A.; Mrowiec, A.; Insausti, C.L.; García-Vizcaíno, E.M.; Ruiz-Canada, C.; López-Martínez, C.; Moraleda, J.M.; Nicolás, F.J. Autocrine TGF-β Induces Epithelial to Mesenchymal Transition in Human Amniotic Epithelial Cells. *Cell Transplant.* **2013**, *22*, 1351–1367. [CrossRef]

82. Halliwell, B. Oxidative stress in cell culture: An under-appreciated problem? *FEBS Lett.* **2003**, *540*, 3–6. [CrossRef]

83. Stolzing, A.; Sethe, S.; Scutt, A.M. Stressed Stem Cells: Temperature Response in Aged Mesenchymal Stem Cells. *Stem Cells Dev.* **2006**, *15*, 478–487. [CrossRef]

84. Terada, S.; Matsuura, K.; Enosawa, S.; Miki, M.; Hoshika, A.; Suzuki, S.; Sakuragawa, N. Inducing Proliferation of Human Amniotic Epithelial (HAE) Cells for Cell Therapy. *Cell Transplant.* **2000**, *9*, 701–704. [CrossRef] [PubMed]

85. Widbiller, M.; Rothmaier, C.; Saliter, D.; Wölflick, M.; Rosendahl, A.; Buchalla, W.; Schmalz, G.; Spruss, T.; Galler, K.M. Histology of human teeth: Standard and specific staining methods revisited. *Arch. Oral Biol.* **2021**, *127*, 105136. [CrossRef] [PubMed]

86. Galler, K.M.; Schweikl, H.; Thonemann, B.; D'Souza, R.N.; Schmalz, G. Human pulp-derived cells immortalized with Simian Virus 40 T-antigen. *Eur. J. Oral Sci.* **2006**, *114*, 138–146. [CrossRef] [PubMed]

87. Dominici, M.; Le Blanc, K.; Mueller, I.; Slaper-Cortenbach, I.; Marini, F.C.; Krause, D.S.; Deans, R.J.; Keating, A.; Prockop, D.J.; Horwitz, E.M. Minimal criteria for defining multipotent mesenchymal stromal cells. The International Society for Cellular Therapy position statement. *Cytotherapy* **2006**, *8*, 315–317. [CrossRef]

88. Schmittgen, T.D.; Livak, K.J. Analyzing real-time PCR data by the comparative C_T method. *Nat. Protoc.* **2008**, *3*, 1101–1108. [CrossRef]

International Journal of
Molecular Sciences

Article

Bone Differentiation Ability of CD146-Positive Stem Cells from Human Exfoliated Deciduous Teeth

Ryo Kunimatsu [1,*], Kodai Rikitake [1], Yuki Yoshimi [1], Nurul Aisyah Rizky Putranti [1], Yoko Hayashi [2] and Kotaro Tanimoto [1]

[1] Department of Orthodontics and Craniofacial Developmental Biology, Graduate School of Biomedical and Health Sciences, Hiroshima University, 1-2-3 Kasumi, Minami-ku, Hiroshima 734-8553, Japan
[2] Analysis Center of Life Science, Natural Science Center for Basic Research and Development, Hiroshima University, 1-2-3 Kasumi, Minami-ku, Hiroshima 734-8553, Japan
* Correspondence: ryoukunimatu@hiroshima-u.ac.jp; Tel.: +81-82-257-5686; Fax: +81-82-257-5687

Abstract: Regenerative therapy for tissues by mesenchymal stem cell (MSCs) transplantation has received much attention. The cluster of differentiation (CD)146 marker, a surface-antigen of stem cells, is crucial for angiogenic and osseous differentiation abilities. Bone regeneration is accelerated by the transplantation of CD146-positive deciduous dental pulp-derived mesenchymal stem cells contained in stem cells from human exfoliated deciduous teeth (SHED) into a living donor. However, the role of CD146 in SHED remains unclear. This study aimed to compare the effects of CD146 on cell proliferative and substrate metabolic abilities in a population of SHED. SHED was isolated from deciduous teeth, and flow cytometry was used to analyze the expression of MSCs markers. Cell sorting was performed to recover the CD146-positive cell population (CD146+) and CD146-negative cell population (CD146-). CD146 + SHED without cell sorting and CD146-SHED were examined and compared among three groups. To investigate the effect of CD146 on cell proliferation ability, an analysis of cell proliferation ability was performed using BrdU assay and MTS assay. The bone differentiation ability was evaluated using an alkaline phosphatase (ALP) stain after inducing bone differentiation, and the quality of ALP protein expressed was examined. We also performed Alizarin red staining and evaluated the calcified deposits. The gene expression of ALP, bone morphogenetic protein-2 (BMP-2), and osteocalcin (OCN) was analyzed using a real-time polymerase chain reaction. There was no significant difference in cell proliferation among the three groups. The expression of ALP stain, Alizarin red stain, ALP, BMP-2, and OCN was the highest in the CD146+ group. CD146 + SHED had higher osteogenic differentiation potential compared with SHED and CD146-SHED. CD146 contained in SHED may be a valuable population of cells for bone regeneration therapy.

Keywords: stem cells from human exfoliated deciduous teeth; bone regeneration; CD146

Citation: Kunimatsu, R.; Rikitake, K.; Yoshimi, Y.; Putranti, N.A.R.; Hayashi, Y.; Tanimoto, K. Bone Differentiation Ability of CD146-Positive Stem Cells from Human Exfoliated Deciduous Teeth. *Int. J. Mol. Sci.* **2023**, *24*, 4048. https://doi.org/10.3390/ijms24044048

Academic Editors: Kerstin M. Galler and Matthias Widbiller

Received: 12 January 2023
Revised: 4 February 2023
Accepted: 15 February 2023
Published: 17 February 2023

1. Introduction

Regenerative medicine is a medical technology that uses stem cells to regenerate tissues that have become dysfunctional; it was developed as a new therapeutic technology to replace organ and bone transplantation [1–4]. Mesenchymal stem cells (MSCs) were first identified as colony-forming cells with the ability to differentiate into osteoblasts, adipocytes, and chondrocytes within bone marrow organs [5]. MSCs are present in skeletal muscles, adipocytes, placentae, dental pulp, and periodontal ligament, and play roles in preparing for and maintaining homeostasis during the restoration of compromised tissues [6,7]. In addition, since MSCs can be collected from tissues and grown in standardized culture conditions, they are used as a transplanted cell preparation for autoimplantation in the medical field; cell preparation for skin and cartilage regeneration has already been marketed [8,9]. Notably, techniques for the isolation and culture of MSCs from oral tissues have been developed in dentistry, and research and development for the practical application of regenerative treatment of dental pulp and periodontal tissues have been promoted [10–12].

Since the beginning of 2000, a search for MSC sources in intraoral tissues has been undertaken. Studies have demonstrated and identified the presence of tissue stemness in stem cells from human exfoliated deciduous teeth (SHED) and dental pulp stem cells (DPSCs) [13,14]. SHED and DPSCs have a higher proliferative potential than bone marrow-derived MSCs (BMSCs), and their potential to differentiate into osteoblasts, adipocytes, chondrocytes, and neural cells has been demonstrated [15–18]. SHED and DPSCs are involved in the formation of dentin and pulp complexes [11,15] and bone [19,20]. Previous studies reported that the successful isolation rates of SHED and DPSCs were approximately 83% and 70%, respectively [21], and SHED and DPSCs had a bone regeneration capacity equivalent to that of BMSCs when SHED, DPSCs, and BMSCs were seeded in the polylactic acid membrane and transplanted into the parietal bone defect model of immunodeficient mice [22]. Kunimatsu et al. found that SHED had a higher cellular proliferative capacity than DPSCs and BMSCs, and SHED showed significantly higher expression of fibroblast growth factor (bFGF) and bone morphogenetic protein-2 (BMP-2) genes compared with hDPSCs and hBMSCs [23]. Moreover, SHED reportedly have a higher capacity for adipogenic and osteogenic differentiation compared to DPSCs [24]. In addition, SHED have been suggested to be distinctly different in nature from DPSC in recent review articles; they also have a higher osteogenic differentiation potential compared to DPSCs [25]. Kichenbrand et al. reported the following benefits of SHED: 1. Hard tissues cover the pulp; thus, less DNA injury occurs in the pulp, 2. It is easy to harvest MSCs, and the procedure is painless and non-invasive, 3. Several tissue samples can be collected, and 4. Since permanent teeth replace deciduous teeth, no ethical concerns exist [26]. In addition, SHED possess advantages for tissue regeneration, which allows quick in vitro expansion before implantation of the tissue, compared to their DPSCs [27]. Thus, SHED is becoming an attractive source of cells and a potential candidate for tissue regeneration. However, the cell population involved in the bone regeneration mechanism of SHED has not yet been elucidated.

Recently, the surface antigens of MSCs have garnered attention, as many surface antigens of MSCs serve as coreceptors for growth factors and provide valuable benefits for the regeneration of various tissues, such as promoting angiogenesis and osteogenesis [28]. MSCs isolated from tissues are a heterogeneous cell population expressing various surface antigens. The effect of surface antigens on cell properties can be investigated using cells with the surface antigen of interest, isolated by cell sorting [29]. Cluster of differentiation (CD)146, a surface antigen, is expressed on MSCs and the plasma membranes of vascular endothelial cells and vascular pericytes, where it functions as a key cellular adhesion molecule in angiogenesis [30,31]. CD146-positive cells isolated by cell sorting from heterogeneous populations of bone marrow- and umbilical cord-derived MSCs have higher bone regenerative potential than CD146-negative cells, and CD146 is related to bone regenerative potential [30–33]. Therefore, we focused on investigating the expression of CD146 in SHED. In a bone defect immunodeficient mouse model, osteogenesis was promoted by the transplantation of CD146-positive SHED cells [34]. However, a detailed examination of the in vitro effect of CD146 on bone regenerative capacity has not yet been reported in SHED. Accordingly, the present study aimed to compare the cell proliferative and osteogenic potentials of CD146-positive cells, CD146-negative cells, and SHED heterogeneous cells.

2. Results

2.1. Surface-Antigen Analyses of Isolated SHED

The representative results of one of five donors are shown in Figure 1a. Cells expressing CD146 accounted for $70.9 \pm 4.3\%$ of the heterogeneous population of SHED cells isolated from the deciduous pulp. MSC-positive markers, CD90 and CD73, were expressed in all donors, and CD105 was expressed in $99.56 \pm 0.37\%$ of donors. CD14, CD19, CD34a, and

CD45 (MSC-negative markers) were expressed in $0.08 \pm 0.02\%$, 0.27%, $0.05 \pm 0.01\%$, and $2.93 \pm 0.54\%$ of donors, respectively (Figure 1a).

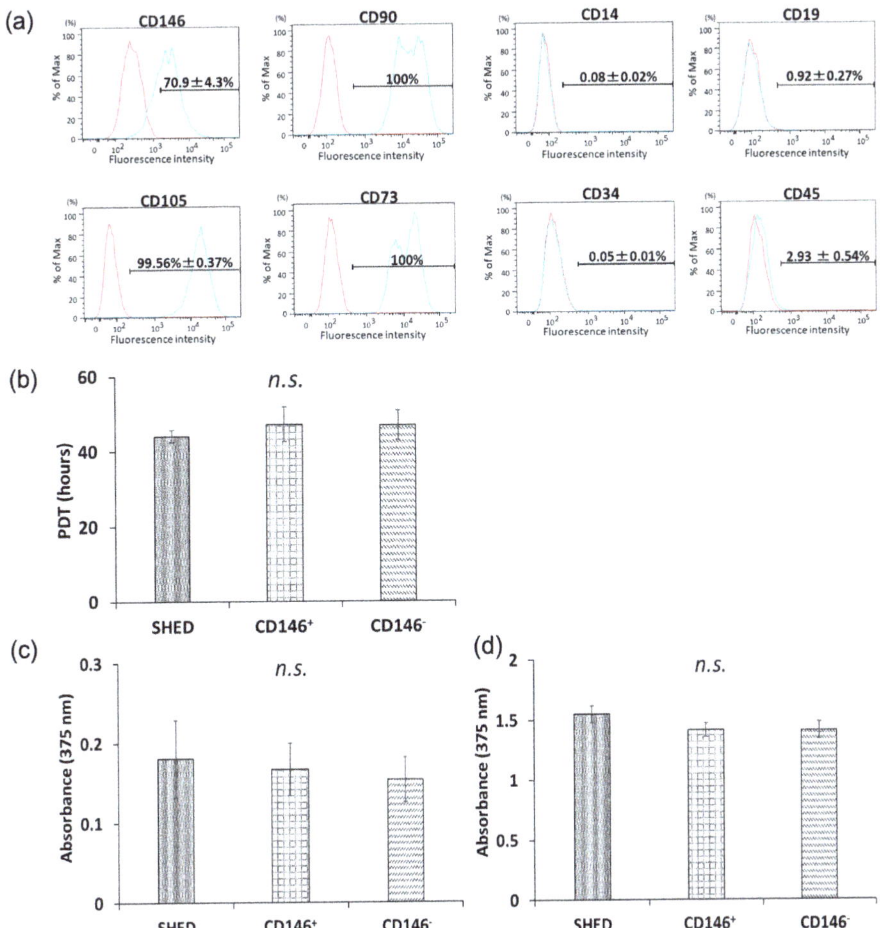

Figure 1. Surface-antigen analysis of SHED and comparative properties of SHED, CD146 + SHED, and CD146-SHED cellular proliferative potential. (**a**) Surface-antigen analysis of SHED. Among the heterogeneous population of SHED isolated from the deciduous pulp, $70.9 \pm 4.3\%$ of cells expressed CD146. MSCs were positive for CD90, CD73, and CD105 but negative for CD14, CD19, CD34, and CD45. (**b**) Comparison of population doubling time (PDT). The log phase was assessed based on the cell growth curve from day 2 to day 6 of incubation, and the PDT was calculated for this period. PDT did not differ significantly among the three groups ($n = 5$, Kruskal–Wallis test, Not significant; N.S.) (**c,d**). Comparison of cell proliferation. (**c**) Two hours after BrdU treatment, DNA synthesized in SHED was slightly higher than that in CD146 + SHED and CD146-SHED, but there were no significant differences among the three groups. (**d**) Twenty-four hours after BrdU treatment, the results were similar ($n = 5$, Kruskal–Wallis method, Not significant; N.S.).

2.2. Cellular Proliferative Capacity of SHED, CD146 + SHED, and CD146-SHED

2.2.1. PDT

PDT was 44.06 h, 46.88 h, and 47.15 h in SHED, CD146-SHED, and CD146 + SHED, respectively. However, there were no significant differences among the three groups (Figure 1b).

2.2.2. BrdU Proliferation Assay

Although the proliferative capacity of SHED was slightly higher than that of CD146 + SHED and CD146-SHED 2 h and 24 h after BrdU administration, there were no significant differences among the three groups (Figure 1c,d).

2.3. Osteogenic Differentiation-Related Gene Expression Analyses in SHED, CD146 + SHED, and CD146-SHED

Before the induction of osteogenic differentiation, the gene expression of ALP, BMP-2, and OCN did not differ significantly among SHED, CD146 + SHED, and CD146-SHED (Figure 2a). However, CD146 + SHED had significantly higher gene expression of ALP, BMP-2, and OCN on days 21 and 28 after osteodifferentiation induction than SHED and CD146-SHED (Figure 2b,c). In addition, SHED showed significantly higher gene expression of ALP, BMP-2, and OCN than CD146-SHED (Figure 2b,c).

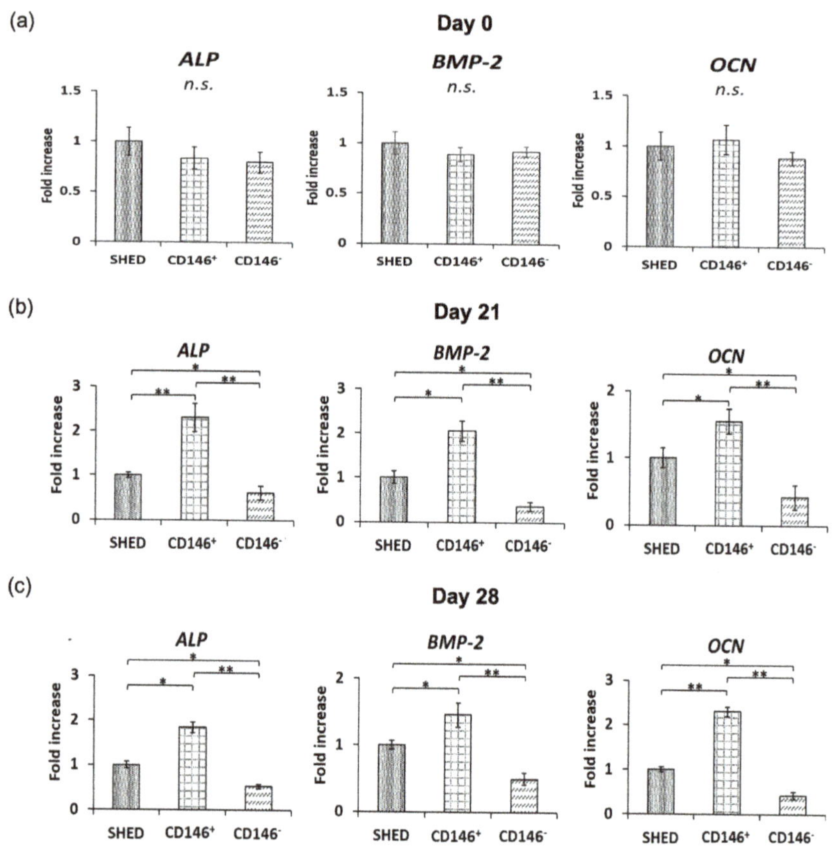

Figure 2. Gene expression in relation to bone differentiation on days 0, 21, and 28 of induction of bone differentiation. (**a**) Uninduced osteogenic differentiation. No significant differences were found among SHED, CD146 + SHED, and CD146-SHED in the gene expression of ALP, BMP-2, and OCN (*n* = 5, Not significant; N.S.). (**b**,**c**). Induced osteogenic differentiation on days 21 and 28. CD146 + SHED showed significantly higher gene expression of ALP, BMP-2, and OCN than SHED and CD146-SHED (*n* = 5, Kruskal–Wallis method, ** *p* < 0.01, * *p* < 0.05).

2.4. Osteogenic Differentiation Potential of SHED, CD146 + SHED, and CD146-SHED

ALP staining performed on day 21 after inductive osteogenic differentiation revealed a decreased order in the intensity of dark staining in CD146 + SHED, SHED, and CD146-

SHED (Figure 3a). In addition, the mean ALP protein levels measured on day 21 after the induction of osteogenic differentiation were 14.28 ± 0.745 μg/mL, 10.27 ± 0.636 μg/mL, and 6.96 ± 0.573 μg/mL in CD146 + SHED, SHED, and CD146-SHED, respectively. CD146 + SHED showed significantly higher ALP protein levels than SHED and CD146-SHED (Figure 3b); SHED showed significantly higher ALP protein levels than CD146-SHED (Figure 3b).

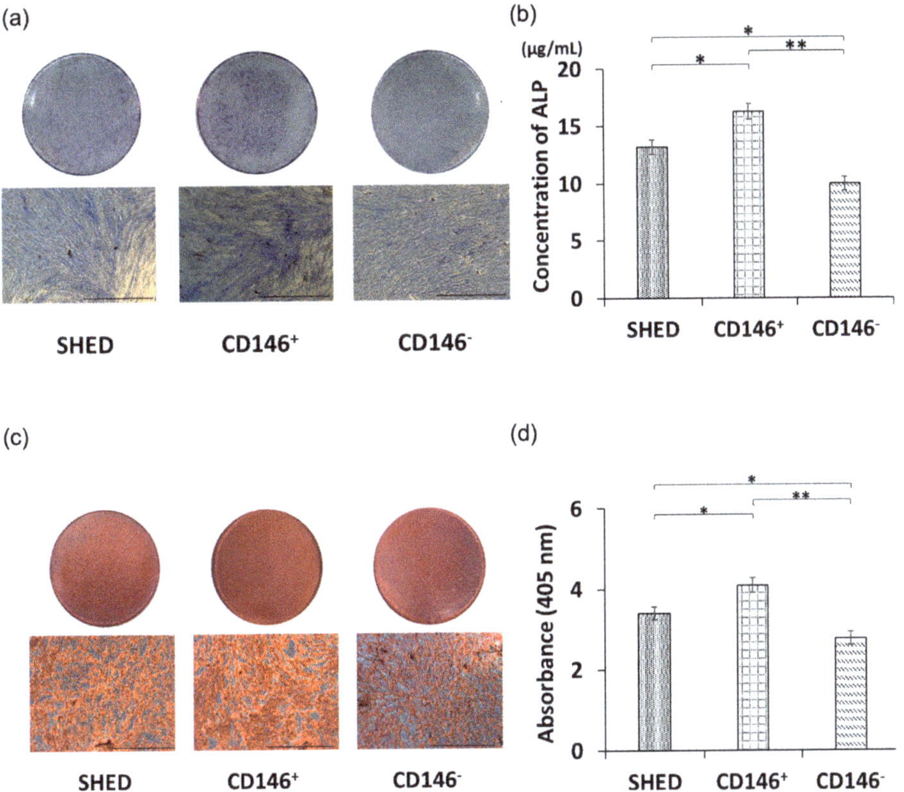

Figure 3. Osteogenic differentiation potential of SHED, CD146 + SHED, and CD146-SHED. (**a**) Osteogenic differentiation assessed by ALP staining on day 21. ALP staining was confirmed in all cells. The ALP staining intensity decreased in the following order: CD146 + SHED, SHED, and CD146-SHED (Scale bar = 500 μm). (**b**) Determination of ALP protein levels by ELISA. ALP protein levels were significantly higher in CD146 + SHED than those in SHED and CD146-SHED. The same was true for SHED compared to CD146-SHED ($n = 5$, Kruskal–Wallis method, ** $p < 0.01$, * $p < 0.05$). (**c**) Induced osteogenic differentiation. Osteogenic differentiation assessed by Alizarin red staining on day 28 decreased in the following order: CD146 + SHED, SHED, and CD146-SHED (Scale bar = 500 μm). (**d**) CD146 + SHED showed significantly higher absorbance values than SHED and CD146-SHED ($n = 5$, Kruskal–Wallis method, ** $p < 0.01$, * $p < 0.05$).

2.5. Comparative Analysis of Calcium Deposition among SHED, CD146 + SHED, CD146-SHED

Alizarin red staining was dense in CD146 + SHED, SHED, and CD146-SHED on day 28 after the induction of osteogenic differentiation (Figure 3c), with a decreased staining intensity observed in CD146 + SHED, SHED, and CD146-SHED, in that order. In addition, the absorbance assay showed that CD146 + SHED had significantly higher levels than SHED and CD146-SHED, indicating an increased calcified deposit in CD146 + SHED (Figure 3d).

3. Discussion

MSCs are isolated from various tissues, such as the bone marrow, adipose tissue, umbilical cord, and dental pulp, and implanted in defective tissues to promote tissue restoration and regeneration. Since deciduous teeth are shed spontaneously with the permanent tooth replacing them, harvesting SHED is less invasive than harvesting BMSCs, adipose-derived MSCs, and umbilical cord-derived MSCs. Since SHED and BMSCs have a similar bone regeneration capacity, the present study investigated SHED and BMSCs as sources of cells for transplantation into bone defects.

In tissues where abundant blood vessels are present, such as the pulp, there is a microenvironment around the blood vessels called the MSC niche [35,36], which comprises MSCs, hematopoietic stem cells, mesenchymal progenitors, fibroblasts, and pericytes (vascular pericytes) [37]. CD146 is expressed in MSCs and the plasma membrane of vascular endothelial cells and pericytes, and is associated with cell-to-cell or cell-to-extracellular matrix [38,39]. The surface antigen CD146 expressed on MSCs is a receptor for growth factors, such as Netrin-1, Wnt-1, and vascular endothelial growth factor (VEGF)-C [40], and it reportedly acts as a coreceptor for VEGF receptor-2 (VEGFR2) and platelet-derived growth factor receptor beta, thereby contributing to angiogenesis and vascular maintenance [41,42]. Since MSCs and pericytes are present in MSC niches and express some similar surface antigens, including CD146, pericytes are considered the origin of MSCs, and CD146 + MSCs may be very close to pericytes [43–45].

In the present study, the surface-antigen analyses of SHED-like cells revealed that more than 99.5% of the heterogeneous cells expressed CD105, CD73, and CD90. However, less than 3% of these cells expressed CD14, CD19, CD34, and CD45. SHED isolated from the pulp of deciduous teeth met the requirements for defining MSCs established by the International Society for Cellular Therapy [46]. CD105, CD73, and CD90 are expressed in these cells under standard culture conditions; however, CD45, CD34, CD14, CD11b, CD79a, CD19, and HLA-DR are not. Although CD105, CD73, and CD90 are MSC-positive markers, they are expressed in MSCs, hematopoietic stem cells, and blood cells [47]. However, MSC-negative markers, i.e., CD14, CD19, CD34, and CD45, are specific markers for hematopoietic stem cells and hemocytes [47]. In the present study, cells isolated from the deciduous pulp were CD105-, CD73-, CD90-, CD14-, CD19-, CD34-, and CD45-negative, and they were very likely to be SHED. Differentiation of SHED (which were isolated in a similar procedure) into osteoblasts, adipocytes, and chondrocytes in vitro systems was confirmed by Miura et al. [13] and Nakajima et al. [23]. In this study, the analysis of CD146 expression rates by flow cytometry showed that CD146 was expressed in approximately 70.9% of SHED heterogeneous populations. In contrast, CD146 was expressed in 48.39–66.3% of SHED heterogeneous populations [48,49]. In this study, adequate numbers of CD146 + SHED were isolated from SHED by cell sorting, which might contribute to slightly higher outcomes than previously reported (70.9% vs. 48.39–66.3%). These findings demonstrate the potential of CD146 + SHED as a valuable source of cells for future clinical applications.

The properties of SHED, CD146 + SHED, and CD146-SHED were investigated in cellular studies. Based on the BrdU cell proliferation assay and PDT analysis, there were no significant differences in proliferative capacity among the three groups, although SHED had a slightly higher proliferative capacity. CD146 + MSCs exhibited significantly higher cell proliferative capacity than CD146-MSCs in human endometrium-derived MSCs and periodontal ligament-derived MSCs [50,51]. However, some papers have demonstrated that CD146-MSCs have higher proliferative potential than CD146 + MSCs [42]. Paduano et al. reported that the proliferative capacity of CD146- MSCs and CD146+ MSCs, cells varied in each study [52]. In the present study, there were no significant differences in the proliferative ability among CD146-SHED, CD146 + SHED, and SHED. Factors accounting for the difference between the reports may include the type of MSC, performance of cellular conditioning or sorting devices used, and effect of the procedure. In this study, cells were passaged after cell sorting, cultured to confluence, and seeded in 96-well and 12-well plates for cell growth testing. However, CD146 + SHED and CD146-SHED grew more slowly immediately after cell sorting than SHED without cell sorting. The longer operating times after releasing cells from the dish, and several centrifugations during the cell sorting procedure, might have minor adverse effects on CD146 + SHED and CD146-SHED. Further investigations with different conditions, such as decreasing the number of passages after cell sorting, are warranted.

Comparative analysis of bone differentiation potential showed that CD146 + SHED had a higher bone differentiation potential than SHED and CD146-SHED based on ALP staining, quantitative ALP analysis, and alizarin red staining of SHED cultured in an osteogenic differentiation-inducing medium. During the osteogenic differentiation-inducing process, the progenitor cells are differentiated into osteoblasts and express ALP from the early- to mid-stage, followed by the expression of bone-specific OCN and, ultimately, the production of hydroxyapatite and collagen I with a crystallographic architecture present in the bone matrix in vivo [53–55].

MSC-differentiated osteoblasts are known to secrete VEGF [56]. As an autocrine factor, VEGF secreted by osteoblasts differentiated from SHED may have promoted the osteogenesis of SHED. VEGFR2 is expressed in CD146-SHED [57] and promotes the expression of bone morphogenetic protein-2 (BMP-2) and Runx2, and CD146 acts as a coreceptor for VEGFR2 in CD146 + SHED. In addition, DPSC bone differentiation is promoted by VEGF in vitro [58]. Moreover, the genetic analysis on days 21 and 28 after the initiation of osteogenic differentiation showed that CD146 + SHED had significantly higher gene expression of ALP, OCN, and BMP-2 than SHED and CD146-SHED, indicating that CD146 + SHED has a higher osteogenic differentiation potential. However, to confirm the relationship between VEGF and CD146 + SHED osteogenic differentiation potential, it is necessary to further examine the signaling pathway; expression analysis of genes (such as VEGF, VEGFR2, Runx2, and Osterix) in osteodifferentiated SHED, CD146 + SHED, and CD146-SHED; and osteogenic differentiation potential by treatment with VEGF and anti-VEGF neutralizing antibodies. Previously, it has been reported that hMSC-CD146(+) cells exhibited greater chemotactic attraction in a transwell migration assay, and when injected intravenously into immune-deficient mice following closed femoral fracture, exhibited wider tissue distribution and significantly increased migration ability as demonstrated by bioluminescence imaging [30]. Therefore, CD146 defines a subpopulation of hMSCs capable of bone formation and in vivo trans-endothelial migration and thus represents a population of hMSCs suitable for use in clinical protocols of bone tissue regeneration [30]. Moreover, newly formed bone matrix with embedded osteocytes of donor origin was reportedly observed upon transplantation of CD146(+) human umbilical cord perivascular cells-Gelfoam-alginate 3D complexes in severe combined immunodeficiency (SCID) mice [32]. In addition, a high expression of CD146 in MSCs from bone marrow reportedly correlates with their robust osteogenic differentiation potential [33]. This study suggested CD146 + SHED had superior bone regeneration potential compared to SHED and CD146-SHED in vitro. Based on the present and previous findings on CD146 + BMSCs [30,32,33,39,41,42,52,59], CD146 + SHED may have the following properties: "They have very close properties to pericytes," and "the binding of VEGF and VEGFR2 further enhances the pathway to promote the expression of bFGF, BMP-2, Runx2, and Osterix as CD146 acts as a coreceptor of VEGFR2." These factors may have promoted angiogenesis and bone regeneration. A previous study on BMSCs with single-cell sorting of BMSC populations revealed osteo-, adipo-, and chondroid differentiation of as many as 100 cells in clonal culture, and only 50% of these cells differentiated into these three lineages. Furthermore, 80% of BMSCs differentiated into the three lineages expressing CD146, and 40% of the cells differentiated into only one or two lineages expressing CD146 [30]. Therefore, even if a population of CD146-positive cells is isolated, there may be a heterogeneous presence of cells within that population, leading to different differentiation potentials [30]. Thus, it is conceivable that CD146 + SHED and CD146-SHED cell populations isolated from heterogeneous SHED cell populations in this study also differ in function and nature, and may be heterogeneous cell populations. At present, it is difficult to investigate the functional heterogeneity of MSC populations in-depth, and much remains to be elucidated [60]. However, single-cell sorting and clonal culture should also be performed in SHED, CD146 + SHED, and CD146-SHED, and the extent of osteodifferentiation and the ability to differentiate into the three lineages in CD146 + SHED and CD146-SHED populations used in this study should be examined in detail in future studies.

4. Materials and Methods

4.1. SHED Isolation and Culture

Pulp tissues were collected from deciduous teeth extracted from five healthy patients (Average 9 years 8 months, $\pm$ 2 years 4.8 months) who provided informed consent at the Department of Orthodontics, Hiroshima University Hospital. SHED were isolated and cultured using a previously described procedure [21–23,34], with reference to the methods of Miura et al. [13] and Gronthos et al. [14]. The following text elaborates on the isolation of SHED from the pulp. A mixture of α-MEM (Sigma-Aldrich, St. Louis, MO, USA), 4 mg/mL collagenase (Thermo Fisher Scientific, Waltham, MA, USA), and 3 mg/mL dispase (Godo Shusei, Tokyo) was prepared. The pulp tissue was immersed in the solution and minced with a scalpel. Adequate dental pulp tissue slices were transferred to a 10-mL tube and then incubated at 37° C under 5% CO_2 for 20 min with shaking. Cell aggregates were eliminated using a 70-μm cell strainer (CORNING, Corning, NY, USA), and the filtered solution was diluted with α-MEM and centrifuged at 1500 rpm for 5 min. The supernatant was aspirated and added to 20% fetal bovine serum (FBS) (Daiichi Kagaku, Tokyo), 0.24 μL/mL kanamycin (Meiji Seika Pharma Co., Ltd., Tokyo), 0.5 μL/mL penicillin (Meiji Seika Pharma), and 1 μL/mL. After being suspended in α-MEM containing mL amphotericin (MP Biomedicals), they were seeded in a cell culture petri dish (CORNING) with a diameter of 35 mm, cultivated at 37°C and 5% CO_2, and the cells were detached from the petri dish using PBS containing 0.25% trypsin (Nacalai Tesque, Kyoto) and 1 mM EDTA (Wako Pure Chemical Industries, Osaka) when confluent and passaged. After the first passage (P1), the cells were cultured in α-MEM containing 10% FBS (Daiichi Kagaku, Tokyo) and the abovementioned antibiotics at 37°C under 5% CO_2. This study was conducted in accordance with the Regulations for Epidemiological Studies of Hiroshima University Hospital (approval no. E-20-2). Cells were independently isolated from the deciduous teeth obtained from the five patients, and the cells were cultured separately.

4.2. Fluorescence-Activated Cell Sorting

MSCs were isolated from the pulp of deciduous teeth, and surface-antigen analysis was performed to confirm the presence of CD146. Each SHED collected from 5 patients was cultured and passaged to P3. Flow cytometry was subsequently performed using one of the 10 cm dishes in which the cells of each donor were cultured (five in total), and the targeted surface antigens were analyzed. The targeted surface antigens were CD146 and MSC-positive (CD73, CD90, and CD105) and MSC-negative (CD14, CD19, CD34, and CD45) markers, as defined by the International Society for Cellular Therapy. The cultured cells were detached using phosphate-buffered saline (PBS) containing 0.25% trypsin and 1 mM ethylenediaminetetraacetic acid. The cell suspension was centrifuged at 1800 rpm for 5 min. After aspirating the supernatant, the cells were washed with PBS containing 2% fetal bovine serum (FBS). Two PBS solutions (2% FBS) containing 1×10^6 cells were prepared for each antibody detection.

One of the prepared solutions was supplemented with 30 μL of PE mouse anti-human CD146, PE mouse anti-human CD90, FITC mouse anti-human CD105, PE-CTM 7 mouse anti-human CD14, APC-H7 mouse anti-human CD19 (BD Pharmingen, San Jose, CA, USA), Brilliant Violet™ 421 mouse anti-human CD73, FITC mouse anti-human CD34, and APC-H7 mouse anti-human CD45 (Becton Dickinson, San Jose, CA, USA). In addition, 5 μL of the corresponding kappa isotype control was added to the other prepared solution and incubated at 4 °C, protected from light, for 20 min. Thereafter, the mixture was washed twice with PBS containing 2% FBS and 3 μL of 7-amino-actinomycin (7-AAD; BD

Pharmingen) was added. FLOWJO software (Tomy Digital Biology Co., Tokyo, Japan) was used to analyze the surface antigens. The cells were sorted based on the surface antigen analysis to separate CD146-positive (CD146 + SHED) and negative (CD146-SHED) cells using FACS Aria II Cell Sorter (BD Biosciences, San Jose, CA, USA). The separated cells were cultured in α- minimum essential medium (α-MEM) at 37 °C in 5% CO_2.

The remaining cells were not used for flow cytometry and were used for subsequent experiments as an unsorted SHED group. Following the analysis of these surface antigens by flow cytometry, the cells were merely sorted for CD146, only into CD146-positive and CD146-negative SHEDs. Therefore, surface antigens other than CD146 were only analyzed and not sorted. In subsequent experiments, three groups of CD146-positive and CD146-negative SHEDs isolated by cell sorting and unsorted SHEDs without flow cytometry were used. The resulting SHED cells were cultured separately. Thereafter, we examined the proliferative and osteogenic differentiation activities of each of the five samples of SHED of an individual.

4.3. Properties of SHED, CD146 + SHED, and CD146-SHED

4.3.1. Cellular Proliferative Capacity of SHED, CD146 + SHED, and CD146-SHED

The following variables were examined to compare and examine the cellular proliferative abilities of SHED, CD146 + SHED, and CD146-SHED.

Population Doubling Time

SHED isolated from deciduous pulp, and CD146 + SHED and CD146-SHED isolated by cell sorting, were cultured in corresponding media, and passage-4 cells were seeded in 24-well plates (CORNING Inc., Corning, NY, USA; 1.0×10^4 cells/well) and cultured in 5% CO_2 at 37 °C. Dead cells were stained with 0.4% trypan blue (MP Biomedicals, Santa Ana, CA, USA), and the number of live cells was counted daily from day 1 to day 10 of culture using a hemocytometer. Subsequently, a cell growth curve was generated, and the logarithmic growth phase was defined as days 2–6. Population doubling time (PDT) was calculated using the following equation [23].

$$\mathrm{PDT} = (t - t0)\ \log 2/\log N - \log N0$$

where t0 indicates the time taken for the cell count and the number of cells at N, N0:t, t0.

Bromodeoxyuridine Cell Proliferation Assay

Cell growth ELISA and Cell Proliferation ELISA Bromodeoxyuridine (BrdU) kits (Roche Diagnostics, Basel, Switzerland) were used. SHED, CD146 + SHED, and CD146-SHED were seeded in 96-well plates (CORNING; 3×10^3 cells/well) and cultured at 37°C in 5% CO_2. After 48 h of growth, the cells were incubated with BrdU for 24 h at 37 °C in 5% CO_2. The absorbance was measured at the wavelength of 375 nm using a microplate reader (MultiskanTM FC; Thermo Fisher Scientific, Waltham, MA, USA).

4.3.2. Induction of Osteogenic Differentiation of SHED, CD146 + SHED, and CD146-SHED

SHED isolated from primary dental pulp, and CD146 + SHED and CD146-SHED isolated by cell sorting, were cultured in corresponding media, and passage-4 cells were seeded in 24-well plates (CORNING) coated with type I collagen (Nippon Ham, Osaka, Japan; 1.0×10^4 cells/well) and cultured at 37 °C in a 5% CO_2 incubator. After the cells reached 80% confluence, 100 nM dexamethasone (Sigma-Aldrich), 0.2 mM ascorbate-phosphate (Sigma-Aldrich), and 10 mM β-glycerolphosphate (Tokyo Kasei Kogyo, Tokyo, Japan) were added to the cell culture media (D-MEM; Sigma-Aldrich). The osteogenic differentiation-inducing media were changed every 2 days.

4.4. Quantitative Real-Time Polymerase Chain Reaction Analysis

SHED, CD146 + SHED, and CD146-SHED were cultured at 37°C under 5% CO_2. After the cells reached 80% confluence, induction of differentiation was initiated with the osteodifferentiation induction medium described above. The cells were harvested before induction, and at 21 and 28 days after induction. The mRNA expression levels of alkaline phosphatase (ALP), osteocalcin (OCN, a bone transcription factor), and bone morphogenetic protein-2 (BMP-2) were determined using quantitative real-time polymerase chain reaction (RT-PCR) analysis with QuantiTect SYBR Green PCR master mix (Qiagen, Valencia, CA, USA) with a LightCycler® 480 II instrument (Roche Diagnostics). Total RNA was extracted from cells using an RNeasy Mini kit (Qiagen) and quantified using a NanoDrop One/Onec spectrophotometer (Thermo Fisher Scientific, Inc., Waltham, MA, USA). RNA purity was also assessed using this instrument based on the OD 260/OD 280 ratio; only samples with an A260/A280 ratio of 1.5–2.0 were used for further analysis. Subsequently, 1 µg of purified total RNA was reverse-transcribed to cDNA using a ReverTra Ace first-strand cDNA synthesis kit (Toyobo, Osaka, Japan). RT-PCR was performed using Thunderbird SYBR qPCR mix (Toyobo) with specific primer sets (Table 1).

Table 1. The sequence of each primer.

Gene		Sequence (5′→3′)
GAPDH	Forward	CCA CTC CTC CAC CTT TGA
	Reverse	CAC CAC CCT GTT GCT GTA
ALP	Forward	ATG GTG GAC TGC TCA CAA C
	Reverse	GAC GTA GTT CTG CTC GTG GA
BMP-2	Forward	AAC ACT GTG CGC AGC TTC C
	Reverse	CTC CGG GTT GTT TTC CCA C
OCN	Forward	GCA GAG TCC AGG AAA GGG TG
	Reverse	GTC AGC AAC TCG TCA CAG

4.5. ALP Staining and Determination of ALP Activity

The cells were fixed on days 3, 7, 14, 21, and 28 after incubation with an osteogenic differentiation medium, and ALP staining was performed using the following methods. After fixation with 4% paraformaldehyde PBS (Fujifilm Wako Pure Chemical, Osaka, Japan) for 10 min, the cells were incubated with PBS containing 0.05% Tween-20 (Roche Diagnostics) (washing buffer). After removing the washing buffer, the cells were incubated with ALP staining solution (Fujifilm Wako Pure Chemical Industries, Ltd.) at room temperature under light-resistant conditions for 10 min.

For the quantitative testing of ALP, the cells were harvested on days 3, 7, 14, 21, and 28 using the pNPP Phosphatase Assay Kit (AnaSpec, Fremont, CA, USA) after incubation with an osteogenic differentiation induction medium. The harvested cells were homogenized using a Sonic Vibra Cell (Sonic & Materials, Newtown, CT, USA), and the supernatant was collected and used as the sample. The sample was mixed with 50 µL of the pNPP substrate solution in a 96-well plate (CORNING), and the absorbance was determined at a wavelength of 405 nm using a microplate reader MultiskanTM FC (Thermo Fisher Scientific).

4.6. Calcium Deposition Analyses (Alizarin Red Staining)

After the cells were cultured with an osteogenic differentiation-inducing medium for 28 days, they were washed with PBS and 10 mM Tris-HCl (pH 7.5) with 0.9% NaCl. The cells were fixed with 4% paraformaldehyde and stained with 1% Alizarin Red S (Kshida Chemical, Osaka, Japan). The stained tissue sections were photographed and observed using a BZ-X810 microscope (Keyence, Osaka, Japan). In addition, for quantification analyses, the cells were incubated with a mixed solution of 10% acetic acid and 20%

methanol at room temperature for 15 min after staining to elute the stained dye. The eluate was added to a 96-well plate (CORNING), and the absorbance was determined at a wavelength of 405 nm using a microplate reader Multiskan™FC (Thermo Fisher Scientific).

4.7. Statistical Analysis

All data are presented as mean ± standard deviation. The Kruskal–Wallis test, a nonparametric test, was performed to analyze significant differences between the groups using the software BellCurve® for Excel (SSRI; Tokyo, Japan). $p < 0.05$ and < 0.01 were considered statistically significant.

5. Conclusions

In conclusion, the results indicated that CD146 + SHED has superior bone regeneration ability compared with SHED and CD146-SHED. Moreover, CD146 affects the bone regeneration ability of SHED, and CD146 + SHED might be helpful for bone regeneration treatment. Further studies are needed to elucidate the detailed mechanism of bone regeneration in CD146 + SHED for the clinical application of SHED.

Author Contributions: Conceptualization, R.K., K.R., Y.Y. and K.T.; methodology, R.K., K.R., Y.Y., Y.H. and K.T.; software, R.K., K.R. and Y.H.; validation, K.R., N.A.R.P. and Y.H. formal analysis, K.R., N.A.R.P., Y.H. and R.K.; investigation, R.K., K.R., Y.Y. and K.T.; resources, R.K., K.R., Y.Y., N.A.R.P., Y.Y. and K.T; data curation, R.K., K.R. and Y.Y.; writing—original draft preparation, R.K. and K.R.; writing—review and editing, K.R., Y.Y., N.A.R.P., Y.H. and K.T.; visualization, R.K. and K.R.; supervision, K.T.; project administration, R.K. and K.T.; funding acquisition, R.K., K.R., Y.Y. and K.T. All authors have read and agreed to the published version of the manuscript.

Funding: This work was supported by Grants-in-Aid for Research Activity Start Up [grant number 22K21062] and Grants-in-Aid for Scientific Research (C) [grant number 21K10186 and 19K10384] from the Japan Society for the Promotion of Science. This study was supported by Hiroshima University's step-up support system and the 2021 Satake Foundation Research Grant.

Institutional Review Board Statement: All procedures performed in this study involving human participants were in accordance with the ethical standards of the institutional research committee and with the 1964 Helsinki Declaration and its later amendments or comparable ethical standards. This study involving live animals followed the recommendations of the ARRIVE (Animal Research: Reporting of In Vivo Experiments) guidelines. The ethical number for this study is E20-2, date approved 20 December 2018.

Informed Consent Statement: Informed consent was obtained from all individual participants included in the study.

Data Availability Statement: The data supporting the findings of this study are available from the corresponding author upon reasonable request.

Acknowledgments: We are grateful to Kengo Nakajima and Tomoka Hiraki for providing advice pertaining to this study.

Conflicts of Interest: The authors declare no competing interests.

References

1. Qi, J.; Yu, T.; Hu, B.; Wu, H.; Ouyang, H. Current Biomaterial-Based Bone Tissue Engineering and Translational Medicine. *Int. J. Mol. Sci.* **2021**, *22*, 10233. [CrossRef]
2. Walmsley, G.G.; Ransom, R.C.; Zielins, E.R.; Leavitt, T.; Flacco, J.S.; Hu, M.S.; Lee, A.S.; Longaker, M.T.; Wan, D.C. Stem Cells in Bone Regeneration. *Stem Cell Rev.* **2016**, *12*, 524–529. [CrossRef]
3. Fu, J.; Wang, Y.; Jiang, Y.; Du, J.; Xu, J.; Liu, Y. Systemic Therapy of MSCs in Bone Regeneration: A Systematic Review and Meta-Analysis. *Stem Cell Res. Ther.* **2021**, *12*, 377. [CrossRef]
4. Edgar, L.; Pu, T.; Porter, B.; Aziz, J.M.; La Pointe, C.; Asthana, A.; Orlando, G. Regenerative Medicine, Organ Bioengineering and Transplantation. *Br. J. Surg.* **2020**, *107*, 793–800. [CrossRef]
5. Caplan, A.I. Mesenchymal Stem Cells. *J. Orthop. Res.* **1991**, *9*, 641–650. [CrossRef]
6. Zuk, P.A.; Zhu, M.; Mizuno, H.; Huang, J.; Futrell, J.W.; Katz, A.J.; Benhaim, P.; Lorenz, H.P.; Hedrick, M.H. Multilineage Cells from Human Adipose Tissue: Implications for Cell-Based Therapies. *Tissue Eng.* **2001**, *7*, 211–228. [CrossRef]

7. Brown, C.; McKee, C.; Bakshi, S.; Walker, K.; Hakman, E.; Halassy, S.; Svinarich, D.; Dodds, R.; Govind, C.K.; Chaudhry, G.R. Mesenchymal Stem Cells: Cell Therapy and Regeneration Potential. *J. Tissue Eng. Regen. Med.* **2019**, *13*, 1738–1755. [CrossRef]
8. Xiang, X.-N.; Zhu, S.-Y.; He, H.-C.; Yu, X.; Xu, Y.; He, C.-Q. Mesenchymal Stromal Cell-Based Therapy for Cartilage Regeneration in Knee Osteoarthritis. *Stem. Cell Res. Ther.* **2022**, *13*, 14. [CrossRef]
9. Shimizu, Y.; Ntege, E.H.; Sunami, H. Current Regenerative Medicine-Based Approaches for Skin Regeneration: A Review of Literature and a Report on Clinical Applications in Japan. *Regen. Ther.* **2022**, *21*, 73–80. [CrossRef]
10. Venkataiah, V.S.; Handa, K.; Njuguna, M.M.; Hasegawa, T.; Maruyama, K.; Nemoto, E.; Yamada, S.; Sugawara, S.; Lu, L.; Takedachi, M.; et al. Periodontal Regeneration by Allogeneic Transplantation of Adipose Tissue Derived Multi-Lineage Progenitor Stem Cells in Vivo. *Sci. Rep.* **2019**, *9*, 921. [CrossRef]
11. Sui, B.; Chen, C.; Kou, X.; Li, B.; Xuan, K.; Shi, S.; Jin, Y. Pulp Stem Cell-Mediated Functional Pulp Regeneration. *J. Dent. Res.* **2019**, *98*, 27–35. [CrossRef]
12. Tassi, S.A.; Sergio, N.Z.; Misawa, M.Y.O.; Villar, C.C. Efficacy of Stem Cells on Periodontal Regeneration: Systematic Review of Pre-Clinical Studies. *J. Periodontal Res.* **2017**, *52*, 793–812. [CrossRef]
13. Miura, M.; Gronthos, S.; Zhao, M.; Lu, B.; Fisher, L.W.; Robey, P.G.; Shi, S. SHED: Stem Cells from Human Exfoliated Deciduous Teeth. *Proc. Natl. Acad. Sci. USA* **2003**, *100*, 5807–5812. [CrossRef]
14. Gronthos, S.; Mankani, M.; Brahim, J.; Robey, P.G.; Shi, S. Postnatal Human Dental Pulp Stem Cells (DPSCs) in Vitro and in Vivo. *Proc. Natl. Acad. Sci. USA* **2000**, *97*, 13625–13630. [CrossRef]
15. Gronthos, S.; Brahim, J.; Li, W.; Fisher, L.W.; Cherman, N.; Boyde, A.; DenBesten, P.; Robey, P.G.; Shi, S. Stem Cell Properties of Human Dental Pulp Stem Cells. *J. Dent. Res.* **2002**, *81*, 531–535. [CrossRef]
16. Graziano, A.; d'Aquino, R.; Laino, G.; Papaccio, G. Dental Pulp Stem Cells: A Promising Tool for Bone Regeneration. *Stem Cell Rev.* **2008**, *4*, 21–26. [CrossRef]
17. Király, M.; Porcsalmy, B.; Pataki, A.; Kádár, K.; Jelitai, M.; Molnár, B.; Hermann, P.; Gera, I.; Grimm, W.-D.; Ganss, B.; et al. Simultaneous PKC and cAMP Activation Induces Differentiation of Human Dental Pulp Stem Cells into Functionally Active Neurons. *Neurochem. Int.* **2009**, *55*, 323–332. [CrossRef]
18. La Noce, M.; Stellavato, A.; Vassallo, V.; Cammarota, M.; Laino, L.; Desiderio, V.; Del Vecchio, V.; Nicoletti, G.F.; Tirino, V.; Papaccio, G.; et al. Hyaluronan-Based Gel Promotes Human Dental Pulp Stem Cells Bone Differentiation by Activating YAP/TAZ Pathway. *Cells.* **2021**, *26*, 2899. [CrossRef]
19. Seo, B.M.; Sonoyama, W.; Yamaza, T.; Coppe, C.; Kikuiri, T.; Akiyama, K.; Lee, J.S.; Shi, S. SHED Repair Critical-Size Calvarial Defects in Mice. *Oral Dis.* **2008**, *14*, 428–434. [CrossRef]
20. d'Aquino, R.; De Rosa, A.; Lanza, V.; Tirino, V.; Laino, L.; Graziano, A.; Desiderio, V.; Laino, G.; Papaccio, G. Human Mandible Bone Defect Repair by the Grafting of Dental Pulp Stem/Progenitor Cells and Collagen Sponge Biocomplexes. *Eur. Cell. Mater.* **2009**, *18*, 75–83. [CrossRef]
21. Nakajima, K.; Kunimatsu, R.; Ando, K.; Hiraki, T.; Rikitake, K.; Tsuka, Y.; Abe, T.; Tanimoto, K. Success Rates in Isolating Mesenchymal Stem Cells from Permanent and Deciduous Teeth. *Sci. Rep.* **2019**, *9*, 16764. [CrossRef]
22. Nakajima, K.; Kunimatsu, R.; Ando, K.; Ando, T.; Hayashi, Y.; Kihara, T.; Hiraki, T.; Tsuka, Y.; Abe, T.; Kaku, M.; et al. Comparison of the Bone Regeneration Ability between Stem Cells from Human Exfoliated Deciduous Teeth, Human Dental Pulp Stem Cells and Human Bone Marrow Mesenchymal Stem Cells. Biochem. *Biophys. Res. Commun.* **2018**, *497*, 876–882. [CrossRef]
23. Kunimatsu, R.; Nakajima, K.; Awada, T.; Tsuka, Y.; Abe, T.; Ando, K.; Hiraki, T.; Kimura, A.; Tanimoto, K. Comparative Characterization of Stem Cells from Human Exfoliated Deciduous Teeth, Dental Pulp, and Bone Marrow–Derived Mesenchymal Stem Cells. Biochem. Biophys. *Res. Commun.* **2018**, *501*, 193–198. [CrossRef]
24. Wang, X.; Sha, X.J.; Li, G.H.; Yang, F.S.; Ji, K.; Wen, L.Y.; Liu, S.Y.; Chen, L.; Ding, Y.; Xuan, K. Comparative characterization of stem cells from human exfoliated deciduous teeth and dental pulp stem cells. *Arch Oral Biol.* **2012**, *57*, 1231–1240. [CrossRef]
25. Ching, H.S.; Luddin, N.; Rahman, I.A.; Ponnuraj, K.T. Expression of odontogenic and osteogenic markers in DPSCs and SHED: A review. *Curr. Stem. Cell Res. Ther.* **2017**, *12*, 71–79. [CrossRef]
26. Kichenbrand, C.; Velot, E.; Menu, P.; Moby, V. Dental Pulp Stem Cell-Derived Conditioned Medium: An Attractive Alternative for Regenerative Therapy. *Tissue Eng. Part B Rev.* **2019**, *25*, 78–88. [CrossRef]
27. Lee, S.H.; Looi, C.Y.; Chong, P.P.; Foo, J.B.; Looi, Q.H.; Ng, C.X.; Ibrahim, Z. Comparison of isolation, expansion and cryopreservation techniques to produce stem cells from human exfoliated deciduous teeth (SHED) with better regenerative potential. *Curr. Stem. Cell Res. Ther.* **2021**, *16*, 551–562. [CrossRef]
28. Uder, C.; Brückner, S.; Winkler, S.; Tautenhahn, H.-M.; Christ, B. Mammalian MSC from Selected Species: Features and Applications: Cross-Species MSC. *Cytometry A* **2018**, *93*, 32–49. [CrossRef]
29. Johnson, K.W.; Dooner, M.; Quesenberry, P.J. Fluorescence Activated Cell Sorting: A Window on the Stem Cell. *Curr. Pharm. Biotechnol.* **2007**, *8*, 133–139. [CrossRef]
30. Harkness, L.; Zaher, W.; Ditzel, N.; Isa, A.; Kassem, M. CD146/MCAM Defines Functionality of Human Bone Marrow Stromal Stem Cell Populations. *Stem Cell Res. Ther.* **2016**, *7*, 4. [CrossRef]
31. Wang, Z.; Yan, X. CD146, a Multi-Functional Molecule beyond Adhesion. *Cancer Lett.* **2013**, *330*, 150–162. [CrossRef]
32. Tsang, W.P.; Shu, Y.; Kwok, P.L.; Zhang, F.; Lee, K.K.H.; Tang, M.K.; Li, G.; Chan, K.M.; Chan, W.-Y.; Wan, C. CD146+ Human Umbilical Cord Perivascular Cells Maintain Stemness under Hypoxia and as a Cell Source for Skeletal Regeneration. *PLoS ONE* **2013**, *8*, e76153. [CrossRef]

33. Ulrich, C.; Abruzzese, T.; Maerz, J.K.; Ruh, M.; Amend, B.; Benz, K.; Rolauffs, B.; Abele, H.; Hart, M.L.; Aicher, W.K. Human Placenta-Derived CD146-Positive Mesenchymal Stromal Cells Display a Distinct Osteogenic Differentiation Potential. *Stem. Cells Dev.* **2015**, *24*, 1558–1569. [CrossRef]

34. Rikitake, K.; Kunimatsu, R.; Yoshimi, Y.; Nakajima, K.; Hiraki, T.; Aisyah Rizky Putranti, N.; Tsuka, Y.; Abe, T.; Ando, K.; Hayashi, Y.; et al. Effect of CD146+ SHED on Bone Regeneration in a Mouse Calvaria Defect Model. *Oral Dis.* **2021**, *29*, 725–734. [CrossRef]

35. Oh, M.; Nör, J.E. The Perivascular Niche and Self-Renewal of Stem Cells. *Front. Physiol.* **2015**, *6*, 367. [CrossRef]

36. Oh, M.; Zhang, Z.; Mantesso, A.; Oklejas, A.E.; Nör, J.E. Endothelial-Initiated Crosstalk Regulates Dental Pulp Stem Cell Self-Renewal. *J. Dent. Res.* **2020**, *99*, 1102–1111. [CrossRef]

37. Kfoury, Y.; Scadden, D.T. Mesenchymal Cell Contributions to the Stem Cell Niche. *Cell Stem. Cell* **2015**, *16*, 239–253. [CrossRef]

38. Leroyer, A.S.; Blin, M.G.; Bachelier, R.; Bardin, N.; Blot-Chabaud, M.; Dignat-George, F. CD146 (Cluster of Differentiation 146): An Adhesion Molecule Involved in Vessel Homeostasis. *Arterioscler. Thromb. Vasc. Biol.* **2019**, *39*, 1026–1033. [CrossRef]

39. Wang, Z.; Xu, Q.; Zhang, N.; Du, X.; Xu, G.; Yan, X. CD146, from a Melanoma Cell Adhesion Molecule to a Signaling Receptor. *Signal Transduct. Target. Ther.* **2020**, *5*, 148. [CrossRef]

40. Zeng, Q.; Wu, Z.; Duan, H.; Jiang, X.; Tu, T.; Lu, D.; Luo, Y.; Wang, P.; Song, L.; Feng, J.; et al. Impaired Tumor Angiogenesis and VEGF-Induced Pathway in Endothelial CD146 Knockout Mice. *Protein Cell* **2014**, *5*, 445–456. [CrossRef]

41. Jiang, T.; Zhuang, J.; Duan, H.; Luo, Y.; Zeng, Q.; Fan, K.; Yan, H.; Lu, D.; Ye, Z.; Hao, J.; et al. CD146 Is a Coreceptor for VEGFR-2 in Tumor Angiogenesis. *Blood* **2012**, *120*, 2330–2339. [CrossRef] [PubMed]

42. Espagnolle, N.; Guilloton, F.; Deschaseaux, F.; Gadelorge, M.; Sensébé, L.; Bourin, P. CD146 Expression on Mesenchymal Stem Cells Is Associated with Their Vascular Smooth Muscle Commitment. *J. Cell. Mol. Med.* **2014**, *18*, 104–114. [CrossRef] [PubMed]

43. Bianco, P. "Mesenchymal" Stem Cells. *Annu. Rev. Cell Dev. Biol.* **2014**, *30*, 677–704. [CrossRef]

44. Crisan, M.; Yap, S.; Casteilla, L.; Chen, C.-W.; Corselli, M.; Park, T.S.; Andriolo, G.; Sun, B.; Zheng, B.; Zhang, L.; et al. A Perivascular Origin for Mesenchymal Stem Cells in Multiple Human Organs. *Cell Stem. Cell* **2008**, *3*, 301–313. [CrossRef]

45. Covas, D.T.; Panepucci, R.A.; Fontes, A.M.; Silva, W.A., Jr.; Orellana, M.D.; Freitas, M.C.C.; Neder, L.; Santos, A.R.D.; Peres, L.C.; Jamur, M.C.; et al. Multipotent Mesenchymal Stromal Cells Obtained from Diverse Human Tissues Share Functional Properties and Gene-Expression Profile with CD146+ Perivascular Cells and Fibroblasts. *Exp. Hematol.* **2008**, *36*, 642–654. [CrossRef]

46. Dominici, M.; Le Blanc, K.; Mueller, I.; Slaper-Cortenbach, I.; Marini, F.; Krause, D.; Deans, R.; Keating, A.; Prockop, D.; Horwitz, E. Minimal Criteria for Defining Multipotent Mesenchymal Stromal Cells. The International Society for Cellular Therapy Position Statement. *Cytotherapy* **2006**, *8*, 315–317. [CrossRef]

47. Ullah, I.; Subbarao, R.B.; Rho, G.J. Human Mesenchymal Stem Cells-Current Trends and Future Prospective. *Biosci. Rep.* **2015**, *35*, e00191. [CrossRef]

48. Nourbakhsh, N.; Soleimani, M.; Taghipour, Z.; Karbalaie, K.; Mousavi, S.-B.; Talebi, A.; Nadali, F.; Tanhaei, S.; Kiyani, G.-A.; Nematollahi, M.; et al. Induced in Vitro Differentiation of Neural-like Cells from Human Exfoliated Deciduous Teeth-Derived Stem Cells. *Int. J. Dev. Biol.* **2011**, *55*, 189–195. [CrossRef]

49. Kerkis, I.; Caplan, A.I. Stem Cells in Dental Pulp of Deciduous Teeth. *Tissue Eng. Part B Rev.* **2012**, *18*, 129–138. [CrossRef]

50. Fayazi, M.; Salehnia, M.; Ziaei, S. Department of Medical Sciences, Najafabad Branch, Islamic Azad University, Najafabad, Iran; Department of Anatomy, Faculty of Medical Sciences, Tarbiat Modares University, Tehran, Iran; Department of Midwifery, Faculty of Medical Sciences Tarbiat Modares University, Tehran, Iran The Effect of Stem Cell Factor on Proliferation of Human Endometrial CD146+ Cells. *Int. J. Reprod. Biomed. (Yazd)* **2016**, *14*, 437–442. [CrossRef]

51. Zhu, W.; Tan, Y.; Qiu, Q.; Li, X.; Huang, Z.; Fu, Y.; Liang, M. Comparison of the Properties of Human CD146+ and CD146− Periodontal Ligament Cells in Response to Stimulation with Tumour Necrosis Factor α. *Arch. Oral Biol.* **2013**, *58*, 1791–1803. [CrossRef]

52. Paduano, F.; Marrelli, M.; Palmieri, F.; Tatullo, M. CD146 Expression Influences Periapical Cyst Mesenchymal Stem Cell Properties. *Stem Cell Rev.* **2016**, *12*, 592–603. [CrossRef] [PubMed]

53. Endo, I.; Mastumoto, T. Bone and Stem Cells. Regulatory mechanism of mesenchymal stem cell differentiation to osteoblasts. *Clin. Calcium* **2014**, *24*, 555–564. [PubMed]

54. Ohgushi, H.; Tamai, S.; Dohi, Y.; Katuda, T.; Tabata, S.; Suwa, Y. In Vitro Bone Formation by Rat Marrow Cell Culture. *J. Biomed. Mater. Res.* **1996**, *32*, 333–340. [CrossRef]

55. Fakhry, M. Molecular Mechanisms of Mesenchymal Stem Cell Differentiation towards Osteoblasts. *World J. Stem Cells* **2013**, *5*, 136. [CrossRef] [PubMed]

56. Hu, K.; Olsen, B.R. Osteoblast-Derived VEGF Regulates Osteoblast Differentiation and Bone Formation during Bone Repair. *J. Clin. Invest.* **2016**, *126*, 509–526. [CrossRef]

57. Hu, K.; Olsen, B.R. The Roles of Vascular Endothelial Growth Factor in Bone Repair and Regeneration. *Bone* **2016**, *91*, 30–38. [CrossRef]

58. D' Alimonte, I.; Nargi, E.; Mastrangelo, F.; Falco, G.; Lanuti, P.; Marchisio, M.; Miscia, S.; Robuffo, I.; Capogreco, M.; Buccella, S.; et al. Vascular Endothelial Growth Factor Enhances in Vitro Proliferation and Osteogenic Differentiation of Human Dental Pulp Stem Cells. *J. Biol. Regul. Homeost. Agents* **2011**, *25*, 57–69.

59. Bai, Y.; Li, P.; Yin, G.; Huang, Z.; Liao, X.; Chen, X.; Yao, Y. BMP-2, VEGF and BFGF Synergistically Promote the Osteogenic Differentiation of Rat Bone Marrow-Derived Mesenchymal Stem Cells. *Biotechnol. Lett.* **2013**, *35*, 301–308. [CrossRef]
60. Muraglia, A.; Cancedda, R.; Quarto, R. Clonal Mesenchymal Progenitors from Human Bone Marrow Differentiate in Vitro According to a Hierarchical Model. *J. Cell Sci.* **2000**, *113*, 1161–1166. [CrossRef]

International Journal of
Molecular Sciences

Article

Transcriptome Analysis Reveals Modulation of Human Stem Cells from the Apical Papilla by Species Associated with Dental Root Canal Infection

Yelyzaveta Razghonova [1,†], Valeriia Zymovets [2,*,†], Philip Wadelius [3], Olena Rakhimova [2], Lokeshwaran Manoharan [4], Malin Brundin [2], Peyman Kelk [5] and Nelly Romani Vestman [2,6]

[1] Department of Microbiology, Virology and Biotechnology, Mechnikov National University, 65000 Odesa, Ukraine
[2] Department of Odontology, Umeå University, 90187 Umeå, Sweden
[3] Department of Endodontics, Region of Västerbotten, 90189 Umeå, Sweden
[4] National Bioinformatics Infrastructure Sweden (NBIS), Lund University, 22362 Lund, Sweden
[5] Section for Anatomy, Department of Integrative Medical Biology (IMB), Umeå University, 90187 Umeå, Sweden
[6] Wallenberg Centre for Molecular Medicine, Umeå University, 90187 Umeå, Sweden
* Correspondence: valeriia.zymovets@umu.se
† These authors contributed equally to this work.

Citation: Razghonova, Y.; Zymovets, V.; Wadelius, P.; Rakhimova, O.; Manoharan, L.; Brundin, M.; Kelk, P.; Romani Vestman, N. Transcriptome Analysis Reveals Modulation of Human Stem Cells from the Apical Papilla by Species Associated with Dental Root Canal Infection. *Int. J. Mol. Sci.* **2022**, *23*, 14420. https://doi.org/10.3390/ijms232214420

Academic Editors: Kerstin M. Galler and Matthias Widbiller

Received: 15 October 2022
Accepted: 17 November 2022
Published: 20 November 2022

Publisher's Note: MDPI stays neutral with regard to jurisdictional claims in published maps and institutional affiliations.

Abstract: Interaction of oral bacteria with stem cells from the apical papilla (SCAP) can negatively affect the success of regenerative endodontic treatment (RET). Through RNA-seq transcriptomic analysis, we studied the effect of the oral bacteria *Fusobacterium nucleatum* and *Enterococcus faecalis*, as well as their supernatants enriched by bacterial metabolites, on the osteo- and dentinogenic potential of SCAPs in vitro. We performed bulk RNA-seq, on the basis of which differential expression analysis (DEG) and gene ontology enrichment analysis (GO) were performed. DEG analysis showed that *E. faecalis* supernatant had the greatest effect on SCAPs, whereas *F. nucleatum* supernatant had the least effect (Tanimoto coefficient = 0.05). GO term enrichment analysis indicated that *F. nucleatum* upregulates the immune and inflammatory response of SCAPs, and *E. faecalis* suppresses cell proliferation and cell division processes. SCAP transcriptome profiles showed that under the influence of *E. faecalis* the upregulation of *VEGFA*, *Runx2*, and *TBX3* genes occurred, which may negatively affect the SCAP's osteo- and odontogenic differentiation. *F. nucleatum* downregulates the expression of *WDR5* and *TBX2* and upregulates the expression of *TBX3* and *NFIL3* in SCAPs, the upregulation of which may be detrimental for SCAPs' differentiation potential. In conclusion, the present study shows that in vitro, *F. nucleatum*, *E. faecalis*, and their metabolites are capable of up- or downregulating the expression of genes that are necessary for dentinogenic and osteogenic processes to varying degrees, which eventually may result in unsuccessful RET outcomes. Transposition to the clinical context merits some reservations, which should be approached with caution.

Keywords: stem cells from the apical papilla (SCAP); regenerative endodontic treatment (RET); osteogenesis; dentinogenesis; *Enterococcus faecalis*; *Fusobacterium nucleatum*; transcriptome analysis; differential gene expression analysis (DEG)

1. Introduction

The oral cavity is an ecological niche for a diverse range of microorganisms, including more than 700 bacterial species [1]. Microorganisms and their host live in a flexible balance that risks disturbance if traumatic dental injuries (TDIs) or oral diseases such as dental caries occur. Accordingly, if microorganisms penetrate dental structures and disturb host cell homeostasis, it can lead to infection (pulpal necrosis), periapical inflammation, and bone resorption (apical periodontitis) [2].

Pulpal necrosis in immature necrotic teeth demands challenging clinical procedures that pose a risk to teeth's long-term survival. Conventional treatment of immature necrotic teeth includes apexification, which consists either of a long-term application of calcium hydroxide (CH) [3] paste to induce an apical barrier or placement of mineral trioxide aggregate (MTA) [4] to achieve closure of the root apex. These procedures, however, do not result in promoting root development [5] and may increase susceptibility to cervical fractures [6].

A currently promising therapeutic option—regenerative endodontic treatment (RET)—encourages tissue regeneration by promoting continued root development [7]. RETs are capable of regenerating vascular dental pulp tissue in animal models and human patients by ex vivo expanded autologous tooth stem cells implanted from deciduous teeth [8]. In this context, stem cells found at the apical papilla of the tooth (SCAP) are a unique group of dental stem cells characterized by their plasticity, potency, and versatility [9]. SCAPs have been shown to differentiate into osteo/odontoblast-like cells, adipocytes, and chondroblasts [10]. Moreover, SCAPs have been proven to take part in the processes of neurite outgrowth and axonal targeting, both in vitro and in vivo [11]. In comparison with dental pulp stem cells (DPSCs) and bone marrow mesenchymal stromal cells (BMMSCs), SCAPs show a higher proliferation rate and osteo/dentinogenic potential, and express multiple osteogenic markers including dentin sialophosphoprotein (DSPP), osteocalcin, and alkaline phosphatase (ALP) [12].

The presence of bacteria such as *Streptococcus oralis* and *Actinomyces naeslundii* and their residuals in the root canal, even after disinfection methods [13], can jeopardize RET success. In fact, micro-environmental conditions, such as pre-existing infection, seem to influence the viability, proliferation, and mineralization capacity of SCAPs [14,15]. Previous studies have confirmed that products from *S. oralis* J22 and *A. naeslundii* T14V-J1 inhibited mineralization of human SCAPs [13]. Moreover, lipopolysaccharides (LPS) extracted from *Porphyromonas gingivalis* resulted in a pronounced osteogenic response, since it significantly upregulates bone sialoprotein gene expression [16].

In a previous study, we reported that key species in dental root canal infection, namely *A. gerensceriae*, *S. exigua*, *F. nucleatum* and *E. faecalis* were able to modulate SCAPs under oxygen-free conditions in a species-dependent fashion. *Moreover, E. faecalis* and *F. nucleatum* reported the strongest binding capacity and significantly reduced SCAP proliferation [15]. As a diverse commensal and opportunistic bacterium, *F. nucleatum* participates in a variety of interactions with other bacteria and human cells, and these interactions can vary from helpful to damaging [17]. *F. nucleatum* is considered to be a key species in a biofilm formation, supporting primary colonizers such as Streptococcus species, and providing a low-oxygen microenvironment in the root canal, thereby protecting a secondary colonizer such as *P. gingivalis* [18]. *F. nucleatum* has been highly correlated with traumatized teeth [19], oral and extraoral human diseases such as periodontitis [20], endodontic infection [21], inflammatory bowel disease [22], and colorectal cancer [23]. Recently, it was shown that *F. nucleatum*, when exposed to immortalized primary colonic epithelium and vascular endothelial cells, upregulated genes related to inflammation, downregulated genes related to histone modification, and significantly remodeled chromatin states [24].

Enterococcus faecalis is associated with failed endodontic treatment [25] and is known for its survival capacity even in harsh conditions thanks to its virulence factors (enterococcal surface protein (esp), gelatinase (gelE), aggregation substance (asa1), cytolysin B (cylB) etc.) and ability to form biofilms [17,26]. It was reported that *E. faecalis* biofilm downregulated dentinogenic genes and upregulated osteoblastic genes in SCAPs [27]. *E. faecalis* is the most frequently isolated bacterial species from symptomatic root canal-treated teeth, accounting for up to 90% of cases [28]. Due to its ability to persist in harsh conditions with nutrient destitution and high alkalinity and despite the presence of intracanal medication, *E faecalis* is commonly found in secondary or chronic cases [29]. Forming a biofilm that is 1000-fold more impervious than planktonic bacteria to the action of anti-microbials, *E. faecalis* represents a particular pathogenicity and eradication problem [17]. Moreover, *E. faecalis* promotes the differentiation of murine bone marrow stem cells into CD11c-positive dendritic cells with aberrant immune functions while retaining the ability to induce proinflammatory cytokines [30].

Transcriptome profiling by RNA sequencing (RNA-seq) has been widely used to provide far higher coverage and greater resolution of the dynamic nature of the transcriptome. Unlike the genome (which is usually the same for all cells of the same lineage), the transcriptome can vary greatly depending on environmental conditions [31].

In the present study, we used RNA-seq transcriptomic analysis to reveal the alteration of gene expression across the SCAP genome in the case of *F. nucleatum* and *E. faecalis* stimulation, which lays a foundation for understanding the cellular changes induced by bacterial stimulation. Furthermore, we searched for osteogenic- and dentinogenic-associated genes expressed by SCAPs upon *F. nucleatum* and *E. faecalis* stimulation.

2. Results

2.1. Mapping and Quantifying SCAP Transcriptomes by RNA-Seq

A total of 109.7 GB of raw sequence data was generated from all fifteen samples (Table 1). There were on average 49 million paired-end reads (2 × 150 bp) for each sample.

Table 1. RNA-seq data statistics. DI—SCAP donor I; DII—SCAP donor II; DIII SCAP donor III; NB—non-treated SCAP; F.n_B—SCAP co-cultured with *F. nucleatum*; F.n._S—SCAP cultured with *F. nucleatum* supernatant; E.f._B—SCAP co-cultured with *E. faecalis*; E.f._S—SCAP cultured with *E. faecalis* supernatant.

Sample	Type of Treatment	Raw Data (Read Count)	Clean Data (Read Count)	Total Mapped Reads	Total Mapping Rate (%)	Percent of Genome Regions (%)		
						Exon	Intron	Intergenic
1	DI NB	43,633,856	43,606,890	43,210,852	99.1	93.0	2.5	4.5
2	DI F.n_B	54,476,736	54,444,824	53,964,109	99.1	93.3	2.4	4.3
3	DI F.n._S	42,956,122	42,933,526	42,576,137	99.2	93.7	2.0	4.3
4	DI E.f_B	55,629,374	55,600,608	54,824,199	98.6	88.9	5.9	5.2
5	DI E.f_S	55,430,062	55,401,924	54,885,257	99.1	92.3	3.2	4.5
6	DII NB	47,027,536	47,002,190	46,541,473	99.0	93.5	2.2	4.3
7	DII F.n_B	48,806,760	48,784,160	48,246,591	98.9	91.6	2.9	5.5
8	DII F.n._S	46,427,206	46,406,140	45,986,861	99.1	92.1	2.7	5.2
9	DII E.f_B	39,262,262	39,242,992	38,806,871	98.9	90.9	4.2	4.9
10	DII E.f_S	49,064,034	49,042,722	48,616,751	99.1	91.9	3.3	4.8
11	DIII NB	48,797,068	48,773,676	48,361,861	99.1	93.7	2.1	4.2
12	DIII F.n_B	52,910,564	52,883,784	52,349,588	98.9	93.2	2.5	4.3
13	DIII F.n._S	52,127,896	52,105,306	51,686,329	99.2	91.9	2.7	5.4
14	DIII E.f_B	46,191,244	46,171,130	45,806,664	99.2	90.8	4.2	5.0
15	DIII E.f_S	49,193,684	49,173,978	48,823,399	99.3	93.4	2.3	4.3

2.2. SCAP Transcriptome Elicits Distinct Profiles Based on Bacterial Stimulation

Principal component analysis (PCA) of the transcriptome profile of all 15 samples revealed four major clusters based on treatment variants. PCA is a dimensionality-reduction method, which helps to visualize differences between expressed gene profiles in controls and in treated samples, and to find the patterns in a dataset. Different clusters are formed by the transcriptomic profiles of SCAPs (donors I–III) treated by viable *E. faecalis* (planktonic stage), *E. faecalis* (supernatant), and viable *F. nucleatum* (planktonic stage). However, the unstimulated SCAP (negative control) and SCAPs co-cultured with *F. nucleatum* supernatants clustered together (Figure 1).

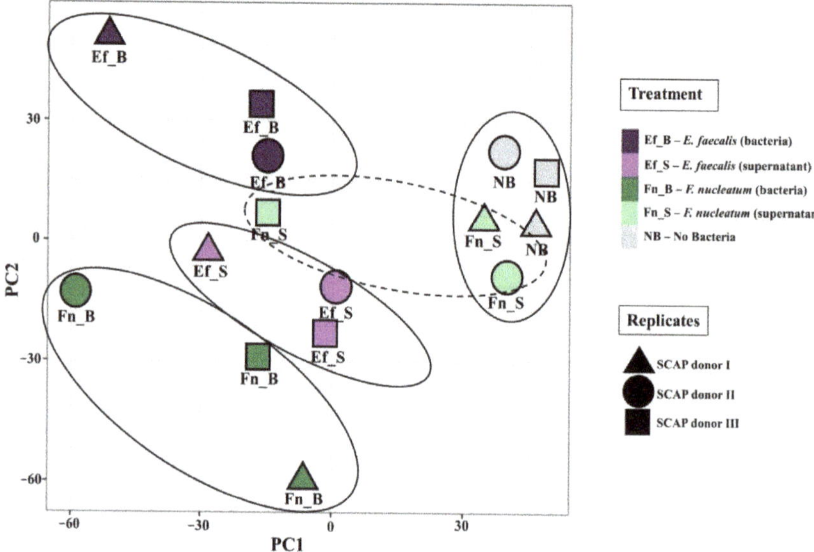

Figure 1. Principal component analysis of various transcriptomic profiles of SCAPs under different treatment conditions. SCAP donors are marked by shapes, and treatment variants are denoted by colors. Suggested clusters are circled. The first principal component (PC1) shown on the *x*-axis represents the most variation in the data, and the second principal component (PC2) on the *y*-axis represents the second-highest level of data variation. The axes are ranked in order of importance: differences along the PC1 axis are more important than differences along the PC2 axis.

2.3. E. faecalis Supernatant Has the Strongest Influence on SCAPs Whereas Supernatant of F. nucleatum Has a Mild Effect

The transcriptome profile of uninfected versus infected SCAPs was compared using differentially expressed genes (DEG) analysis.

As a result of DEG analysis of transcriptomic profiles of SCAPs under the influence of bacteria or their supernatants, direct bacterial contact with *E. faecalis* resulted in differential expression of 1350 genes, and its supernatant resulted in the differential expression of 1453 genes. Moreover, 1252 genes were identified as DEGs via treatment with viable *F. nucleatum*, and only 135 genes by *F. nucleatum* supernatant infection (Figure 2).

Furthermore, differentially expressed genes (DEGs, *p* < 0.05 and log$_2$ fold change > 1.5) were used to compare SCAP gene expression patterns by different treatment variants. For each comparison, the number of unique and shared genes was investigated, as shown in Figure 3. First, a comparison of up- and downregulated genes after SCAP co-cultivation with viable *F. nucleatum* and *E. faecalis* showed a similar number of co-upregulated genes; 191 genes in total. There were 207 genes which were commonly downregulated for direct bacterial treatment with *F. nucleatum* and *E. faecalis*. A comparison of bacterial supernatant treatments showed that the number of individually downregulated expressed genes was much higher under the influence of *E. faecalis* supernatant (808) than after *F. nucleatum* supernatant treatment (at 16). A difference in individually upregulated genes was also observed: 563 genes for *E. faecalis's* supernatant and 37 genes for *F. nucleatum's* supernatant. A higher number of differentially expressed genes (both up- or downregulated) would have a higher influence on the SCAPs. In this context, *E. faecalis's* supernatant has the strongest influence on SCAPs. In contrast, *F. nucleatum's* supernatant was characterized as having a very mild effect on SCAPs.

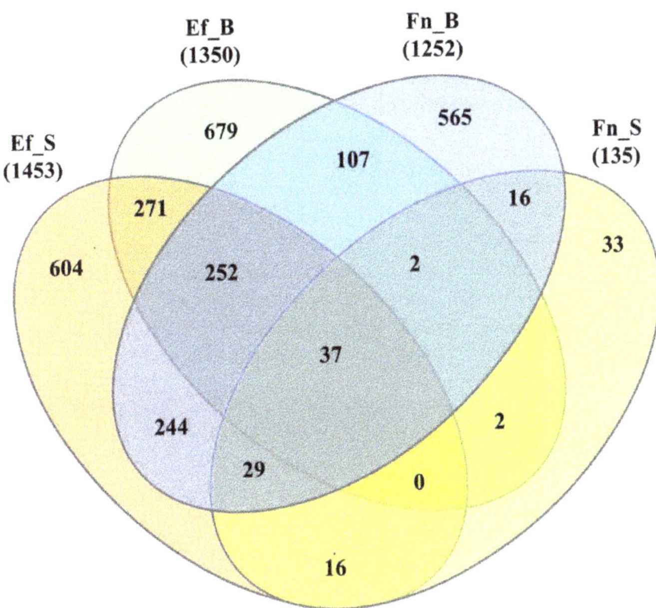

Figure 2. Venn diagram of SCAPs' differentially expressed genes, by treatment variant. Data is shown in absolute numbers of genes, where the differentially expressed genes (in relation to control) in each treatment are compared with each other. Pooled data for SCAP (donors I–III): F.n_B: SCAP co-cultured with *F. nucleatum*; F.n._S: SCAP co-cultured with *F. nucleatum* supernatant; E.f._B: SCAP co-cultured with *E. faecalis*; E.f._S: SCAP co-cultured with *E. faecalis* supernatant.

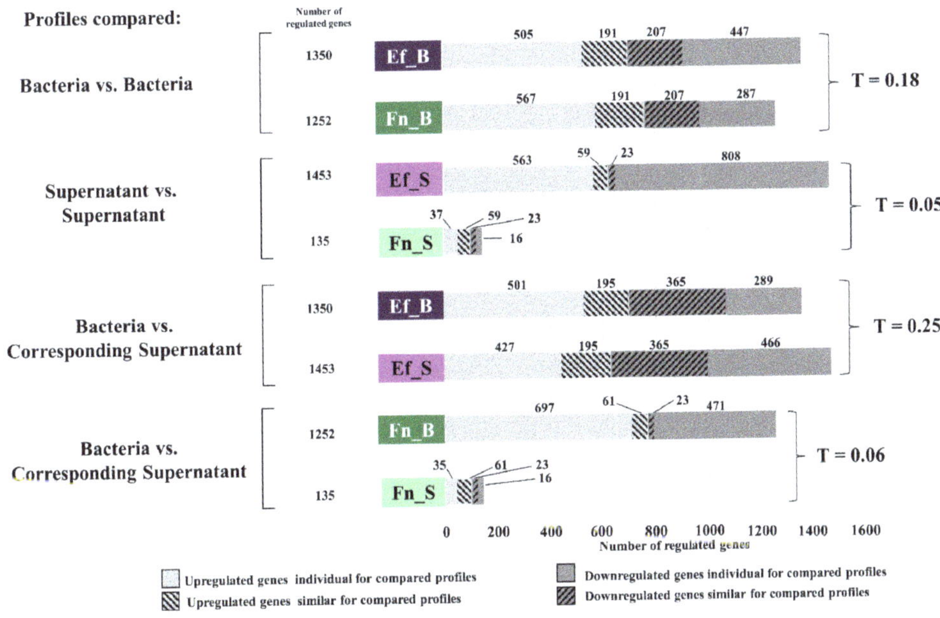

Figure 3. Bar graph of pairwise comparison of differentially expressed genes (DEGs) in absolute numbers. F.n_B: SCAP co-cultured with *F. nucleatum*; F.n._S: SCAP cultured with *F. nucleatum* supernatant;

E.f._B: SCAP co-cultured with *E. faecalis*; E.f._S: SCAP cultured with *E. faecalis* supernatant. DEG degree of similarity was analyzed based on the Tanimoto coefficient (T; highest possible similarity = 1.0). The genes that were up- or downregulated exceptionally for one of the compared profiles are named as individual genes; shared genes that were up- or downregulated in both types of treatment are named as similar genes.

The Tanimoto coefficient was used to assess the degree of similarity between each compared pair of DEG profiles. A Tanimoto coefficient value of = 1.0 represents the highest degree of similarity between the two sets of elements. Comparing different pairs of DEG profiles, the highest Tanimoto coefficient, i.e., the greatest similarity between the two DEG profiles, was found for a pair of *E. faecalis* bacterial and *E. faecalis* supernatant treatments (T = 0.25). Accordingly, as the highest level of Tanimoto coefficient was 0.25 between treatment variants, it is suggested that the investigated treatments influence SCAPs co-equally.

2.4. F. nucleatum Upregulates Immune and Inflammatory Response whereas E. faecalis Downregulates Cell Division and Proliferation in SCAP

To investigate the function of the DEGs, gene ontology (GO) term enrichment analysis was performed using the Database for Annotation, Visualization and Integrated Discovery (DAVID). The GO analysis presented the top GO terms for upregulated and downregulated DEGs sorted by *p*-value and categorized into biological processes as a functional group (Figure 4).

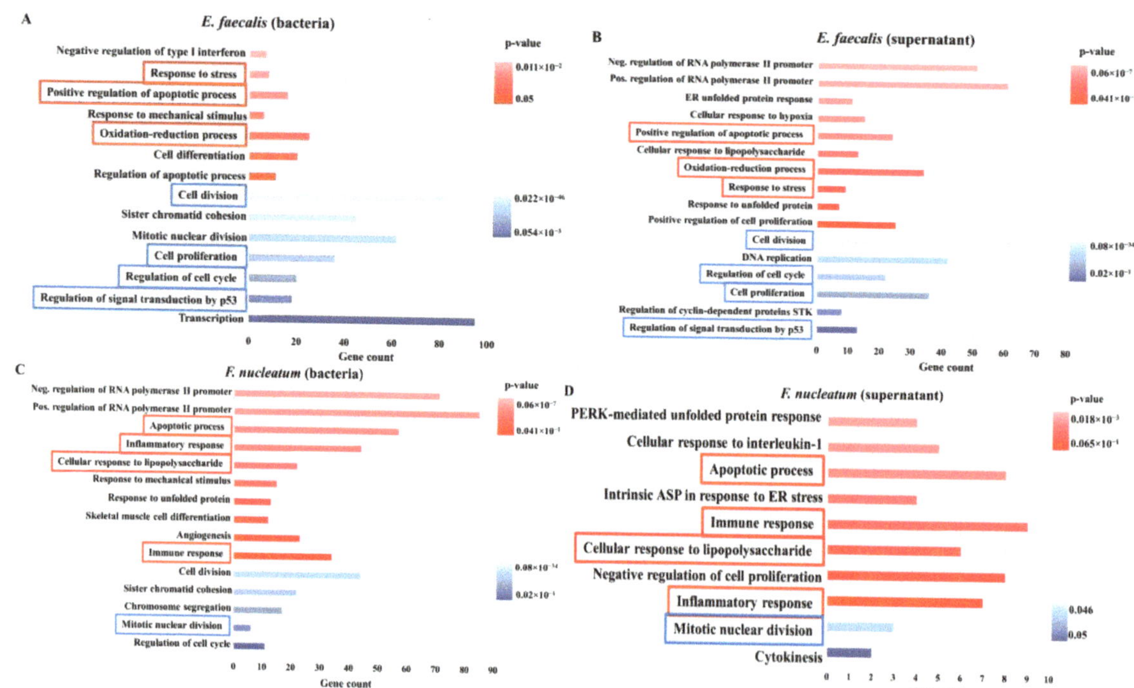

Figure 4. Gene ontology annotation. Upregulated biological processes (red bars) and downregulated processes (blue bars) sorted by *p*-value. Biological processes with the highest *p*-value are placed on the bottom, with the smallest *p*-value on the top (for both up- and downregulated processes): (**A**) SCAPs co-cultured with viable *E. faecalis*; (**B**) SCAPs co-cultured with *E. faecalis* supernatant; (**C**) SCAPs co-cultured with viable *F. nucleatum*; and (**D**) SCAPs co-cultured with *F. nucleatum* supernatants. Biological processes

(upregulated or downregulated) that were similar between treatment with planktonic bacteria and corresponding supernatants are marked by red or blue frames, respectively. ER: endoplasmic reticulum, STK: serine/threonine kinase, PERK: PKR-like ER kinase, ASP: apoptotic signaling pathway.

Co-culture of SCAPs with *E. faecalis* (planktonic and supernatant) leads to activation/downregulation of several analogous processes in SCAPs (Figure 4A,B). In this context, the identical upregulated biological processes were responses to stress, positive regulation of apoptotic processes, and oxidation-reduction processes (Figure 4A). Cell division, cell cycle regulation, cell proliferation, and regulation of signal transduction by p53 were commonly downregulated biological processes in SCAPs co-cultured with viable *E. faecalis* and its supernatant (Figure 4A,B).

Similarly, a set of identical biological processes was upregulated/downregulated in SCAPs co-cultured with viable *F. nucleatum* and *F. nucleatum* supernatants. The identical upregulated processes were apoptotic processes, inflammatory responses, cellular response to lipopolysaccharides, and immune responses; whereas mitotic nuclear division was the only commonly downregulated biological process (Figure 4C,D). It is worth mentioning that despite the common regulated biological processes between direct and indirect bacterial contact, there was both up- and downregulation of biological processes that were unique to each type of treatment. For example, only the *F. nucleatum* supernatant led to the upregulation of the negative regulation of cell proliferation and downregulation of cytokinesis processes (Figure 4D). Conversely, the supernatant of *E. faecalis* upregulated the processes of positive regulation of cell proliferation, the cellular response to lipopolysaccharides, the cellular response to hypoxia, etc.—processes that were not stimulated under the direct *E. faecalis* stimuli (Figure 4B). Overall, *F. nucleatum* modulated SCAP through upregulated genes which were mainly involved in immune and inflammatory responses, and downregulated genes mainly involved in the process of cell cycle regulation. Interestingly, the effect of *F. nucleatum* supernatants on SCAPs was very mild and was evidenced by the small gene count, as an exception among other treatment cases.

2.5. Osteogenic/Odontogenic Genes in SCAPs Are Strongly Influenced by F. nucleatum and E. faecalis Associated with Endodontic Infections

Several genes were chosen for their role in the mineralization process and were sorted into the following categories: dentinogenic, osteogenic cell surface, or osteogenic intracellular and osteogenic secreted genes (Table 2). Accordingly, Figure 5 showed SCAP transcriptomic analysis and corresponding genes that were up- or downregulated.

Table 2. Genes which were examined in the SCAP transcriptome, sorted by their function in osteogenesis and dentinogenesis.

	Gene	Function	Reference
Dentinogenic genes	Vascular endothelial growth factor A (*VEGFA*)	Inducing proliferation and differentiation of hDPSCs into odontoblasts	Matsushita et al., 2000 [32]
	Fibroblast growth factor 2 (*FGF2*)	Potent regulator of mineralization	Roberts-Clark and Smith, 2000; Madan and Kramer, 2005; Cooper et al., 2010; Miraoui and Marie, 2010; Marie et al., 2012; Smith et al., 2012 [33–38]
	Platelet-derived growth factor C (*PDGFC*)	Enhancement of DPSCs proliferation, odontoblast differentiation, and regeneration of dentin–pulp complex	Tsutsui, 2020 [39]
	Transforming growth factor β-1 (*TGF-β1*)	Importance in regulating reparative dentinogenesis	Toyono et al., 1997; Piatelli et al., 2004; Unterbrink et al., 2002 [40–42]

Table 2. *Cont.*

	Gene	Function	Reference
Osteogenic cell surface markers	Gamma-aminobutyric acid B receptor 1 (*GABABR1*)	Negative regulation of osteoblastogenesis	Takahata et al., 2011 [43]
	Gamma-aminobutyric acid B receptor 2 (*GABABR2*)		
	Parathyroid hormone 1 receptor (*PTH1R*)	Committing MSCs to the osteoblast lineage and promoting bone formation	Yu et al., 2012 [44]
	Receptor activator of nuclear factor-κB (*RANK*) or *TNFRSF11A*	Suppression of osteoblast differentiation	Chen et al., 2018 [45]
	Integrin alpha-V (*ITGAV*)	Osteoblast differentiation promotion	Cheng et al., 2001 [46]
	Osteoclast-associated receptor (*OSCAR*)	Regulator of osteoclast differentiation	Barrow et al., 2011 [47]
	Activated leukocyte cell adhesion molecule (*ALCAM/CD166*)	Immature osteoblast marker, promotes osteoblast differentiation	Hooker et al., 2015 [48]
Osteogenic intracellular markers	Nuclear factor interleukin-3-regulated (*NFIL3*)	Transcriptional repressor in osteoblasts	Hariri et al., 2020 [49]
	Runt-related transcription factor 2 (*RUNX2*)	Essential for initial commitment of MSCs to the osteoblastic lineage	Camilleri et al., 2006 [50]
	T cell immune regulator 1 (*TCIRG1*)	Osteoclastogenesis regulation	Zhang et al., 2020 [51]
	WD repeat domain 5 protein (*WDR5*)	Critical for MSCs osteogenic differentiation	Zhu et al., 2016 [52]
	T-box 2 (*TBX2*)	Positive regulation of osteogenic differentiation	Govoni et al., 2009; Abrahams et al., 2010 [53,54]
	T-box 3 (*TBX3*)		
	Distal-less homeobox 5 (*DLX-5*)	Transcriptional regulation of osteoblast differentiation	Hassan et al., 2004 [55]
	Early B cell factor 2 (*EBF2*)	Inhibition of osteoblast differentiation	Kieslinger et al., 2005 [56]
Osteogenic secreted markers	Signal peptide, CUB, and EGF-like domain-containing protein 3 (*SCUBE3*)	Controlling growth, morphogenesis, and bone and teeth development	Lin et al., 2021 [57]
	Type 1 collagen A (*COL1A1*)	Early marker of osteoblast	Kannan et al., 2020 [58]
	Insulin-like growth factor binding protein-3 (*IGFBP-3*)	Osteoblasts differentiation suppression	Li et al., 2013 [59]
	Insulin-like growth factor binding protein-4 (*IGFBP-4*)	Osteoclastogenesis regulation	Maridas et al., 2017 [60]
	Secreted protein acidic and rich in cysteine (*SPARC*)	Regulation of bone remodeling and bone mass maintenance	Rosset et al., 2016 [61]
	SPARC-related modular calcium-binding protein 1 (*SMOC1*)	Increases the expression of osteoblast differentiation-related genes in BMSCs	Choi et al., 2010 [62]
	Acid phosphatase 5 (*ACP5*)	Promotion of odontoblast differentiation and mineralization during tooth development	Choi et al., 2016 [63]
	Biorientation of chromosomes in cell division1-like 1(*BOD1L1*)	Positive regulation of osteoblasts differentiation	Okamura et al., 2017 [64]; NCBI [65]
	Aggrecan (*ACAN*)	Bone tissue formation	Viti et al., 2016 [66]
	Tissue-nonspecific alkaline phosphatase (*ALPL*)	Essential for bone mineralization; osteoblast marker	Nakamura et al., 2020 [67]
	Biglycan (*BGN*)	Modulation of osteoblast differentiation	Parisuthiman et al., 2005 [68]
	Decorin (*DCN*)	Key marker of odontoblasts	Matsuura et al., 2001 [69]
	Fibronectin 1 (*FN1*)	Essential for osteoblast differentiation and mineralization	Globus et al., 1998 [70]
	Bone morphogenetic protein (*BMP2*)	Important in osteoblast differentiation	Yang et al., 2012 [71]

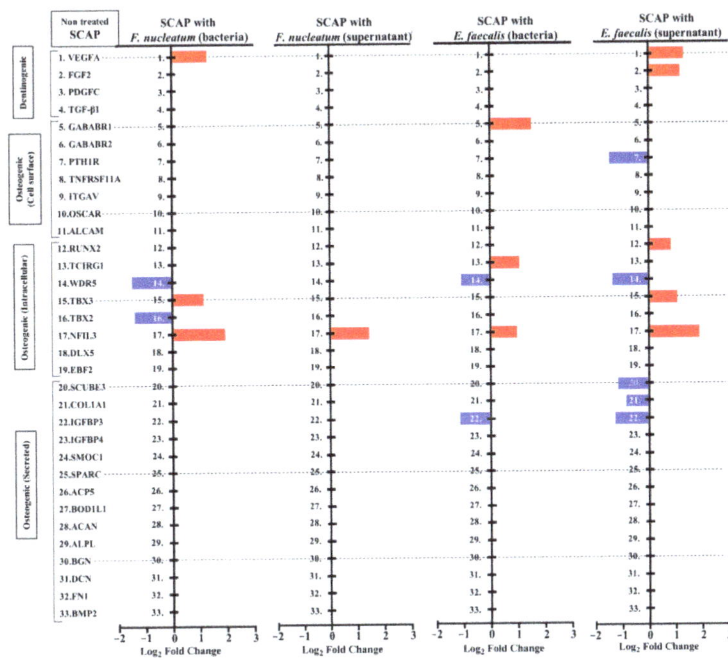

Figure 5. Up- and downregulation of genes associated with bone and dentin formation detected in SCAP culture variants. Transcriptome analysis of SCAP mRNA from three donors was performed. Gene expression is shown in log₂ fold change. Blue boxes represent downregulation. Red boxes represent upregulation.

The largest number of dentinogenic/osteogenic-associated genes were regulated when SCAPs were exposed to *E. faecalis* supernatants. Vascular endothelial growth factor A (*VEGFA*), fibroblast growth factors (*FGF2*), runt-related transcription factor 2 (*RUNX2*), T-box transcription factor 3 (*TBX3*), and nuclear factor interleukin 3 (*NFIL3*) were upregulated in comparison with non-treated SCAPs. In contrast, the parathyroid hormone 1 receptor (*PTH1R*); WD repeat-containing protein 5 (*WDR5*); signal peptides; CUB domain and EGF-like domain (Epidermal growth factor) containing 3 (*SCUBE3*); insulin-like growth factor-binding protein 3 (*IGFBP3*); and collagen, type I, alpha 1 (*COL1A1*) were downregulated in comparison with non-treated SCAPs.

Co-culture of SCAPs with viable *E. faecalis* led to the regulation of five osteogenic genes. *NFIL3*, the T cell immune regulator gene 1 (*TCIRG1*), and gamma-aminobutyric acid B receptor 1 (*GABABR1*) were upregulated, and *WDR5* and *IGFBP3* were downregulated. In addition, five dentinogenic/osteogenic genes were regulated when SCAPs were exposed to *F. nucleatum*. *VEGFA*, *TBX3*, and *NFIL3* were upregulated, and *WDR5*, as well as T-box transcription factor 2 (*TBX2*), were downregulated. Only one osteogenic-associated gene was upregulated when SCAP was co-cultured with *F. nucleatum* supernatant (*NFIL3*). Interestingly, *NFIL3* was upregulated in all treatment variants (Figure 5).

3. Discussion

A transcriptome is the whole set of cell transcripts, and their quantity for a specific stage of cell development or physiological state. Comprehension of cell transcripts as transcriptomes is critical for development and disease conception, as well as for elucidation of the functional aspects of the genome and in disclosing the molecular elements of cells and tissues [72]. Through RNA-seq transcriptomic analysis, we managed to obtain transcriptomic profiles of healthy SCAPs under the exposure of opportunistic bacteria of the oral cavity and their metabolites. In this study, with the help of transcriptomic analysis, we established that the *E. faecalis* supernatant showed a distinguished stimulant effect on the expression of osteo- and dentinogenic genes in SCAPs, compared with other treatment types. In contrast, cells treated by *F. nucleatum* supernatants had the smallest effect on the expression of these genes. This work on a cellular model of healthy SCAP modulated by oral bacteria in vitro is an initial step towards understanding the complex processes potentially occurring in the in vivo ecosystem of SCAP and oral bacteria.

Moreover, PCA analysis of the transcriptome profiles of SCAPs formed four clusters founded by treatment variants. Three of these clusters were formed strictly by the treatment variant: SCAPs under the direct bacterial influence of *F. nucleatum*; SCAPs treated with the planktonic stage of *E. faecalis*; and SCAPs treated with *E. faecalis* supernatants. However, it is noteworthy that SCAPs treated by *F. nucleatum* supernatant (donors I–II) and untreated SCAPs (donors I–III) were grouped in one cluster.

In addition, differential gene expression analysis (DEG) has shown similarities between up- and downregulated genes in *E. faecalis* and *F. nucleatum* in direct bacterial treatment. However, when comparing the action of the supernatants of these bacteria, the number of DEGs under the influence of the supernatant of *E. faecalis* was ten times higher than DEGs under the influence of *F. nucleatum* supernatants. The similarity between the two types of cell treatment was evaluated using the Tanimoto coefficient, which in this case was the smallest (0.05). The low value of Tanimoto coefficient (0.06) in comparing direct and indirect bacterial contact by DEG number confirmed the meager influence of *F. nucleatum* supernatants on SCAP. The strength of the treatment's influence on stem cells from the strongest to the weakest was as follows: *E. faecalis* supernatant > *E. faecalis* bacteria > *F. nucleatum* bacteria > *F. nucleatum* supernatant.

According to the search of gene ontology annotation, in between upregulated biological processes, apoptotic processes were upregulated in all treatment variants: through both direct contact with bacteria in the planktonic stage and through supernatants saturated by bacterial metabolites. This means that the presence of bacteria in the planktonic stage or secreted bacterial metabolites in the tooth at the investigated concentrations can possibly lead to root aberration and failure of regeneration treatments, as was already shown in several studies using other cells or models [73,74]. The key differences between the influence of *F. nucleatum* (bacteria as well as supernatant) in comparison with *E. faecalis* (bacteria as well as supernatant) on SCAPs were the upregulation of the immune and inflammatory responses occurring only in the case of SCAP treatment with *F. nucleatum* bacteria/supernatant. Intensification of inflammatory processes would lead to dentist attention and improve the probability of improving the regenerative process [75], whereas the effect of *E. faecalis* bacteria in either the planktonic stage or bacterial metabolites may not be detected as quickly because it remains hidden from the immune response [76]. Chong et al. demonstrated this on a mouse model with non-healing wounds infected with *E. faecalis* that led to the suppression of the inflammatory cytokines, notwithstanding the immune cell infiltration [77]. It was shown that the ability of bacteria to reduce proinflammatory cytokine secretion is an intrinsic feature that reflects the degree of bacterial virulence [78].

Interestingly, the number of gene counts in the biological processes that were identified via the gene ontology tool were shifted to the side of downregulated processes in the case of SCAP co-culture with *E. faecalis*. If SCAPs were cultured in *E. faecalis* supernatant, biological processes were balanced between up- and downregulation. Upregulated biological processes predominated if SCAPs were treated by *F. nucleatum* bacteria or its supernatant. The decreased number of gene counts for upregulated genes could be an indication of cell infection with pathogenic bacteria, as was reported by Stekel and co-authors on a model of human intestinal epithelial cells [79].

It is noteworthy that the regulated biological processes found via the GO tool for SCAPs co-cultured with bacteria and for SCAPs cultured in the supernatant from the corresponding bacteria were not only similar, but also unique in their upregulated and downregulated processes. This very likely indicates that the effect of bacteria on cells is mediated not only by direct contact with microbial-associated molecular patterns (MAMP) [80], but also, possibly, by secreted bacterial metabolites [81].

Recent findings considered SCAPs to be a new and promising source of stem cells for regenerative endodontic treatments [9,82,83] which is why the key issue in the present study was to examine the influence of two specified bacterial species, isolated from the root canal, on the differentiation directions of SCAPs. To this end, 33 dentinogenic- and osteogenic-related genes that were selected according to their role in mineralization processes (Table 2) were examined in the transcriptomics/gene expression profile of non-infected SCAPs, after which the expression of these genes was compared with gene expression profiles in the treated groups. The selected genes were divided into two subgroups according to the genesis of either dentin or bone (osteogenic or dentinogenic markers), and the subgroup of osteogenic markers was further divided according to the marker's location in the cell: on the cell surface, intracellular, or secreted.

This study showed that direct bacterial treatment with SCAPs and the bacterial metabolites of *F. nucleatum* affect the expression of dentinogenic and osteogenic intracellular-associated genes in various ways. One important gene for dental pulp repair and reparative dentine formation—*VEGFA*—was upregulated only in the group that underwent direct bacterial treatment with *F. nucleatum*. Mendes et al. showed that *F. nucleatum* was able to induce an immune response in endothelial cells and increase *VEGF* cell secretion, but the expression of this gene at the mRNA level was lower [84]. Immunohistochemical data of cells from the inflammatory infiltrate of irreversible pulpitis showed strong positive *VEGF* expression [85]. Despite the fact that *VEGF* is important for pulp healing and angiogenesis in general [86], stimulation of the expression of this gene by *F. nucleatum* may indicate an inflammatory process in SCAPs, since angiogenesis may in fact increase the severity of the inflammatory process [85].

We have furthermore shown that direct bacterial contact of SCAPs with *F. nucleatum* leads to upregulated expression of two other osteogenic intracellular markers: *TBX3* and *NFIL3*. A previous study conducted by Govoni et al. had demonstrated that *TBX3* inhibits mineralization of osteoblast cells on mouse pre-osteoblast cells [53]. Another upregulated gene—*NFIL3*—was found to be upregulated in all variants of SCAP treatment and the only osteogenic gene that was regulated in the case of SCAP culture in *F. nucleatum* supernatant. Besides the involvement of *NFIL3* in the development of the innate immunity cells in a mouse model [87], it has been shown that *NFIL3* can act as a transcriptional repressor in osteoblasts while exhibiting differential activity as an activator in osteocytes on mouse osteoblastic cell lines [49].

However, direct bacterial contact of *F. nucleatum* with SCAPs resulted in the downregulation of two genes, *WDR5* and *TBX2*. It should be noted that downregulation of the expression of these genes was also observed under direct and indirect bacterial exposure to *E. faecalis*. Studies show that the expression of these genes works as a stimulator of the canonical Wnt pathway, and therefore acts as a stimulus for differentiation of odonto- and osteoblasts [52,88].

Concerning the other treatment group, *E. faecalis*'s bacterial suspension and supernatant, it was shown that the supernatant of these bacteria modulated SCAPs to a greater extent. Direct bacterial stimulation of SCAPs with *E. faecalis* resulted in the upregulation of three genes: *GABABR1*, *TCIRG1* and *NFIL3*. To date, there is no data on how the expression of these genes affects the osteo/odontogenic potential of SCAP, but it is known that the secretion of these factors may indicate an inflammatory process and inhibition of osteoblastogenesis [43,49,89].

Interestingly, it was the indirect effect of *E. faecalis* on SCAPs (via the bacterial supernatant) that resulted in the greatest influence on the transcriptomic SCAP profile. Only in the case of indirect bacterial contact was the expression of *VEGF* and *FGF2* observed. As mentioned before, *VEGF* is important both in angiogenesis and osteogenesis; however, recent studies show that overexpression of *VEGF* by dental pulp cells exposed to gram-positive bacterial toxins may, in the long run, lead to pulp necrosis owing to intra-pulpal pressure caused by *VEGF* [90]. The upregulated expression of *RUNX2*, *TBX3*, and *NFIL3* indicate that under the influence of *E. faecalis*, SCAPs' differentiation potential may be affected negatively [91]. We strongly emphasize the observation that only this treatment option caused the upregulation of *RUNX2*. *RUNX2* could be activated by the mitogen-activated protein kinase (MAPK) pathway, and in turn this pathway could be stimulated by cell treatment with the osteogenic growth factor, FGF2 [92]. Since in the present study *FGF2* expression was found to be upregulated only in the *E. faecalis* supernatant treatment option, we can assume that *RUNX2* is being activated and it can induce differentiation to osteoblasts instead of odontoblasts [91]. In addition, with *E. faecalis* supernatant treatment, the downregulation of *PTH1R*, *SCUBE3*, *COL1A1m*, which are important for osteogenic differentiation, was shown [57,93–95]. Furthermore, it is worth noting that the expression of the *IGFBP3* gene was downregulated in this type of SCAP treatment, and a recent study by Aizawa et al. in a mouse model showed that *IGFBP3* is required for pre-odontoblast differentiation [96].

Additionally, the transcriptomic analysis is very specific, sensitive, and reproducible, and allows for the assay of multiple samples simultaneously [72].

SCAPs are a promising source of local stem cells, which could be used for regenerative endodontic treatment (RET) of traumatic dental injuries to immature teeth [97]. Traumatic dental injuries create the possibility for oral bacteria, which exist in the planktonic stage, to come into direct contact with SCAP cells [98] through the surface components of bacteria (flagella, pili, surface layer proteins, capsular polysaccharides, lipoteichoic acid, lipopolysaccharides) via SCAPs' toll-like receptors (TLR). This then regulates several signaling pathways, such as nuclear factor B (NF-κB) and mitogen-activated protein kinases (MAPK), as was shown on epithelial cells [99,100]. The dysregulation of the latter pathway leads to diminishment of the mineralization processes in human dental pulp stem cells (DPSCs) [101,102].

However, oral bacteria exist in most cases in the oral cavity in the form of biofilms and can continuously secrete metabolites produced by bacteria, such as vesicles, extracellular proteins, organic acids, indoles, etc. Those metabolites can easily reach the SCAPs in the case of traumatic dental injury and, probably, stimulate signaling cascades and the production of specific cytokines and chemokines, which then up- or downregulate inflammation, resulting in a changed microenvironment which can influence regeneration [27,103–105]. The success of regenerative endodontic treatment depends on a better understanding of the consequences of SCAPs' and live bacteria's direct interaction in comparison with SCAP stimulation by bacterial metabolites alone.

Our results presume rigorous disinfection procedures in the case of traumatic dental injuries to be one of the keys to successful regenerative treatment. One limitation of this study is the short duration of cell exposure to different treatment options (with a longer exposure, the processes of cell death began to prevail due to live bacteria), which prevented us from seeing the development and the result of the differentiation process.

Undoubtedly, the influence on healthy SCAPs by bacteria, which are most often isolated from infected root canals, is improperly perceived as a model of an infected injured tooth; however, interest in and feasibility of the total elimination of the two proposed opportunistic bacteria merits further discussion on the one hand, and on the other hand so do the interactions of these two bacteria with other species of the oral microbiota. Thus, from a translational perspective, our investigations suggest that further studies investigating not only one species but a cell–microbiota interaction may be warranted in order to mimic the complex clinical situation.

Furthermore, extrapolating the results to mimic the situation of a chronic process should be done with care, as we are using an in vitro system. Nevertheless, our method still provides useful information to increase understanding of in vivo processes.

In summary, the differentiation of tissue-specific MSCs, such as SCAPs to osteoblasts, is a very complex process that is finely orchestrated by a number of cytokines, chemokines, signaling molecules, and mechanical stimuli. Such a stimulus should be applied at specific time points of the differentiation process [106,107]. In the previous study, we showed that *F. nucleatum* stimulates the inflammatory response of SCAP, whereas *E. faecalis* reduces it and decreases the level of the key proteins of Wnt/β-Catenin and NF-κB signaling pathways that are important for bone formation [98]. In the present study, stem cells were exposed to bacteria at the planktonic stage or to the corresponding bacterial supernatant over a 24 h period in vitro. Direct and indirect bacterial stimulation of cells led to a change in the transcriptomic profile of SCAPs, namely up- and downregulation of genes, changes in the levels of which may adversely affect the processes of osteo- and dentinogenesis.

4. Materials and Methods

4.1. Cell Isolation and Culture

In this study, we used three clinical isolates obtained from impacted human teeth ($n = 3$; two lower jaw third molars and one upper jaw canine) from three healthy patients (one male and two females with mean age of 17 years and a range 11–20 years), due to retention and/or lack of space in orthodontic treatment [10,108]. The authenticity of the multipotent stromal cells was confirmed by the presence of CD73, CD90, CD105, and CD146, and the absence of CD11b, CD19, CD34, CD45, and HLA-DR. Detection was performed with PE-conjugated antibodies against the above-mentioned markers by flow cytometry (FCM, Becton Dickinson, Accuri C6), and analyzed with FlowJo Software V9. The multipotency and stemness of isolated SCAPs from these donors, in a step toward adipogenic and osteogenic differentiation, were previously published [15]. Collection, culture, storage, and usage of all cell lines were approved by the local research ethics committee at Umeå University (Reg. no. 2013-276-31M).

The SCAPs we used in this study were isolated 4–5 years ago and cryo-preserved in cryomedium (90% FBS and 10% DMSO). SCAPs were brought back from cryopreservation and grown until 95% confluency in cell culture medium α-MEM, GlutaMAX™ GIBCO (ThermoFisher (Lifetech), Waltham, MA, USA, #32561029), supplemented by a 10% FBS and 1% Penicillin-Streptomycin solution (Merck (Sigma-Aldrich, St. Louis, MO, USA) # P0781). Cells were harvested using trypsin/EDTA solution (Merck (Sigma-Aldrich) #T3924), counted by Countess II Automated Cell Counters (ThermoFisher Scientific) according to the manufacture's protocol, and seeded at a cell density of 2×10^4 cells/mL into 10 cm cell culture dishes, and kept in the cell incubator at +37 °C with 5% CO_2 overnight until adherence. SCAP cells (4th passage) were used for the experiments.

4.2. Bacterial Strains and Culture Conditions

We used clinical isolates of *F. nucleatum subsp. polymorphum* and *E. faecalis* (Table S1) obtained from root canal samples of traumatized necrotic teeth of young patients that were referred to the Endodontic Department, Region Västerbotten, Sweden (Reg. no. 2016/520-31) in this study. Sample collection, processing, and characterization of isolates was performed as previously described [19]. Briefly, samples were collected from root canals under strict aseptic conditions. Before entering the pulp space, a rubber dam was applied and the tooth, clamp, and dam were disinfected with hydrogen peroxide (30%) and tincture of iodine (5%). The contents of the root canal were absorbed into sterile paper points. The paper points were then moved to a TE buffer followed by culturing in anaerobic conditions on Fastidious Anaerobe Agar (FAA) (Svenska LabFab) for one week. Colonies with different phenotypic patterns were selected from each plate, amplified by PCR, and sequenced to identify bacterial species [13]. Sequences were compared with the eHOMD database (Expanded Human Oral Microbiome Database, Forsyth, (http://www.ehomd.org (accessed on 9 February 2017) for detection at the species level with >98.5% sequence similarity with regard to their 16S rRNA genes.

The cryostocks of the designated species were preserved in 20% sterile skimmed milk and stored at −80 °C until needed for experiments. Bacteria were removed from the cryostocks and passaged on Fastidious Anaerobe Agar (FAA) (Svenska LabFab, Söderhamn, Sweden) medium supplemented with 5% citrated bovine blood (Svenska LabFab FIE

34200500) and 16 µg/L of vitamin K (Sigma-Aldrich M5850) in an anaerobic atmosphere (10% CO_2, 10% H_2, 80% N_2) at +37 °C for 5–7 days.

Bacteria were then harvested and resuspended in cell culture medium MEM-α enriched with 10% FBS. Optical density of each bacterial suspension was adjusted by spectrophotometer at 600 nm to 1.0 (corresponds to 1×10^8 CFU/mL). Bacterial suspensions were used directly in the co-culture experiment with SCAPs or for preparation of bacterial supernatants. Bacterial strains were quantitatively inoculated in MEM-α supplemented with 10% FBS in order to have the equivalent of multiplicity of infection (MOI) equal to 100. Each strain was grown individually over 24 h in anaerobic conditions at +37 °C. Bacteria were then pelleted by centrifugation at $10.000 \times g$ for 10 min at 4 °C and supernatants enriched by bacterial metabolites (hereafter referred to as 'bacterial supernatants') were filtrated through a sterile syringe filter with 0.22 µm pore size (Fisher Scientific #10268401), aliquoted, and stored at -80 °C until use.

4.3. SCAP Infection by Bacterial Strains: Co-culture Experiments

Species of *E. faecalis* and *F. nucleatum* were used in this study because of their strong binding capacity and proliferation effects on SCAPs [15]. Viable bacteria in the planktonic stage and bacterial supernatants were analyzed in the following treatment variants: (i) SCAPs (donors I–III) co-cultured with viable *F. nucleatum* (planktonic stage); (ii) SCAPs (donors I–III) co-cultured with viable *E. faecalis* (planktonic stage); (iii) SCAPs (donors I–III) co-cultured with *F. nucleatum* (supernatant); (iv) SCAPs (donors I–III) co-cultured with *E. faecalis* (supernatant); and (iv) SCAPs (donors I–III) without bacterial infection, used as controls.

Previous study showed that selected bacterial species modulated SCAP cytokine secretion only after 24 h of co-cultivation [15]. Based on the results of the previous study, SCAPs were cultured in an anaerobic atmosphere (10% CO_2, 10% H_2, 80% N_2) at +37 °C for 24 h. For the co-culture experiments, viable bacteria or their supernatants were resuspended in antibiotic-free cell culture medium and adjusted to MOI 100 on SCAPs as previously described [15]. MOI 100 was determined to be an effective concentration of bacteria per cell, by both the dose response test and the neutral red cytotoxicity test [15].

4.4. RNA Extraction and Quality Evaluation

After bacteria or supernatants were co-cultured with SCAP for 24 h, cell monolayers were washed with PBS and detached via a trypsin/EDTA solution. Collected cells were treated using an RNA stabilizer (RNAprotect Cell Reagent Qiagen, #76526) and kept overnight at +4 °C. The next day, collected cell samples were lysed and homogenized (QIAshreder Qiagen, #79654). RNA extraction was performed using a RNeasy Mini Kit (RNeasy Mini Kit, Qiagen, #74104) according to the manufacturer's protocol with an additional step of on-column treatment by DNase for elimination of residual DNA contamination (DNase I, RNase-free, ThermoFisher Scientific #EN0521). The quality of isolated RNA (yield, purity, and integrity) was assessed using an Agilent 2100 instrument (Agilent Technologies, Santa Clara, CA, USA).

mRNA extraction, conversion to complementary DNA (cDNA), and sequencing library preparation (150 nucleotides) were performed by Novogene Bioinformatics Technology Co., Ltd. (Beijing, China). The RNA integrity number (RIN)—an important tool in conducting valid gene expression measurement experiments—satisfied the quality requirements for transcriptomic analysis (mean 9.13) (Table 3). Thus, good RNA quality assessment is considered one of the most critical elements in obtaining meaningful gene expression data via transcriptomics [109]. Agarose gel electrophoresis confirmed that none of the samples were contaminated by DNA or protein and that the RNA was intact and showed two sharp 28S and 18S rRNA bands (Figure S1).

Table 3. Characteristics of total RNA extracted from the designated sample. The 'No Bacteria' type of treatment should be considered a control.

Sample	SCAP Source	Treatment	Concentration, pg/µL	Absorbance Ratio 260/280	RNA Integrity Number
1	Donor I	No Bacteria	4405	2.086	9.9
2	Donor I	*F. nucleatum* B *	3737	2.113	9.4
3	Donor I	*F. nucleatum* S **	3016	2.088	9.1
4	Donor I	*E. faecalis* B	3977	2.089	6.4
5	Donor I	*E. faecalis* S	3078	2.102	9.8
6	Donor II	No Bacteria	4791	2.097	9.8
7	Donor II	*F. nucleatum* B	6566	2.123	9.2
8	Donor II	*F. nucleatum* S	5288	2.096	10.0
9	Donor II	*E. faecalis* B	5208	2.087	8.0
10	Donor II	*E. faecalis* S	4465	2.080	9.5
11	Donor III	No Bacteria	4193	2.068	10.0
12	Donor III	*F. nucleatum* B	4929	2.058	8.7
13	Donor III	*F. nucleatum* S	4673	2.064	9.5
14	Donor III	*E. faecalis* B	4871	2.068	7.6
15	Donor III	*E. faecalis* S	4089	2.069	10.0

* B: bacteria, ** S: supernatants.

4.5. Transcriptomic Analysis, Data Preprocessing and Bioinformatics

The paired-end reads obtained from NovaSeq were checked for quality using FastQC [110]. Initially, we pre-processed the RNA-seq data from our fifteen samples using Cutadapt (v3.1) [111] by trimming reads containing adapter and poly-N sequences and low-quality raw data reads. All downstream analyses were based on clean data of high quality. The trimmed reads were aligned separately to the human genome by HISAT2 (v2.2.1) [112] using default parameters. The genome sequences and the human annotations (GRCh38.p13) were obtained from the NCBI genome database (https://www.ncbi.nlm.nih.gov/genome, accessed on 5 March 2021). Post-alignment QC metrics were generated using RSeQC (v2.6.4) [113], which detected junctions, read distribution, experiment type, and ribosomal contamination [114].

For quantitation of the mapped read numbers of each gene we used FeatureCounts (subread v2.0.0). Differential expression analysis was performed using DESeq2 (v1.32) [115]. Differentially expressed genes were those whose expression changed at least 1.5 times at the $\log_2$ level compared with the expression of genes in the control group, and those who had a level of statistical significance of $p < 0.05$. For clarity, DEGs exhibiting higher levels of expression in treated samples compared with controls were designated as 'highly regulated', whereas those exhibiting the opposite ratio were designated as 'reduced'. Unwanted variation was removed using the removeBatchEffects function from Limma (v3.48.3) through technical heterogeneity [116]. DESeq2 was used to compute a VST (variance stabilizing transformation) of the original count data for visualization in principal component analysis (PCA) using R (4.0.3). Variability due to the three replicate data sets being higher (Figure S2), referred to as 'batch effects', was corrected using the 'Limma' R package [116] prior to further analysis to avoid introducing biologically irrelevant signals into the high-throughput data and misleading conclusions [117].

The Database for Annotation, Visualization and Integrated Discovery (DAVID; version 6.8; david.ncifcrf.gov/ (accessed on 20 April 2021)) was used to perform GO (www.geneontology.org (accessed on 20 April 2021)) enrichment; $p < 0.05$ was considered to indicate a statistically significant difference [118–120].

A Tanimoto coefficient was used to assess the degree of similarity between each compared pair of DEG profiles. A Tanimoto coefficient ranges from 0 (no similarity) to 1 (high similarity), where values greater than 0.85 reflect a high probability of similarity between the two sets of elements [121].

Supplementary Materials: The following supporting information can be downloaded at: https://www.mdpi.com/article/10.3390/ijms232214420/s1.

Author Contributions: Conceptualization, N.R.V. and O.R.; methodology, P.W., Y.R. and V.Z.; formal analysis, L.M. and Y.R.; data curation, P.W., Y.R. and V.Z.; writing—original draft preparation, Y.R., V.Z. and O.R.; writing—review and editing, N.R.V., M.B., and P.K.; supervision, N.R.V.; funding acquisition, N.R.V., M.B. and P.K. All authors have read and agreed to the published version of the manuscript.

Funding: This research was funded by the Knut and Alice Wallenberg Foundation (project number 7003503, the Region of Vasterbotten (Sweden) via ALF grant number 7004361, Kempestiftelserna Kempe SMK-1966 and the research fund of the County Council of Vasterbotten (Project Number: 7003459 and 7003589).

Institutional Review Board Statement: The study was conducted in accordance with the Declaration of Helsinki, and approved by the Ethics Committee at Umeå University (Reg. no. 2013-276-31M and Reg. no. 2016/520-31).

Informed Consent Statement: Informed consent was obtained from all subjects involved in the study.

Data Availability Statement: The raw data supporting the conclusions of this article will be made available by the authors, without undue reservation.

Conflicts of Interest: The authors declare no conflict of interest. The funders had no role in the design of the study; in the collection, analyses, or interpretation of data; in the writing of the manuscript, or in the decision to publish the results.

References

1. Kilian, M.; Chapple, I.L.C.; Hannig, M.; Marsh, P.D.; Meuric, V.; Pedersen, A.M.L.; Tonetti, M.S.; Wade, W.G.; Zaura, E. The oral microbiome—An update for oral healthcare professionals. *Br. Dent. J.* **2016**, *221*, 657–666. [CrossRef] [PubMed]
2. Fouad, A.F. Microbiological aspects of traumatic injuries. *Dent. Traumatol.* **2019**, *35*, 324–332. [CrossRef] [PubMed]
3. Farhad, A.; Mohammadi, Z. Calcium hydroxide: A review. *Int. Dent. J.* **2005**, *55*, 293–301. [CrossRef] [PubMed]
4. Torabinejad, M.; Chivian, N. Clinical applications of mineral trioxide aggregate. *J. Endod.* **1999**, *25*, 197–205. [CrossRef]
5. Lin, J.; Zeng, Q.; Wei, X.; Zhao, W.; Cui, M.; Gu, J.; Lu, J.; Yang, M.; Ling, J. Regenerative Endodontics Versus Apexification in Immature Permanent Teeth with Apical Periodontitis: A Prospective Randomized Controlled Study. *J. Endod.* **2017**, *43*, 1821–1827. [CrossRef]
6. Andreasen, J.O.; Farik, B.; Munksgaard, E.C. Long-term calcium hydroxide as a root canal dressing may increase risk of root fracture. *Dent. Traumatol.* **2002**, *18*, 134–137. [CrossRef]
7. Galler, K.M.; Krastl, G.; Simon, S.; Van Gorp, G.; Meschi, N.; Vahedi, B.; Lambrechts, P. European Society of Endodontology position statement: Revitalization procedures. *Int. Endod. J.* **2016**, *49*, 717–723. [CrossRef]
8. Xuan, K.; Li, B.; Guo, H.; Sun, W.; Kou, X.; He, X.; Zhang, Y.; Sun, J.; Liu, A.; Liao, L.; et al. Deciduous autologous tooth stem cells regenerate dental pulp after implantation into injured teeth. *Sci. Transl. Med.* **2018**, *10*, eaaf3227. [CrossRef]
9. Kang, J.; Fan, W.; Deng, Q.; He, H.; Huang, F. Stem Cells from the Apical Papilla: A Promising Source for Stem Cell-Based Therapy. *Biomed Res. Int.* **2019**, *2019*, 6104738. [CrossRef]
10. Pettersson, L.F.; Kingham, P.J.; Wiberg, M.; Kelk, P. In Vitro Osteogenic Differentiation of Human Mesenchymal Stem Cells from Jawbone Compared with Dental Tissue. *Tissue Eng. Regen. Med.* **2017**, *14*, 763–774. [CrossRef]
11. de Almeida, J.F.A.; Chen, P.; Henry, M.A.; Diogenes, A. Stem Cells of the Apical Papilla Regulate Trigeminal Neurite Outgrowth and Targeting through a BDNF-Dependent Mechanism. *Tissue Eng. Part A* **2014**, *20*, 3089–3100. [CrossRef]
12. Bakopoulou, A.; Leyhausen, G.; Volk, J.; Tsiftsoglou, A.; Garefis, P.; Koidis, P.; Geurtsen, W. Comparative analysis of in vitro osteo/odontogenic differentiation potential of human dental pulp stem cells (DPSCs) and stem cells from the apical papilla (SCAP). *Arch. Oral Biol.* **2011**, *56*, 709–721. [CrossRef]
13. Petridis, X.; van der Sluis, L.W.M.; Dijkstra, R.J.B.; Brinker, M.G.L.; van der Mei, H.C.; Harmsen, M.C. Secreted products of oral bacteria and biofilms impede mineralization of apical papilla stem cells in TLR-, species-, and culture-dependent fashion. *Sci. Rep.* **2018**, *8*, 12529. [CrossRef]
14. Diogenes, A.; Hargreaves, K.M. Microbial Modulation of Stem Cells and Future Directions in Regenerative Endodontics. *J. Endod.* **2017**, *43*, S95–S101. [CrossRef]
15. Rakhimova, O.; Schmidt, A.; Landström, M.; Johansson, A.; Kelk, P.; Romani Vestman, N. Cytokine Secretion, Viability, and Real-Time Proliferation of Apical-Papilla Stem Cells Upon Exposure to Oral Bacteria. *Front. Cell. Infect. Microbiol.* **2021**, *10*, 620801. [CrossRef]
16. Brennan, C.A.; Garrett, W.S. Fusobacterium nucleatum—Symbiont, opportunist and oncobacterium. *Nat. Rev. Microbiol.* **2019**, *17*, 156–166. [CrossRef]

17. Tan, H.C.; Cheung, G.S.P.; Chang, J.W.W.; Zhang, C.; Lee, A.H.C. Enterococcus faecalis Shields Porphyromonas gingivalis in Dual-Species Biofilm in Oxic Condition. *Microorganisms* **2022**, *10*, 1729. [CrossRef]

18. Lertchirakarn, V.; Aguilar, P. Effects of Lipopolysaccharide on the Proliferation and Osteogenic Differentiation of Stem Cells from the Apical Papilla. *J. Endod.* **2017**, *43*, 1835–1840. [CrossRef]

19. Manoharan, L.; Brundin, M.; Rakhimova, O.; Chávez de Paz, L.; Romani Vestman, N. New Insights into the Microbial Profiles of Infected Root Canals in Traumatized Teeth. *J. Clin. Med.* **2020**, *9*, 3877. [CrossRef]

20. Moore, W.E.C.; Moore, L.V.H. The bacteria of periodontal diseases. *Periodontol. 2000* **1994**, *5*, 66–77. [CrossRef]

21. Topcuoglu, N.; Bozdoğan, E.; Aktoren, O.; Kulekci, G. Presence of Oral Bacterial Species in Primary Endodontic Infections of Primary Teeth. *J. Clin. Pediatr. Dent.* **2013**, *38*, 155–160. [CrossRef] [PubMed]

22. Despins, C.A.; Brown, S.D.; Robinson, A.V.; Mungall, A.J.; Allen-Vercoe, E.; Holt, R.A. Modulation of the Host Cell Transcriptome and Epigenome by Fusobacterium nucleatum. *Mbio* **2021**, *12*, e02062-21. [CrossRef] [PubMed]

23. Castellarin, M.; Warren, R.L.; Freeman, J.D.; Dreolini, L.; Krzywinski, M.; Strauss, J.; Barnes, R.; Watson, P.; Allen-Vercoe, E.; Moore, R.A.; et al. Fusobacterium nucleatum infection is prevalent in human colorectal carcinoma. *Genome Res.* **2012**, *22*, 299–306. [CrossRef] [PubMed]

24. Kang, W.; Sun, T.; Tang, D.; Zhou, J.; Feng, Q. Time-Course Transcriptome Analysis of Gingiva-Derived Mesenchymal Stem Cells Reveals That Fusobacterium nucleatum Triggers Oncogene Expression in the Process of Cell Differentiation. *Front. Cell Dev. Biol.* **2020**, *7*, 359. [CrossRef] [PubMed]

25. Siqueira, J.F.; Rôças, I.N. Uncultivated Phylotypes and Newly Named Species Associated with Primary and Persistent Endodontic Infections. *J. Clin. Microbiol.* **2005**, *43*, 3314–3319. [CrossRef]

26. Endo, M.S.; Ferraz, C.C.R.; Zaia, A.A.; Almeida, J.F.A.; Gomes, B.P.F.A. Quantitative and qualitative analysis of microorganisms in root-filled teeth with persistent infection: Monitoring of the endodontic retreatment. *Eur. J. Dent.* **2013**, *7*, 302–309. [CrossRef]

27. Alghamdi, F.; Shakir, M. The Influence of Enterococcus faecalis as a Dental Root Canal Pathogen on Endodontic Treatment: A Systematic Review. *Cureus* **2020**, *12*, e7257. [CrossRef]

28. Rocas, I.; Siqueirajr, J.; Santos, K. Association of Enterococcus faecalis With Different Forms of Periradicular Diseases. *J. Endod.* **2004**, *30*, 315–320. [CrossRef]

29. Vishwanat, L.; Duong, R.; Takimoto, K.; Phillips, L.; Espitia, C.O.; Diogenes, A.; Ruparel, S.B.; Kolodrubetz, D.; Ruparel, N.B. Effect of Bacterial Biofilm on the Osteogenic Differentiation of Stem Cells of Apical Papilla. *J. Endod.* **2017**, *43*, 916–922. [CrossRef]

30. Elashiry, M.M.; Elashiry, M.; Zeitoun, R.; Elsayed, R.; Tian, F.; Saber, S.E.; Elashry, S.H.; Tay, F.R.; Cutler, C.W. Enterococcus faecalis Induces Differentiation of Immune-Aberrant Dendritic Cells from Murine Bone Marrow-Derived Stem Cells. *Infect. Immun.* **2020**, *88*, e00338-20. [CrossRef]

31. Ozsolak, F.; Milos, P.M. RNA sequencing: Advances, challenges and opportunities. *Nat. Rev. Genet.* **2011**, *12*, 87–98. [CrossRef]

32. Matsushita, K.; Motani, R.; Sakutal, T.; Yamaguchi, N.; Koga, T.; Matsuo, K.; Nagaoka, S.; Abeyama, K.; Maruyama, I.; Torii, M. The Role of Vascular Endothelial Growth Factor in Human Dental Pulp Cells: Induction of Chemotaxis, Proliferation, and Differentiation and Activation of the AP-1-dependent Signaling Pathway. *J. Dent. Res.* **2000**, *79*, 1596–1603. [CrossRef]

33. Roberts-Clark, D.; Smith, A. Angiogenic growth factors in human dentine matrix. *Arch. Oral Biol.* **2000**, *45*, 1013–1016. [CrossRef]

34. Madan, A.K.; Kramer, B. Immunolocalization of fibroblast growth factor-2 (FGF-2) in the developing root and supporting structures of the murine tooth. *J. Mol. Histol.* **2005**, *36*, 171–178. [CrossRef]

35. Cooper, P.R.; Takahashi, Y.; Graham, L.W.; Simon, S.; Imazato, S.; Smith, A.J. Inflammation–regeneration interplay in the dentine–pulp complex. *J. Dent.* **2010**, *38*, 687–697. [CrossRef]

36. Miraoui, H.; Marie, P.J. Fibroblast Growth Factor Receptor Signaling Crosstalk in Skeletogenesis. *Sci. Signal.* **2010**, *3*, re9. [CrossRef]

37. Marie, P.J.; Miraoui, H.; Sévère, N. FGF/FGFR signaling in bone formation: Progress and perspectives. *Growth Factors* **2012**, *30*, 117–123. [CrossRef]

38. Smith, A.J.; Scheven, B.A.; Takahashi, Y.; Ferracane, J.L.; Shelton, R.M.; Cooper, P.R. Dentine as a bioactive extracellular matrix. *Arch. Oral Biol.* **2012**, *57*, 109–121. [CrossRef]

39. Tsutsui, T. Dental Pulp Stem Cells: Advances to Applications. *Stem Cells Cloning Adv. Appl.* **2020**, *13*, 33–42. [CrossRef]

40. Toyono, T.; Nakashima, M.; Kuhara, S.; Akamine, A. Expression of TGF-β Superfamily Receptors in Dental Pulp. *J. Dent. Res.* **1997**, *76*, 1555–1560. [CrossRef]

41. Piattelli, A.; Rubini, C.; Fioroni, M.; Tripodi, D.; Strocchi, R. Transforming Growth Factor-beta 1 (TGF-beta 1) expression in normal healthy pulps and in those with irreversible pulpitis. *Int. Endod. J.* **2004**, *37*, 114–119. [CrossRef] [PubMed]

42. Unterbrink, A.; O'Sullivan, M.; Chen, S.; MacDougall, M. TGFβ-1 Downregulates DMP-1 and DSPP in Odontoblasts. *Connect. Tissue Res.* **2002**, *43*, 354–358. [CrossRef] [PubMed]

43. Takahata, Y.; Takarada, T.; Hinoi, E.; Nakamura, Y.; Fujita, H.; Yoneda, Y. Osteoblastic γ-Aminobutyric Acid, Type B Receptors Negatively Regulate Osteoblastogenesis toward Disturbance of Osteoclastogenesis Mediated by Receptor Activator of Nuclear Factor κB Ligand in Mouse Bone. *J. Biol. Chem.* **2011**, *286*, 32906–32917. [CrossRef] [PubMed]

44. Yu, B.; Zhao, X.; Yang, C.; Crane, J.; Xian, L.; Lu, W.; Wan, M.; Cao, X. Parathyroid hormone induces differentiation of mesenchymal stromal/stem cells by enhancing bone morphogenetic protein signaling. *J. Bone Miner. Res.* **2012**, *27*, 2001–2014. [CrossRef]

45. Chen, X.; Zhi, X.; Wang, J.; Su, J. RANKL signaling in bone marrow mesenchymal stem cells negatively regulates osteoblastic bone formation. *Bone Res.* **2018**, *6*, 34. [CrossRef]

46. Cheng, S.-L.; Lai, C.-F.; Blystone, S.D.; Avioli, L.V. Bone Mineralization and Osteoblast Differentiation Are Negatively Modulated by Integrin αvβ3. *J. Bone Miner. Res.* **2001**, *16*, 277–288. [CrossRef]

47. Barrow, A.D.; Raynal, N.; Andersen, T.L.; Slatter, D.A.; Bihan, D.; Pugh, N.; Cella, M.; Kim, T.; Rho, J.; Negishi-Koga, T.; et al. OSCAR is a collagen receptor that costimulates osteoclastogenesis in DAP12-deficient humans and mice. *J. Clin. Investig.* **2011**, *121*, 3505–3516. [CrossRef]

48. Hooker, R.A.; Chitteti, B.R.; Egan, P.H.; Cheng, Y.H.; Himes, E.R.; Meijome, T.; Srour, E.F.; Fuchs, R.K.; Kacena, M.A. Activated leukocyte cell adhesion molecule (ALCAM or CD166) modulates bone phenotype and hematopoiesis. *J. Musculoskelet. Neuronal Interact.* **2015**, *15*, 83–94.

49. Hariri, H.; Pellicelli, M.; St-Arnaud, R. Nfil3, a target of the NACA transcriptional coregulator, affects osteoblast and osteocyte gene expression differentially. *Bone* **2020**, *141*, 115624. [CrossRef]

50. Camilleri, S.; McDonald, F. Runx2 and dental development. *Eur. J. Oral Sci.* **2006**, *114*, 361–373. [CrossRef]

51. Zhang, D.; Lin, L.; Yang, B.; Meng, Z.; Zhang, B. Knockdown of Tcirg1 inhibits large-osteoclast generation by down-regulating NFATc1 and IP3R2 expression. *PLoS ONE* **2020**, *15*, e0237354. [CrossRef]

52. Zhu, E.D.; Demay, M.B.; Gori, F. Wdr5 Is Essential for Osteoblast Differentiation. *J. Biol. Chem.* **2008**, *283*, 7361–7367. [CrossRef]

53. Govoni, K.E.; Linares, G.R.; Chen, S.-T.; Pourteymoor, S.; Mohan, S. T-box 3 negatively regulates osteoblast differentiation by inhibiting expression of osterix and runx2. *J. Cell. Biochem.* **2009**, *106*, 482–490. [CrossRef]

54. Abrahams, A.; Parker, M.I.; Prince, S. The T-box transcription factor Tbx2: Its role in development and possible implication in cancer. *IUBMB Life* **2009**, *62*, 92–102. [CrossRef]

55. Hassan, M.Q.; Javed, A.; Morasso, M.I.; Karlin, J.; Montecino, M.; van Wijnen, A.J.; Stein, G.S.; Stein, J.L.; Lian, J.B. Dlx3 transcriptional regulation of osteoblast differentiation: Temporal recruitment of Msx2, Dlx3, and Dlx5 homeodomain proteins to chromatin of the osteocalcin gene. *Mol. Cell. Biol.* **2004**, *24*, 9248–9261. [CrossRef]

56. Kieslinger, M.; Folberth, S.; Dobreva, G.; Dorn, T.; Croci, L.; Erben, R.; Consalez, G.G.; Grosschedl, R. EBF2 Regulates Osteoblast-Dependent Differentiation of Osteoclasts. *Dev. Cell* **2005**, *9*, 757–767. [CrossRef]

57. Lin, Y.-C.; Niceta, M.; Muto, V.; Vona, B.; Pagnamenta, A.T.; Maroofian, R.; Beetz, C.; van Duyvenvoorde, H.; Dentici, M.L.; Lauffer, P.; et al. SCUBE3 loss-of-function causes a recognizable recessive developmental disorder due to defective bone morphogenetic protein signaling. *Am. J. Hum. Genet.* **2021**, *108*, 115–133. [CrossRef]

58. Kannan, S.; Ghosh, J.; Dhara, S.K. Osteogenic differentiation potential of porcine bone marrow mesenchymal stem cell subpopulations selected in different basal media. *Biol. Open* **2020**, *9*, bio053280. [CrossRef]

59. Li, J.; Jin, D.; Fu, S.; Mei, G.; Zhou, J.; Lei, L.; Yu, B.; Wang, G. Insulin-like growth factor binding protein-3 modulates osteoblast differentiation via interaction with vitamin D receptor. *Biochem. Biophys. Res. Commun.* **2013**, *436*, 632–637. [CrossRef]

60. Maridas, D.E.; DeMambro, V.E.; Le, P.T.; Nagano, K.; Baron, R.; Mohan, S.; Rosen, C.J. IGFBP-4 regulates adult skeletal growth in a sex-specific manner. *J. Endocrinol.* **2017**, *233*, 131–144. [CrossRef]

61. Rosset, E.M.; Bradshaw, A.D. SPARC/osteonectin in mineralized tissue. *Matrix Biol.* **2016**, *52–54*, 78–87. [CrossRef] [PubMed]

62. Choi, Y.-A.; Lim, J.; Kim, K.M.; Acharya, B.; Cho, J.-Y.; Bae, Y.-C.; Shin, H.-I.; Kim, S.-Y.; Park, E.K. Secretome Analysis of Human BMSCs and Identification of SMOC1 as an Important ECM Protein in Osteoblast Differentiation. *J. Proteome Res.* **2010**, *9*, 2946–2956. [CrossRef] [PubMed]

63. Choi, H.; Kim, T.-H.; Yun, C.-Y.; Kim, J.-W.; Cho, E.-S. Testicular acid phosphatase induces odontoblast differentiation and mineralization. *Cell Tissue Res.* **2016**, *364*, 95–103. [CrossRef] [PubMed]

64. Okamura, H.; Yoshida, K.; Morimoto, H.; Teramachi, J.; Ochiai, K.; Haneji, T.; Yamamoto, A. Role of Protein Phosphatase 2A in Osteoblast Differentiation and Function. *J. Clin. Med.* **2017**, *6*, 23. [CrossRef] [PubMed]

65. National Center for Biotechnology Information (NCBI). Bethesda (MD): National Library of Medicine (US), National Center for Biotechnology Information. 1988. Available online: https://www.ncbi.nlm.nih.gov/gene/259282#bibliog (accessed on 7 May 2022).

66. Viti, F.; Landini, M.; Mezzelani, A.; Petecchia, L.; Milanesi, L.; Scaglione, S. Osteogenic Differentiation of MSC through Calcium Signaling Activation: Transcriptomics and Functional Analysis. *PLoS ONE* **2016**, *11*, e0148173. [CrossRef]

67. Nakamura, T.; Nakamura-Takahashi, A.; Kasahara, M.; Yamaguchi, A.; Azuma, T. Tissue-nonspecific alkaline phosphatase promotes the osteogenic differentiation of osteoprogenitor cells. *Biochem. Biophys. Res. Commun.* **2020**, *524*, 702–709. [CrossRef]

68. Parisuthiman, D.; Mochida, Y.; Duarte, W.R.; Yamauchi, M. Biglycan Modulates Osteoblast Differentiation and Matrix Mineralization. *J. Bone Miner. Res.* **2005**, *20*, 1878–1886. [CrossRef]

69. Matsuura, T.; Duarte, W.R.; Cheng, H.; Uzawa, K.; Yamauchi, M. Differential expression of decorin and biglycan genes during mouse tooth development. *Matrix Biol.* **2001**, *20*, 367–373. [CrossRef]

70. Globus, R.K.; Doty, S.B.; Lull, J.C.; Holmuhamedov, E.; Humphries, M.J.; Damsky, C.H. Fibronectin is a survival factor for differentiated osteoblasts. *J. Cell Sci.* **1998**, *111*, 1385–1393. [CrossRef]

71. Yang, W.; Harris, M.A.; Cui, Y.; Mishina, Y.; Harris, S.E.; Gluhak-Heinrich, J. Bmp2 Is Required for Odontoblast Differentiation and Pulp Vasculogenesis. *J. Dent. Res.* **2012**, *91*, 58–64. [CrossRef]

72. Wang, Z.; Gerstein, M.; Snyder, M. RNA-Seq: A revolutionary tool for transcriptomics. *Nat. Rev. Genet.* **2009**, *10*, 57–63. [CrossRef]

73. Lee, F.J.; Williams, K.B.; Levin, M.; Wolfe, B.E. The Bacterial Metabolite Indole Inhibits Regeneration of the Planarian Flatworm Dugesia japonica. *iScience* **2018**, *10*, 135–148. [CrossRef]

74. Chatzivasileiou, K.; Kriebel, K.; Steinhoff, G.; Kreikemeyer, B.; Lang, H. Do oral bacteria alter the regenerative potential of stem cells? A concise review. *J. Cell. Mol. Med.* **2015**, *19*, 2067–2074. [CrossRef]

75. Kinane, D.F.; Lappin, D.F. Immune Processes in Periodontal Disease: A Review. *Ann. Periodontol.* **2002**, *7*, 62–71. [CrossRef]

76. Tien, B.Y.Q.; Goh, H.M.S.; Chong, K.K.L.; Bhaduri-Tagore, S.; Holec, S.; Dress, R.; Ginhoux, F.; Ingersoll, M.A.; Williams, R.B.H.; Kline, K.A. Enterococcus faecalis Promotes Innate Immune Suppression and Polymicrobial Catheter-Associated Urinary Tract Infection. *Infect. Immun.* **2017**, *85*, e00378-17. [CrossRef]

77. Chong, K.K.L.; Tay, W.H.; Janela, B.; Yong, A.M.H.; Liew, T.H.; Madden, L.; Keogh, D.; Barkham, T.M.S.; Ginhoux, F.; Becker, D.L.; et al. Enterococcus faecalis Modulates Immune Activation and Slows Healing During Wound Infection. *J. Infect. Dis.* **2017**, *216*, 1644–1654. [CrossRef]

78. Tajima, A.; Seki, K.; Shinji, H.; Masuda, S. Inhibition of interleukin-8 production in human endothelial cells by Staphylococcus aureus supernatant. *Clin. Exp. Immunol.* **2006**, *147*, 148–154. [CrossRef]

79. Stekel, D.J. Analysis of host response to bacterial infection using error model based gene expression microarray experiments. *Nucleic Acids Res.* **2005**, *33*, e53. [CrossRef]

80. Christmas, P. Toll-like receptors: Sensors that detect infection. *Nat. Educ.* **2010**, *3*, 85.

81. Sharma, A.K.; Dhasmana, N.; Dubey, N.; Kumar, N.; Gangwal, A.; Gupta, M.; Singh, Y. Bacterial Virulence Factors: Secreted for Survival. *Indian J. Microbiol.* **2017**, *57*, 1–10. [CrossRef]

82. Nada, O.A.; El Backly, R.M. Stem Cells From the Apical Papilla (SCAP) as a Tool for Endogenous Tissue Regeneration. *Front. Bioeng. Biotechnol.* **2018**, *6*, 103. [CrossRef] [PubMed]

83. Chandra, P.K.; Soker, S.; Atala, A. Tissue engineering: Current status and future perspectives. In *Principles of Tissue Engineering*; Elsevier: Amsterdam, The Netherlands, 2020; pp. 1–35.

84. Mendes, R.T.; Nguyen, D.; Stephens, D.; Pamuk, F.; Fernandes, D.; Van Dyke, T.E.; Kantarci, A. Endothelial Cell Response to Fusobacterium nucleatum. *Infect. Immun.* **2016**, *84*, 2141–2148. [CrossRef] [PubMed]

85. Artese, L.; Rubini, C.; Ferero, G.; Fioroni, M.; Santinelli, A.; Piatelli, A. Vascular Endothelial Growth Factor (VEGF) Expression in Healthy and Inflamed Human Dental Pulps. *J. Endod.* **2002**, *28*, 20–23. [CrossRef] [PubMed]

86. Lambrichts, I.; Driesen, R.B.; Dillen, Y.; Gervois, P.; Ratajczak, J.; Vangansewinkel, T.; Wolfs, E.; Bronckaers, A.; Hilkens, P. Dental Pulp Stem Cells: Their Potential in Reinnervation and Angiogenesis by Using Scaffolds. *J. Endod.* **2017**, *43*, S12–S16. [CrossRef] [PubMed]

87. Geiger, T.L.; Abt, M.C.; Gasteiger, G.; Firth, M.A.; O'Connor, M.H.; Geary, C.D.; O'Sullivan, T.E.; van den Brink, M.R.; Pamer, E.G.; Hanash, A.M.; et al. Nfil3 is crucial for development of innate lymphoid cells and host protection against intestinal pathogens. *J. Exp. Med.* **2014**, *211*, 1723–1731. [CrossRef]

88. Kawata, K.; Narita, K.; Washio, A.; Kitamura, C.; Nishihara, T.; Kubota, S.; Takeda, S. Odontoblast differentiation is regulated by an interplay between primary cilia and the canonical Wnt pathway. *Bone* **2021**, *150*, 116001. [CrossRef]

89. Todd, W.M.; Kafrawy, A.H.; Newton, C.W.; Brown, C.E. Immunohistochemical study of gamma-aminobutyric acid and bombesin/gastrin releasing peptide in human dental pulp. *J. Endod.* **1997**, *23*, 152–157. [CrossRef]

90. Telles, P.D.S.; Hanks, C.T.; Machado, M.A.A.M.; Nör, J.E. Lipoteichoic Acid Up-regulates VEGF Expression in Macrophages and Pulp Cells. *J. Dent. Res.* **2003**, *82*, 466–470. [CrossRef]

91. Li, S.; Kong, H.; Yao, N.; Yu, Q.; Wang, P.; Lin, Y.; Wang, J.; Kuang, R.; Zhao, X.; Xu, J.; et al. The role of runt-related transcription factor 2 (Runx2) in the late stage of odontoblast differentiation and dentin formation. *Biochem. Biophys. Res. Commun.* **2011**, *410*, 698–704. [CrossRef]

92. Franceschi, R.T.; Xiao, G.; Jiang, D.; Gopalakrishnan, R.; Yang, S.; Reith, E. Multiple signaling pathways converge on the Cbfa1/Runx2 transcription factor to regulate osteoblast differentiation. *Connect. Tissue Res.* **2003**, *44* (Suppl. S1), 109–116. [CrossRef]

93. Rutkovskiy, A.; Stensløkken, K.-O.; Vaage, I.J. Osteoblast Differentiation at a Glance. *Med. Sci. Monit. Basic Res.* **2016**, *22*, 95–106. [CrossRef]

94. Tu, C.-F.; Tsao, K.-C.; Lee, S.-J.; Yang, R.-B. SCUBE3 (signal peptide-CUB-EGF domain-containing protein 3) modulates fibroblast growth factor signaling during fast muscle development. *J. Biol. Chem.* **2014**, *289*, 18928–18942. [CrossRef]

95. Makareeva, E.; Leikin, S. Collagen Structure, Folding and Function. In *Osteogenesis Imperfecta*; Elsevier: Amsterdam, The Netherlands, 2014; pp. 71–84.

96. Aizawa, C.; Saito, K.; Ohshima, H. Regulation of IGF-I by IGFBP3 and IGFBP5 during odontoblast differentiation in mice. *J. Oral Biosci.* **2019**, *61*, 157–162. [CrossRef]

97. Zymovets, V.; Razghonova, Y.; Rakhimova, O.; Aripaka, K.; Manoharan, L.; Kelk, P.; Landström, M.; Romani Vestman, N. Combined Transcriptomic and Protein Array Cytokine Profiling of Human Stem Cells from Dental Apical Papilla Modulated by Oral Bacteria. *Int. J. Mol. Sci.* **2022**, *23*, 5098. [CrossRef]

98. Hargreaves, K.M.; Diogenes, A.; Teixeira, F.B. Treatment Options: Biological Basis of Regenerative Endodontic Procedures. *J. Endod.* **2013**, *39*, S30–S43. [CrossRef]

99. Wells, J.M.; Rossi, O.; Meijerink, M.; van Baarlen, P. Epithelial crosstalk at the microbiota–mucosal interface. *Proc. Natl. Acad. Sci. USA* **2011**, *108*, 4607–4614. [CrossRef]

100. Lebeer, S.; Vanderleyden, J.; De Keersmaecker, S.C.J. Host interactions of probiotic bacterial surface molecules: Comparison with commensals and pathogens. *Nat. Rev. Microbiol.* **2010**, *8*, 171–184. [CrossRef]

101. Tom-Kun Yamagishi, V.; Torneck, C.D.; Friedman, S.; Huang, G.T.-J.; Glogauer, M. Blockade of TLR2 Inhibits Porphyromonas gingivalis Suppression of Mineralized Matrix Formation by Human Dental Pulp Stem Cells. *J. Endod.* **2011**, *37*, 812–818. [CrossRef]
102. Abe, S.; Imaizumi, M.; Mikami, Y.; Wada, Y.; Tsuchiya, S.; Irie, S.; Suzuki, S.; Satomura, K.; Ishihara, K.; Honda, M.J. Oral bacterial extracts facilitate early osteogenic/dentinogenic differentiation in human dental pulp–derived cells. *Oral Surg. Oral Med. Oral Pathol. Oral Radiol. Endodontology* **2010**, *109*, 149–154. [CrossRef]
103. Ricucci, D.; Siqueira, J.F. Biofilms and Apical Periodontitis: Study of Prevalence and Association with Clinical and Histopathologic Findings. *J. Endod.* **2010**, *36*, 1277–1288. [CrossRef]
104. Fransson, H.; Petersson, K.; Davies, J.R. Effects of bacterial products on the activity of odontoblast-like cells and their formation of type 1 collagen. *Int. Endod. J.* **2014**, *47*, 397–404. [CrossRef] [PubMed]
105. Rooks, M.G.; Garrett, W.S. Gut microbiota, metabolites and host immunity. *Nat. Rev. Immunol.* **2016**, *16*, 341–352. [CrossRef] [PubMed]
106. Amarasekara, D.S.; Kim, S.; Rho, J. Regulation of Osteoblast Differentiation by Cytokine Networks. *Int. J. Mol. Sci.* **2021**, *22*, 2851. [CrossRef] [PubMed]
107. Huang, W. Signaling and transcriptional regulation in osteoblast commitment and differentiation. *Front. Biosci.* **2007**, *12*, 3068. [CrossRef] [PubMed]
108. Kolar, M.K.; Itte, V.N.; Kingham, P.J.; Novikov, L.N.; Wiberg, M.; Kelk, P. The neurotrophic effects of different human dental mesenchymal stem cells. *Sci. Rep.* **2017**, *7*, 12605. [CrossRef]
109. Schroeder, A.; Mueller, O.; Stocker, S.; Salowsky, R.; Leiber, M.; Gassmann, M.; Lightfoot, S.; Menzel, W.; Granzow, M.; Ragg, T. The RIN: An RNA integrity number for assigning integrity values to RNA measurements. *BMC Mol. Biol.* **2006**, *7*, 3. [CrossRef]
110. Andrews, S. FastQC: A Quality Control Tool for High Throughput Sequence Data. 2010. Available online: https://www.bioinformatics.babraham.ac.uk/projects/fastqc/ (accessed on 5 March 2021).
111. Martin, M. Cutadapt removes adapter sequences from high-throughput sequencing reads. *EMBnet.journal* **2011**, *17*, 10. [CrossRef]
112. Kim, D.; Paggi, J.M.; Park, C.; Bennett, C.; Salzberg, S.L. Graph-based genome alignment and genotyping with HISAT2 and HISAT-genotype. *Nat. Biotechnol.* **2019**, *37*, 907–915. [CrossRef]
113. Wang, L.; Wang, S.; Li, W. RSeQC: Quality control of RNA-seq experiments. *Bioinformatics* **2012**, *28*, 2184–2185. [CrossRef]
114. Gadepalli, V.S.; Ozer, H.G.; Yilmaz, A.S.; Pietrzak, M.; Webb, A. BISR-RNAseq: An efficient and scalable RNAseq analysis workflow with interactive report generation. *BMC Bioinform.* **2019**, *20*, 670. [CrossRef]
115. Love, M.I.; Huber, W.; Anders, S. Moderated estimation of fold change and dispersion for RNA-seq data with DESeq2. *Genome Biol.* **2014**, *15*, 550. [CrossRef]
116. Ritchie, M.E.; Phipson, B.; Wu, D.; Hu, Y.; Law, C.W.; Shi, W.; Smyth, G.K. limma powers differential expression analyses for RNA-sequencing and microarray studies. *Nucleic Acids Res.* **2015**, *43*, e47. [CrossRef]
117. Parker, H.S.; Leek, J.T.; Favorov, A.V.; Considine, M.; Xia, X.; Chavan, S.; Chung, C.H.; Fertig, E.J. Preserving biological heterogeneity with a permuted surrogate variable analysis for genomics batch correction. *Bioinformatics* **2014**, *30*, 2757–2763. [CrossRef]
118. Huang, D.W.; Sherman, B.T.; Lempicki, R.A. Bioinformatics enrichment tools: Paths toward the comprehensive functional analysis of large gene lists. *Nucleic Acids Res.* **2009**, *37*, 1–13. [CrossRef]
119. Huang, D.W.; Sherman, B.T.; Lempicki, R.A. Systematic and integrative analysis of large gene lists using DAVID bioinformatics resources. *Nat. Protoc.* **2009**, *4*, 44–57. [CrossRef]
120. Mi, H.; Muruganujan, A.; Ebert, D.; Huang, X.; Thomas, P.D. PANTHER version 14: More genomes, a new PANTHER GO-slim and improvements in enrichment analysis tools. *Nucleic Acids Res.* **2019**, *47*, D419–D426. [CrossRef]
121. Patterson, D.E.; Cramer, R.D.; Ferguson, A.M.; Clark, R.D.; Weinberger, L.E. Neighborhood Behavior: A Useful Concept for Validation of "Molecular Diversity" Descriptors. *J. Med. Chem.* **1996**, *39*, 3049–3059. [CrossRef]

 International Journal of **Molecular Sciences**

Article

Electrospun Azithromycin-Laden Gelatin Methacryloyl Fibers for Endodontic Infection Control

Afzan A. Ayoub [1,2], Abdel H. Mahmoud [1], Juliana S. Ribeiro [1,3], Arwa Daghrery [1,4], Jinping Xu [1], J. Christopher Fenno [5], Anna Schwendeman [6], Hajime Sasaki [1], Renan Dal-Fabbro [1] and Marco C. Bottino [1,7,*]

[1] Department of Cariology, Restorative Sciences, and Endodontics, School of Dentistry, University of Michigan, Ann Arbor, MI 48103, USA

[2] Comprehensive Care Centre of Studies, Faculty of Dentistry, Universiti Teknologi MARA (UiTM), UiTM Campus Sg Buloh, Jalan Hospital, Sungai Buloh 47000, Malaysia

[3] Department of Clinical Dentistry (Endodontics), School of Dentistry, Federal University of Santa Catarina, Florianopolis 88040-900, Brazil

[4] Department of Restorative Dental Sciences, School of Dentistry, Jazan University, Jazan 45142, Saudi Arabia

[5] Department of Biologic and Materials Sciences & Prosthodontics, University of Michigan School of Dentistry, Ann Arbor, MI 48109, USA

[6] Department of Pharmaceutical Sciences, College of Pharmacy and Biointerfaces Institute, University of Michigan, Ann Arbor, MI 48103, USA

[7] Department of Biomedical Engineering, College of Engineering, University of Michigan, Ann Arbor, MI 48103, USA

* Correspondence: mbottino@umich.edu; Tel.: +1-734-763-2206; Fax: +1-734-936-1597

Citation: Ayoub, A.A.; Mahmoud, A.H.; Ribeiro, J.S.; Daghrery, A.; Xu, J.; Fenno, J.C.; Schwendeman, A.; Sasaki, H.; Dal-Fabbro, R.; Bottino, M.C. Electrospun Azithromycin-Laden Gelatin Methacryloyl Fibers for Endodontic Infection Control. *Int. J. Mol. Sci.* **2022**, *23*, 13761. https://doi.org/10.3390/ijms232213761

Academic Editor: Matthias Widbiller

Received: 11 October 2022
Accepted: 3 November 2022
Published: 9 November 2022

Publisher's Note: MDPI stays neutral with regard to jurisdictional claims in published maps and institutional affiliations.

Abstract: This study was aimed at engineering photocrosslinkable azithromycin (AZ)-laden gelatin methacryloyl fibers via electrospinning to serve as a localized and biodegradable drug delivery system for endodontic infection control. AZ at three distinct amounts was mixed with solubilized gelatin methacryloyl and the photoinitiator to obtain the following fibers: GelMA+5%AZ, GelMA+10%AZ, and GelMA+15%AZ. Fiber morphology, diameter, AZ incorporation, mechanical properties, degradation profile, and antimicrobial action against *Aggregatibacter actinomycetemcomitans* and *Actinomyces naeslundii* were also studied. In vitro compatibility with human-derived dental pulp stem cells and inflammatory response in vivo using a subcutaneous rat model were also determined. A bead-free fibrous microstructure with interconnected pores was observed for all groups. GelMA and GelMA+10%AZ had the highest fiber diameter means. The tensile strength of the GelMA-based fibers was reduced upon AZ addition. A similar pattern was observed for the degradation profile in vitro. GelMA+15%AZ fibers led to the highest bacterial inhibition. The presence of AZ, regardless of the concentration, did not pose significant toxicity. In vivo findings indicated higher blood vessel formation, mild inflammation, and mature and thick well-oriented collagen fibers interweaving with the engineered fibers. Altogether, AZ-laden photocrosslinkable GelMA fibers had adequate mechanical and degradation properties, with 15%AZ displaying significant antimicrobial activity without compromising biocompatibility.

Keywords: antimicrobial; electrospinning; dentistry; drug delivery; endodontics; fibers

1. Introduction

Dental trauma and caries are the most prevalent causes of pulp necrosis, leading to tooth loss if left untreated [1]. It is well recognized that pulpal necrosis in permanent teeth with incomplete apex development presents a severe danger of fracture and therefore decreases the long-term survival [2,3]. To diminish this issue, the indication of the once termed revascularization technique to treat necrotic immature teeth has been documented during the last decade [4]. Briefly, this two-step regenerative-based technique first includes disinfecting the root canal system with irrigation solutions and minimal instrumentation, then placing an intracanal medication containing antibiotics to eliminate and restrict bacterial growth to improve the regenerative outcomes [5–8].

Although regenerative endodontics constitutes a fairly modern clinical domain and has matured over the last two decades, the topic still presents unstudied potential therapeutics for immature permanent teeth with pulp necrosis [9]. For instance, recent advances in electrospinning technology have made the synthesis of natural and/or synthetic polymer scaffolds and drug delivery systems feasible, with significant clinical potential to manage endodontic infection prior to regenerative procedures [7,10,11].

Gelatin methacryloyl (GelMA) is a semi-synthetic biocompatible, degradable, and tunable hydrogel that mimics many essential characteristics of the native extracellular matrix (ECM). Furthermore, GelMA contains cell binding and matrix metalloproteinase (MMP)-sensitive degradation sites and has shown great potential in drug delivery and regenerative medicine [12,13]. From a chemical standpoint, GelMA is synthesized by replacing the amine-containing side groups of gelatins with methacrylamide and methacrylate groups [14,15]. Importantly, due to its hydrophilicity, GelMA can be incorporated with distinct compounds, including but not limited to antibiotics and other clinically relevant therapeutic agents (e.g., dexamethasone-loaded clay nanotubes) [14–16]. Notably, when combined with a photoinitiator, customarily 2-Hydroxy-1-(4-(2-hydroxyethoxy)phenyl)-2-methylpropan-1-one (Irgacure^{TM} 2959) and ultraviolet (UV) light with appropriate intensity and wavelength, the GelMA precursor undergoes radical polymerization to generate a covalently crosslinked fibers network [12]. Regrettably, the use of UV light has been proven to evoke several biological deleterious effects, such as DNA damage by pyrimidine dimerization or the production of reactive oxygen species (ROS), leading to oxidative damage to DNA, accelerated tissue aging, and immunosuppression [17]. Nonetheless, it has been recently established that when using a proper solvent, GelMA can be employed into the fabrication of photocrosslinkable (UV) nanofibers [15]; however, aside from the stated phototoxic effects, UV light penetrates tissues, hydrogels, and fibers only to a limited extent, posing a further barrier to its medical uses [12]. Collectively, these issues have prompted scientists to examine the use of higher-wavelength light, such as visible light, combined with a photoinitiator that can be triggered at those wavelengths [12]. In this way, lithium phenyl(2,4,6-trimethylbenzoyl)phosphinate (LAP), a colorless, single-component initiation system with strong thermal stability and good solubility, which can be triggered by conventional dental curing light, has garnered substantial notice [18].

Azithromycin (AZ) is a semi-synthetic, acid-stable second-generation macrolide antibiotic with a 15-membered azlactone ring, demonstrating a broad spectrum of bacteriostatic action and enhanced pharmacokinetics, and is clinically proven to be effective against Gram-positive, Gram-negative, and atypical infections [19]. Interestingly, in a recent study assessing the effect of AZ on pre-existing experimental periapical lesions, the antibiotic-modulated macrophage polarization from pro-inflammatory (M1) macrophages to pro-resolving (M2) macrophages led to the resolution of periapical inflammation through its immunomodulatory effect [20].

In the present study, we detail for the first time the development of visible-light photocrosslinkable GelMA electrospun fibers loaded with azithromycin as a biodegradable and biocompatible localized drug delivery system for endodontic infection control. Herein, we hypothesized that a combination of LAP-triggered photocrosslinked GelMA fibers with optimal azithromycin content would lead to significant antimicrobial action against endodontic pathogens, with acceptable toxicity to dental pulp stem cells and minimal inflammatory effects upon subcutaneous implantation of the engineered AZ-laden GelMA fibers in rats.

2. Results

2.1. Morphological and Chemical Characteristics

A bead-free fibrous network with interconnected pores was observed for all groups (Figure 1). The diameters of GelMA+5%AZ and GelMA+15%AZ electrospun fibers ranged between 0.66 ± 0.18 µm and 0.83 ± 0.34 µm and were considerably smaller ($p < 0.05$) than pure GelMA fibers (1.08 ± 0.40 µm). At the same time, the fiber diameter of the GelMA+10%AZ fibers (1.01 ± 0.34 µm) did not differ from the control group ($p < 0.05$).

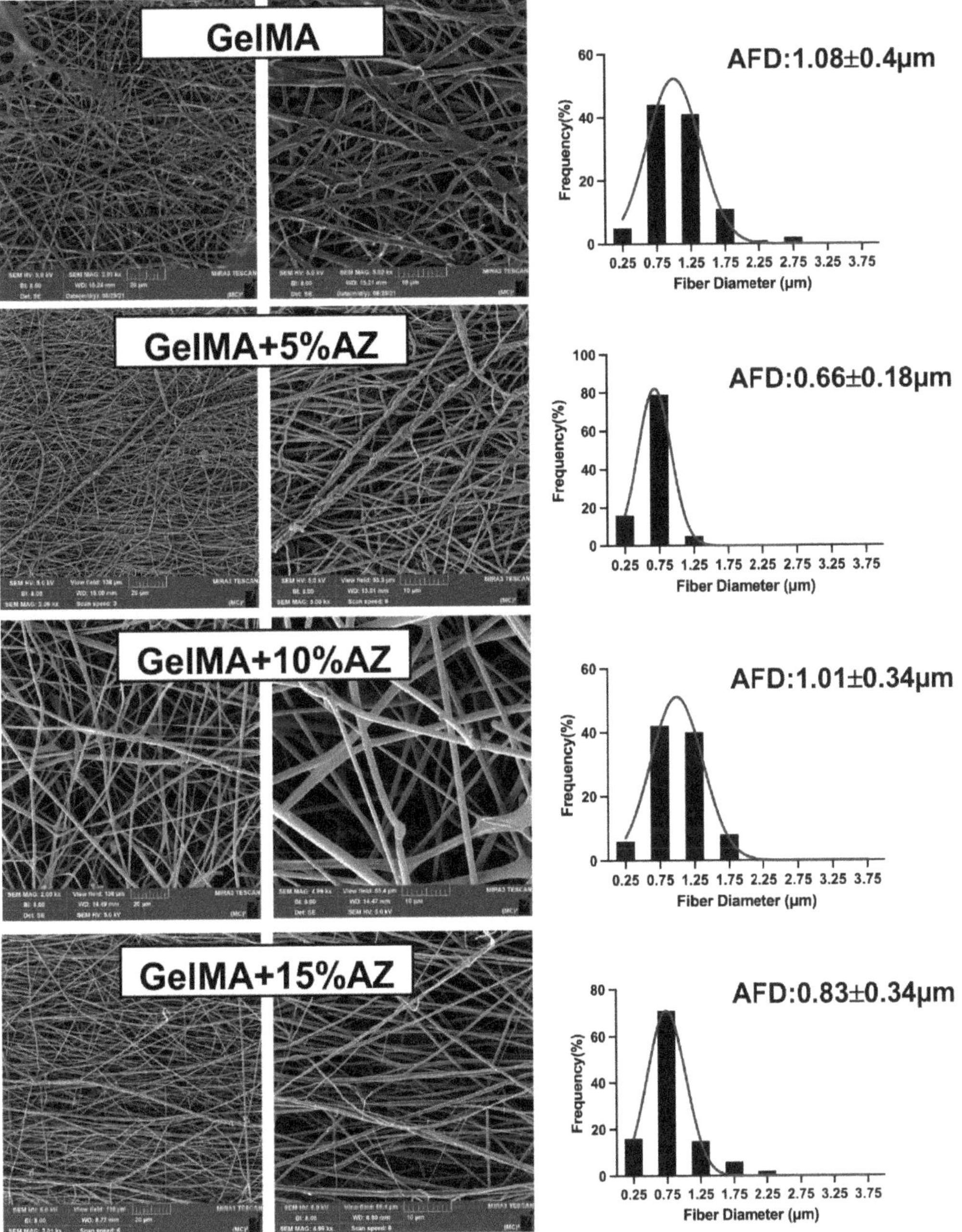

Figure 1. (**Left**) Representative scanning electron microscopy (SEM) images of the electrospun azithromycin-laden gelatin methacryloyl and the pure (antibiotic-free) GelMA fibers. (**Right**) Histograms showing the fiber diameter frequency and average fiber diameter (AFD) with standard deviation.

The FTIR analysis and interpretations of peaks for pure GelMA and different AZ concentrations before and after photocrosslinking are presented in Figure 2. The characteristic peaks of AZ at 3493.03, 1719.9, 1249.85, and 1082.15 cm^{-1} correspond to the hydroxyl group, H-bonded OH, ketone carbonyl compound, aromatic ethers, aryl -O stretch, and organic sulfates [21]. All four similar characteristic peaks were present in uncrosslinked fibers; 3285.51–3307.80, 1645.29–1650.39, 1242.73–1243.73, and 1080.57–10.80.92 cm^{-1}. The spectrum of AZ has characteristic peaks at 2907, 1720, 1363, 1190, 1108 cm^{-1}, and about 1350 corresponded, respectively, to C-H (methyl group) stretching, C=O stretching, C=C bending, C-O-C asymmetrical stretching, C-O-C symmetrical stretching, and C-N stretching.

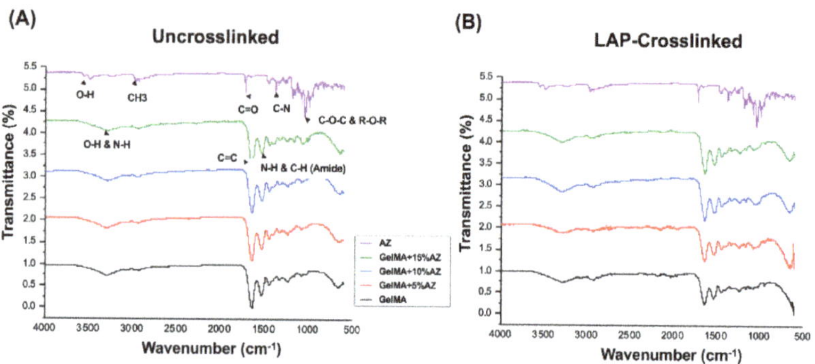

Figure 2. Fourier transform infrared spectroscopy (FTIR) spectra of (**A**) uncrosslinked and (**B**) crosslinked GelMA-based fibers.

Regarding pure GelMA fibers, peaks at 3288 cm^{-1}, 1640 cm^{-1}, 1650 cm^{-1}, and 1530 cm^{-1} correspond to O-H and N-H stretching, C=C stretching of the methacrylate group, C=O stretching of the amide group, and N-H bending coupled to C-H stretching of the amide group, respectively. Peaks for AZ between 100–1300 cm^{-1} are present for the GelMA+AZ groups, particularly for crosslinked groups. Conversely, azithromycin's main peak at 1720 cm^{-1} is unclear in the GelMA+AZ groups, possibly because of the chemical reaction between the methyl group of AZ (CH$_3$) and the methacrylate group of GelMA (C=C).

2.2. Mechanical Properties and Degradation Profile

The tensile strength of the electrospun samples (in MPa) decreased, along with the ascending concentration of the added AZ as follows: 1.17 ± 0.35 (GelMA), 0.76 ± 0.24 (GelMA+5%AZ), 0.66 ± 0.27 (GelMA+10%AZ), and 0.53 ± 0.28 (GelMA+15%AZ). The tensile strength of the GelMA fibers was significantly ($p < 0.05$) higher than GelMA+10%AZ and GelMA+15%AZ. Meanwhile, all the AZ-laden GelMA fiber groups did not show significant differences in strength between them (Figure 3A). Young's modulus was significantly reduced by adding AZ to the GelMA fibers. However, no differences were found comparing the three distinct AZ concentrations (Figure 3B). When the elongation at break was evaluated, pure GelMA and GelMA+5%AZ exhibited similar values and were significantly ($p < 0.05$) higher than GelMA+10%AZ or GelMA+15%AZ (Figure 3C). The degradation profile for the synthesized GelMA-based fibers is shown in Figure 3D. All the groups had a higher mass loss in the first three days of PBS incubation. The greater degradability values were associated with groups loaded with higher AZ concentrations ($p < 0.0001$). Pure GelMA lost ~5% of mass by day 3, and the GelMA+15%AZ fibers lost about 33% of the mass, followed by 25% and 12% mass loss for 10% and 5% AZ-laden GelMA fibers, respectively. It is important to note that all groups remained stable after reaching a plateau from day 3 until day 14 with minimal mass loss variation.

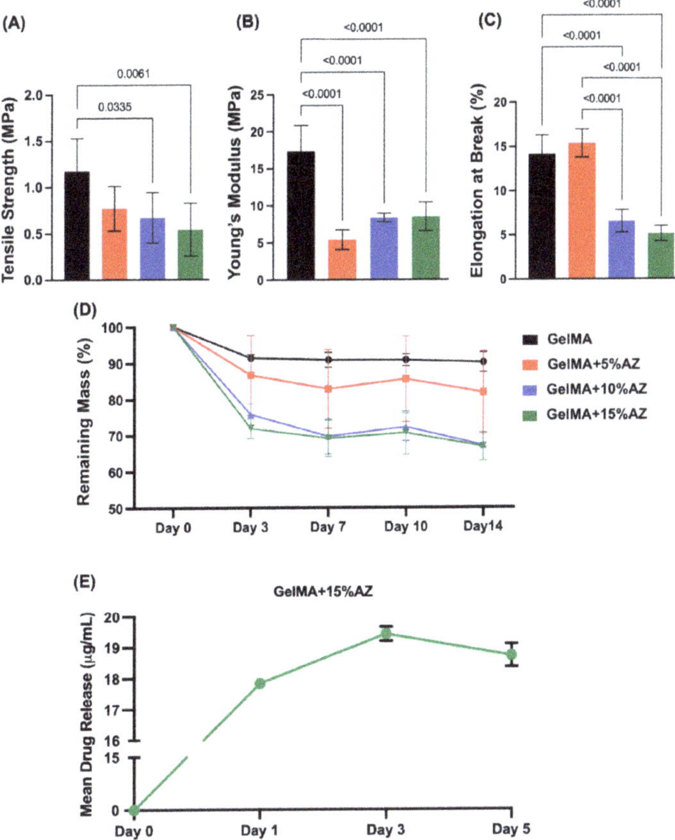

Figure 3. Mechanical analyses of the engineered GelMA-based electrospun fibers laden with 3 distinct amounts of azithromycin. (**A**) Tensile strength in MPa; (**B**) Young's modulus in MPa; (**C**) Elongation at break in %; and (**D**) Degradation profile over 14 days in %. (**E**) Release profile of AZ from GelMA+15%AZ fibers over 5 days. Data are shown as mean and standard deviation.

2.3. Drug Release

The in vitro release profile of azithromycin from the GelMA+15%AZ fibers is shown in Figure 3E. A burst release of AZ was observed in the first 24 h, reaching 17.8 ± 0.04 µg/mL. From day 1 to day 3, an 8% increase in the antibiotic release was noticed, attaining 19.4 ± 0.23 µg/mL. From day 3 to day 5, there was a slight decrease in the AZ release compared to day 3, achieving 18.72 ± 0.36 µg/mL.

2.4. Cytocompatibility

Determining the in vitro cytotoxicity of our novel GelMA-based fibers is essential to supporting its potential application as a localized drug delivery system for endodontic infection control. Here, we investigated the cytocompatibility of the fabricated AZ-laden GelMA-based fibers with hDPSCs. The results are shown in Figure 4A as the viability percentage of each group using the control group (cells seeded but without any aliquots) as 100%. Compared with the pure GelMA group, the number of surviving cells decreased after exposure to the aliquots collected on day 1 for AZ concentrations at 5 and 10%. The same pattern was observed for aliquots collected on day 3, with the 15% AZ group showing the highest decrease ($p = 0.0002$), which was significantly distinct from 10% AZ ($p = 0.0004$). For aliquots collected on day 7, a viability plateau of around 70% was reached for all tested groups, without differences. For aliquots collected on day 14, the AZ-laden fibers demonstrated slightly higher viability than pure GelMA, with the GelMA+15%AZ group displaying significant ($p = 0.0298$) differences.

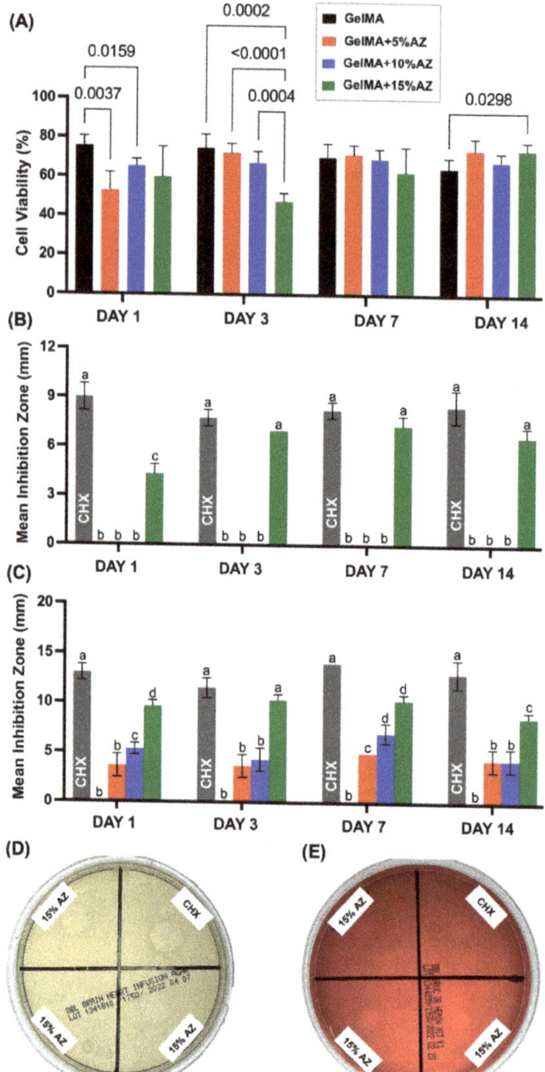

Figure 4. (**A**) Graphic representation of mean and standard deviation of hDPSCs cell viability (%) determined after 72 h using aliquots collected at 1, 3, 7, and 14 d. The percentage of cell viability was normalized by the mean absorbance of hDPSCs cultured on the plate on day 1 (100%). Absorbance was measured at 490 nm. Antimicrobial potential against *Actinomyces naeslundii* (**B**), and *Aggregatibacter actinomycetemcomitans* (**C**), evaluated through the Kirby-Bauer diffusion test (**D**) for *Aa* and (**E**) for *An* using 0.12% CHX as a positive control. Different lowercase letters represent statistical differences between groups compared on the same day ($p < 0.05$).

2.5. Antimicrobial Assessment

The electrospun AZ-laden GelMA-based fibers' antimicrobial effect was assessed through agar diffusion against *Actinomyces naeslundii* and *Aggregatibacter actinomycetemcomitans* performed by an indirect contact assay (i.e., eluates collected over 14 days from sample incubation in PBS). No inhibition was observed in the negative control (PBS). A 12 mm mean inhibition zone was observed for the positive control (0.12% chlorhexidine, CHX) for all time points. Agar diffusion analyses confirmed the antimicrobial properties of the AZ-laden GelMA fibers (Figure 4B–E). Only the highest AZ concentration (15%) had an antimicrobial effect against *An* on all aliquots collected from days 1 to 14 (Figure 4B). For the *Aa*, GelMA fibers containing 15% AZ presented a higher inhibition (~10 mm) than 10% and 5% AZ concentrations for aliquots collected on days 1 and 3. Notably, GelMA+15%AZ aliquot on day 3 led to similar inhibition when compared to the positive control (Figure 4C). For aliquots collected on day 7, 10% and 15% AZ concentrations demonstrated statistically higher inhibition than 5%; and for 14-day aliquots, antibiotic activity was still present, since the 15% AZ concentration demonstrated statistically higher inhibition than the other two AZ-laden groups (Figure 4C).

2.6. In Vivo Biocompatibility

The biocompatibility (H&E) and collagen fiber production (Picrosirius Red) results of AZ-laden GelMA fibers subcutaneously implanted in rats are shown in Figures 5 and 6. After 14 days, the host tissue response at the interface between the implanted material and subcutaneous tissue had an exacerbated inflammatory reaction for pure GelMA fibers when compared to Bio-Gide® (control) and GelMA+15%AZ fibers mainly consisting of polymorphonuclear over the mononuclear inflammatory cells. Blood vessel formation was observed for the three groups, specifically for GelMA+15%AZ with larger diameters. At 28 days post-implantation, the inflammation for all 3 groups slightly decreased over time. Bio-Gide® presented better results, with minimal mononuclear inflammatory infiltrate and initial material resorption. Pure GelMA evoked mild inflammation compared to the control, along with the presence of fatty infiltrates. The GelMA+15%AZ showed slightly decreased inflammation, predominantly consisted of mononuclear inflammatory cells, showing higher vascularity formation and well-organized collagen fibrils' layers when compared to pure GelMA. There were no apparent differences in the level of degradation or fragmentation of the GelMA-based fibrous mats.

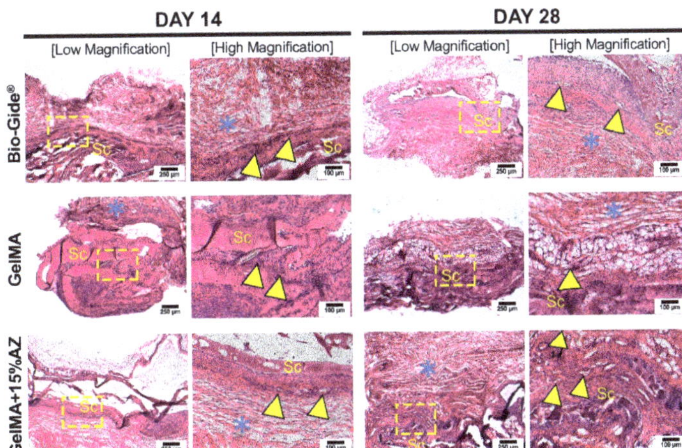

Figure 5. H&E-stained slices for implanted materials in rats' subcutaneous tissue at 14 and 28 d. 4X magnification (scale bar = 250 µm) and 10X magnification (scale bar = 100 µm). A dashed yellow square delimits the area selected for higher magnification. Scaffold: Sc, Blood vessels: yellow arrowhead, Fibroblasts with collagen fibrils: blue asterisk. Note at 14 d, the enhanced inflammatory reaction evoked by pure GelMA, compared to the other two groups, and higher blood vessel formation and well-oriented collagen fibers from the group GelMA+15%AZ at 28 d.

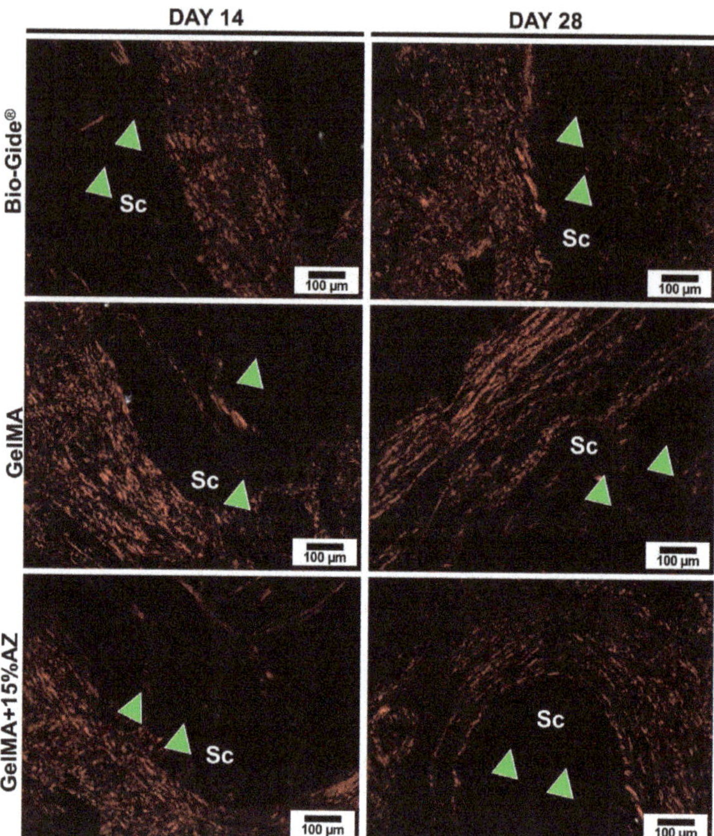

Figure 6. Representative Picrosirius Red stained slices for the implanted scaffolds in rat's subcutaneous tissue at 14 days and 28 days. Original magnification: 10× (scale bar = 100 µm). Scaffold: Sc, Red fibers considered as mature and thick interspersing the scaffold: green arrowhead. On day 28, note the presence of mature collagen fibers interspersing the area initially occupied by the scaffold.

3. Discussion

Among the primary aims of endodontic treatment is to eradicate microbial infection, which is usually a multispecies infection involving aerobic and anaerobic bacteria, with obligate anaerobes being the predominant ones [22]. For this investigation, azithromycin was chosen based on a recently disclosed potential for eliminating pre-existing experimental periapical inflammation [20]. To the best of our knowledge, this is the first report on electrospun LAP-photocrosslinked GelMA-based fibers with and without the incorporation of azithromycin. Our findings indicate that incorporation of AZ into the GelMA matrix does not compromise biocompatibility and yet supports the development of antibiotic-laden GelMA-based fibers as a biodegradable and localized drug delivery method for endodontic therapy.

The GelMA-based fibers produced by electrospinning were observed to be homogeneous and smooth. The addition of AZ was observed to decrease the diameter for the 5% and 15% concentrations. This may be attributed to an enhancement of the electrical conductivity of the electrospinning solution. This higher conductivity suppresses varicose instability and enhances whipping instability, consequently forming finer fibers [23]. The novel GelMA-based fibers were also characterized based on their physicochemical characteristics. First, chemical analysis reveals that the incorporation of AZ did not affect the peaks related to vibrational bands of GelMA in the FTIR spectra, which supports the chemical stability of the fibers. Meanwhile, the AZ's main peak at 1720 cm^{-1} is unclear in the GelMA+AZ groups, probably due to (i) the overlapping peaks of GelMA and AZ, and (ii) the chemical reaction between the methyl group of AZ (CH3) and the methacrylate group of GelMA; consequently, AZ-related peaks were not easy to identify. Nevertheless, the effective incorporation of AZ into GelMA nanofibers was confirmed based on the antimicrobial findings. Next, to demonstrate that the GelMA-based fibers can be successfully utilized as an effective and highly tunable platform for delivering antimicrobial drugs, an in vitro degradation assay was performed. Hydrogel degradation plays an essential role in the controlled drug release, avoiding rapid drug clearance, and leading to a long-term drug release [24]. Based on the presented results, it can be concluded that the degradation of AZ-laden fibers was directly proportional to the concentration of the added antibiotic. However, even with the highest concentration, after significant degradation in the first 24 h (around 30% weight loss), the fibers remained stable until day 14 (about 70% of the original weight). Currently, it is well-established that the use of GelMA at a concentration equal to or less than 15% (*w/v*) has a satisfactory degradation rate for tissue regeneration and is highly biocompatible [25]. In addition, the use of LAP has shown a direct influence on the rate of degradation, as well as on drug release [26]. Therefore, the ability to crosslink GelMA-based fibers ensured greater control of the degradation rate without compromising biocompatibility.

From a mechanical perspective, the presence of AZ led to a reduction in tensile strength, which was inversely proportional to the AZ concentration. Despite Young's modulus being significantly reduced by AZ addition into the GelMA fibers, no differences were found comparing the three distinct concentrations. Notably, since the equilibrium between hydrophobicity and hydrophilicity is essential to producing physical-mechanically stable matrices, we hypothesized that the hydrophobicity of AZ led to the observed reduction in tensile strength, which based on the foreseen clinical applications as drug delivery systems (e.g., intracanal or in deep caries lesions) should not compromise the in vivo performance of the material.

Localized drug delivery systems represent a promising therapeutic strategy for applications in treating oral infections. An indispensable attribute of the present study was to validate the antimicrobial action of the AZ-laden GelMA fibers against endodontic pathogens associated with primary endodontic infection and persistent infection due to failed endodontic treatment [27–29]. In this case, our indirect agar diffusion antimicrobial results revealed that GelMA+15%AZ led to a substantial and prolonged antimicrobial action toward *An* and *Aa* due to the elevated antibiotic concentration, thus proving the release of AZ from the fibers from the first 24 h, keeping the antimicrobial action until day 14. Notably, based on our drug release investigation, the same pattern was observed; from the first day of incubation, the AZ released by the GelMA+15%AZ successfully attained higher values than 1 µg/mL, which represents the amount at which $\geq$90% of the *Aa* population is inhibited (MIC90) [30], and higher than the 0.5 µg/mL MIC for *An* [31]. Due to its dual-base structure, AZ is actively absorbed by various cells and acts by binding to the 23S rRNA of the bacterial 50S ribosomal subunit inhibiting the transpeptidation/translocation step of protein synthesis, resulting in the control of various bacterial infections. Specifically, AZ is known to accumulate inside neutrophils, enhancing phagocytic killing of the *Aa* [32].

In addition to the antimicrobial activity, assessing the in vitro cytotoxicity of the developed GelMA-based fibers is key to confirming their potential application and safety as an auxiliary drug delivery system for endodontic infection management. In the present study, using human dental pulp stem cells, a minor cytotoxic effect was observed for aliquots collected at days 1 and 3, likely due to the more significant release of AZ on the first 3 days. From day 7 and beyond, the incorporation of AZ did not compromise cell viability. Also, it is essential to note that the minor concentration of added LAP did not demonstrate cytotoxicity. These data reinforce safe use of the proposed nanofibers in endodontic infection control protocols and pulp regeneration therapies.

After the promising results from in vitro experiments, we decided to investigate the biocompatibility in vivo. The rat subcutaneous model explored cellular infiltration properties, collagen production, and GelMA-based fibers morphological changes. Although the implantation site diverged from the suggested clinical application, the subcutaneous in vivo biocompatibility test is a well-established model used to represent mechanisms and consequences of tissue-biomaterial interactions [33]. Hematoxylin/eosin, and picrosirius red images of retrieved tissue with implanted AZ-laden GelMA fibers indicated no compromised biocompatibility. Moreover, adding AZ to the newly synthesized GelMA-based fibers enhanced the collagen deposition between the fibers.

The anti-inflammatory properties of AZ can be attributed to its action on macrophages, since mechanistic studies demonstrated immunomodulatory activity through the regulation of cellular processes involved in inflammation through the reduction in NF-kB activation and, consequently, reduction in the up-regulation of pro-inflammatory cytokines [34,35]. Also, AZ exhibits immunomodulatory properties by shifting the inflammatory response toward an M2 macrophage state, characterized by regulation of inflammation and repair [36]. AZ also inhibits the expression of phospholipases A2, an enzyme involved in cell signaling processes that produces arachidonic acid byproducts [37], induced by LPS, an essential component of the outer membrane in Gram-negative bacteria, such as *Aggregatibacter actinomycetemcomitans* [38].

It is important to emphasize that, to the best of our knowledge, this is the first study reporting on the fabrication of visible-light photocrosslinkable GelMA-based fibers loaded with azithromycin as a biodegradable and biocompatible localized drug delivery system for endodontic infection control. Thus, more studies are needed to determine the drug release rate, stability, and durability of the antimicrobial activity. Lastly, since the healing of apical periodontitis is crucial for continuing root development and the overall regenerative outcome, we plan to address in a future investigation the potential benefits of AZ-laden GelMA electrospun fibers as alternative antimicrobial and immunomodulatory therapeutics within the context of regenerative endodontics.

4. Materials and Methods

4.1. Gelatin Methacryloyl Synthesis

GelMA production was performed following formerly reported studies [16,26]. Briefly, on a heating plate at 50 °C, type A gelatin from porcine skin (300 bloom, Sigma-Aldrich, St. Louis, MO, USA) at 10% w/v was solubilized into Dulbecco's phosphate-buffered saline (DPBS, Gibco Invitrogen Corporation, Grand Island, NY, USA). Next, in a dropwise manner, 8 mL of methacrylic acid (MA) was added to the gelatin solution, allowing them to react for 2 h under continuous stirring. The reaction was interrupted by adding an equal quantity of DPBS at 40 °C. Lastly, unreacted monomers and salts were eliminated by dialyzing the mixture in deionized water (DI) and employing a 12–14 kDa dialyze tube at 45 ± 5 °C for 7 d, with DI water changed twice daily. The prepared solution was frozen at −80 °C, lyophilized (Labconco FreeZone 2.5L, Labconco Corporation, Kansas City, MO, USA) for a week, and stored at −80 °C until further use.

4.2. Electrospinning and Nanofibers Preparation

Pure GelMA (i.e., AZ-free) and AZ-laden GelMA fibers were engineered via electrospinning based on formulation of a single antibiotic at three distinct amounts (5%, 10%, and 15%, w/w), hereafter referred to as GelMA+5%AZ, GelMA+10%AZ, and GelMA+15%AZ, respectively. These formulations were based on an initial screening performed by our team to find suitable cytocompatibility and electrospinning ability. Briefly, lyophilized GelMA was dissolved at a concentration of 15% w/v in acetic acid (Sigma-Aldrich). The mixture was left on the hot plate at 50 °C overnight. Next, AZ was individually added at the specified concentrations and left stirring for 2 h. Then, a photoinitiator, i.e., lithium phenyl-2,4,6-trimethylbenzoylphosphinate (LAP, TCI America Inc., Portland, OR, USA), was added at the concentration 0.075% w/v to all groups [16,26]. The prepared GelMA-based solutions were separately loaded into 5 mL plastic syringes capped with a 27-gauge metallic needle. The electrospun fibers were spun using the specifications 2 mL/h, 18 cm distance, and 18 kV on a rotating mandrel overlaid with aluminum foil at 120 rpm in a custom-made electrospinning box. Notably, to protect the GelMA-based solutions from room light exposure during electrospinning, both the plastic syringes and the electrospinning box were wrapped with black paper. The electrospun mats were placed in a desiccator overnight at room temperature (RT) to ensure complete acetic acid evaporation. Lastly, the fibers were peeled from the aluminum foil, cut into the desired size according to the experiment, wetted with 100% amorphous ethanol, gently dried with low-lint wipes (Kimwipes, Kimberly-Clark Corporation, Irving, TX, USA), and light-cured for 5 min on each side using an LED light box (Light Zone II, BesQual-E300N, Meta Dental Corp, Glendale, NY, USA).

4.3. Morphological and Chemical Characteristics

Scanning electron microscopy (SEM, MIRA3, FEG-SEM, Tescan, Czech Republic) was performed to assess fiber morphology and microstructure. Using double-sided carbon adhesive tape, the samples were fixed on Al stubs and sputter-coated with a thin (~5–10 nm) layer of Au-Pd (SPI-Module Carbon/Sputter Coater, SPI Supplies, West Chester, PA, USA) before imaging. The average fiber diameter (AFD) was determined using ImageJ (National Institutes of Health, Bethesda, MD, USA) software. Four SEM images per group were used to measure fiber diameter and the data reported as average $\pm$ standard deviation (n = 25/image/group). To confirm AZ incorporation, Fourier-transform infrared spectroscopy FTIR (ATR-FTIR, Thermo-Nicolet IS-50, Thermo Fischer Scientific, Inc., Waltham, MA, USA) using attenuated total reflection was conducted. The spectra recording was performed between 4000 and 600 cm^{-1} at 4 cm^{-1} resolution.

4.4. Mechanical Properties

The tensile strength of the fibrous mats was evaluated by uniaxial tensile testing (eXpert 5601; ADMET Inc., Norwood, MA, USA) [39]. Testing of rectangular-shaped specimens (25 mm $\times$ 3 mm^2, n = 6/group) was performed at a crosshead speed of 1 mm/min. Three distinct mechanical properties (i.e., tensile strength, Young's modulus, and elongation at break) were recorded or determined from the load-position curves.

4.5. Degradation Profile

In vitro degradation of the fabricated GelMA-based fibers was assessed by incubating them in PBS and recording their weight variation over time. The electrospun fibrous mats were cut into square-shaped samples (20 $\times$ 20 mm), treated with 100% ethanol, and light-cured for 5 min/side. The samples (n = 3) were weighed, immersed into 2 mL sterile PBS, and incubated for 14 days. At prearranged time points, the samples were removed from the incubation medium, gently dried with low-lint wipes (Kimwipes, Kimberly Clark Corporation), and washed twice with DI water and dried at RT for 24 h before the weight was recorded. The degradation profile was calculated by the formula, where Wt denotes the residual weight over time and w0 denotes the original dried weight.

$$\text{Degradation ratio } (\%) = \frac{w_t}{w_0} \times 100$$

4.6. Drug Release

To determine the kinetics of azithromycin release, four square-shaped (10 $\times$ 10 mm^2) GelMA+15%AZ fibrous mats were individually incubated in glass vials containing 10 mL PBS at 37 °C under constant shaking. After 1, 3, and 5 days, 1000 µL aliquots were drawn. Equal amounts of fresh PBS were added back to the incubation following aliquot retrieval. The AZ content was determined by quantifying the absorption of the clear supernatant using a UV-spectrophotometer (SpectraMax iD3, Molecular Devices LLC, San Jose, CA, USA) at 410 nm in triplicate. The AZ concentration at each time point was calculated by comparing it with the established standard curve [14].

4.7. Cytocompatibility

Four square-shaped (15 mm $\times$ 15 mm^2) fibrous mats per group were individually incubated in glass vials containing Glucose Dulbecco Modified Eagle Medium (DMEM, Gibco Invitrogen Corporation) for up to 14 days. 500 μL aliquots were drawn at prearranged time points and identical amounts were added to maintain the extraction volume constant. Using a 0.22 μm nylon membrane under vacuum, the aliquots were filtered before further analysis [16].

Human dental pulp stem cells (hDPSCs) obtained from permanent third molars (passages 6–8) were cultured in DMEM containing 15% fetal bovine serum (FBS; Hyclone Laboratories, Inc., Logan, UT, USA) and 1% penicillin-streptomycin (Sigma-Aldrich) in a humidified incubator at 37 °C with 5% CO_2. The cells were seeded at a density of 3×10^3/well (100 μL cell suspension) in 96-well tissue culture microtiter plates. After 4 h of incubation, the media was removed and replaced by the collected aliquots in triplicate (100 μL), adjusted to 15%FBS and 1% penicillin-streptomycin, and the positive control (0.3% vol phenol solution) [16]. Quadruplicated wells were arranged using medium without cells (blank control) and medium with cells but without any aliquot (reflecting 100% survival) [16]. The microplates were then incubated in a 5% CO_2 chamber. After 3 days, 20 μL CellTiter 96 Aqueous One Solution Reagent (Promega Corporation, Madison, WI, USA) was added to the test wells and incubated for 2 h. The color change reaction was assessed by reading the absorbance at 490 nm in a microplate reader (Spectra iD3; Molecular Devices LLC, San Jose, CA, USA) against blank wells [16].

4.8. Antimicrobial Assessment

GelMA-based AZ-laden fibers were further investigated using agar diffusion against *Actinomyces naeslundii* (*An*, American Type Culture Collection, ATCC 12104, Manassas, VA, USA), and *Aggregatibacter actinomycetemcomitans* (*Aa*, American Type Culture Collection, ATCC 43718). Like the procedure for cell compatibility, square-shaped (n = 3; 15 mm $\times$ 15 mm^2) fibrous mats were first disinfected by UV light (30 min/side) and rinsed twice with sterile PBS. Next, the GelMA-based mats were individually incubated in PBS for 2 weeks. At predetermined time points, 500 μL aliquots were drawn and equal amounts were added to keep the extraction volume unchanged. The aliquots were stored at -20 °C until further use [26].

Broth cultures of bacterial strains were grown for 24 h at 37 °C in BHI broth (Becton, Dickinson & Co. Sparks, MD) in a 5% CO2 atmosphere (*A. actinomycetemcomitans*) or anaerobically in 5% CO_2, 10% H_2 in N_2 atmosphere (*A. naeslundii*). Bacterial cultures on agar plates were grown under the same conditions, using tryptic soy agar with 5% *v/v* sheep blood (Hardy Diagnostics, Santa Maria, CA, USA) for *A. actinomycetemcomitans* and Brucella blood agar with hemin and vitamin K (Hardy Diagnostics) for *A. naeslundii*. One hundred μL of each broth culture was spread onto appropriate agar plates to create a lawn of bacteria. Four individual zones receiving 10 μL of eluates from AZ-laden GelMA-based fibers (days 1, 3, 7, and 14) were created on each plate. Chlorhexidine (0.12%) and sterile PBS were positive and negative controls, respectively. The plates were incubated according to the bacteria strain. After 2 days of incubation, the diameters (in mm) of the clear growth inhibition zones were calculated.

4.9. In Vivo Biocompatibility

All animal procedures followed the ARRIVE guidelines for reporting animal research and were in accordance with the procedures of the local Institutional Animal Care and Use Committee (PRO00010329). Four 6-week-old male Fischer 344 rats (280–300 g) were used for the experiments (Envigo RMS, Inc., Oxford, MI, USA). All surgical procedures were performed under general anesthesia induced with 50 mg/kg of ketamine (Hospira, Inc., Lake Forest, IL, USA) and 5 mg/kg xylazine (Akorn, Inc., Lake Forest, IL, USA) intraperitoneally. After anesthesia, a 2 cm incision in a head-tail orientation with a size 15 scalpel blade was performed, and subsequently, four small separate subcutaneous pockets were created through tissue divulsion [40].

Square-shaped samples (10 mm height × 10 mm width × 1 mm depth) of electrospun GelMA+15%AZ fibers (scaffolds) were implanted. Antibiotic-free (i.e., pure) GelMA electrospun fibers and Bio-Gide® (Geistlich Pharma North America Inc., Princeton, NJ, USA) were used as controls. After wound closure with Coated Vicryl® polyglactin 910 (Ethicon Endo-Surgery, Inc., Cincinnati, OH, USA), the animals were allowed to recover from anesthesia. At 14- or 28-days post-implantation, the animals were euthanized by CO_2 inhalation, and the implanted biomaterials with surrounding tissue were retrieved, fixed in 10% buffered formalin overnight, embedded in paraffin, cut into 6 µm-thick sections, and stained with hematoxylin and eosin (H&E) to investigate under light microscopy (Nikon E800, Nikon Corporation, Tokyo, Japan) for the presence of inflammatory cells and neovascularization. They were then, stained with picrosirius red (PSR) to analyze under polarized light microscopy the orientation, pattern, and interweaving of the collagen fibers in the implanted material, where greenish-yellow fibers were considered to be immature and thin, while yellowish-red fibers were considered to be mature and thick [41].

4.10. Statistical Analysis

All the analyses were performed with GraphPad Prism 9 software (GraphPad Software, San Diego, CA, USA). The fiber diameter and tensile strength were analyzed using a one-way analysis of variance. Each group's inhibition zone and cytocompatibility data were compared using an analysis of variance that included a random effect to account for correlations within a specimen over time. Tukey's post-hoc test was used to count for differences among groups. The level of significance was set at 5%.

5. Conclusions

The AZ-laden fibers (GelMA+15%AZ) have been shown to be a promising alternative for the sustained delivery of azithromycin for endodontic infection control, particularly within the context of regenerative endodontics due to their potent antimicrobial activity against *Aa* and *An* without compromising the biocompatible properties.

Author Contributions: Conceptualization, M.C.B. and A.A.A.; methodology, A.A.A., A.H.M., A.D., and R.D.-F.; software, A.H.M. and R.D.-F.; validation, J.X. and R.D.-F.; formal analysis, J.X. and R.D.-F.; investigation, A.A.A. and A.H.M.; resources, J.C.F. and M.C.B.; data curation, A.S.; writing—original draft preparation, A.A.A., A.H.M., J.S.R., and A.D.; writing—review and editing, R.D.-F., A.S., H.S., J.C.F., and M.C.B.; visualization, J.X., H.S., and J.C.F.; supervision, J.X., A.S., H.S., and J.C.F.; project administration, M.C.B.; funding acquisition, M.C.B. All authors have read and agreed to the published version of the manuscript.

Funding: M.C.B. acknowledges the National Institutes of Health (National Institute of Dental and Craniofacial Research—Grants R01DE026578 and R01DE031476). The graphical abstract in this manuscript was partially created with Bio Render. The content is solely the responsibility of the authors and does not necessarily represent the official views of the National Institutes of Health.

Institutional Review Board Statement: The animal study protocol was approved by the local Institutional Animal Care and Use Committee of the University of Michigan (protocol code: PRO00010329 and date of approval: 08/05/2021).

Informed Consent Statement: Not applicable.

Data Availability Statement: The data presented in this study are available on request from the corresponding author.

Conflicts of Interest: The authors declare no conflict of interest.

References

1. Pitts, N.B.; Zero, D.T.; Marsh, P.D.; Ekstrand, K.; Weintraub, J.A.; Ramos-Gomez, F.; Tagami, J.; Twetman, S.; Tsakos, G.; Ismail, A. Dental caries. *Nat. Rev. Dis. Primers* **2017**, *3*, 17030. [CrossRef] [PubMed]
2. Kim, S.G.; Malek, M.; Sigurdsson, A.; Lin, L.M.; Kahler, B. Regenerative endodontics: A comprehensive review. *Int. Endod. J.* **2018**, *51*, 1367–1388. [CrossRef] [PubMed]

3. Songtrakul, K.; Azarpajouh, T.; Malek, M.; Sigurdsson, A.; Kahler, B.; Lin, L.M. Modified Apexification Procedure for Immature Permanent Teeth with a Necrotic Pulp/Apical Periodontitis: A Case Series. *J. Endod.* **2020**, *46*, 116–123. [CrossRef] [PubMed]

4. Cehreli, Z.C.; Isbitiren, B.; Sara, S.; Erbas, G. Regenerative Endodontic Treatment (Revascularization) of Immature Necrotic Molars Medicated with Calcium Hydroxide: A Case Series. *J. Endod.* **2011**, *37*, 1327–1330. [CrossRef]

5. Smith, A.J.; Cooper, P.R. Regenerative Endodontics: Burning Questions. *J. Endod.* **2017**, *43*, S1–S6. [CrossRef]

6. Verma, P.; Nosrat, A.; Kim, J.R.; Price, J.B.; Wang, P.; Bair, E.; Xu, H.H.; Fouad, A.F. Effect of Residual Bacteria on the Outcome of Pulp Regeneration In Vivo. *J. Dent. Res.* **2017**, *96*, 100–106. [CrossRef]

7. Soares, D.G.; Bordini, E.A.F.; Swanson, W.B.; de Souza Costa, C.A.; Bottino, M.C. Platform technologies for regenerative endodontics from multifunctional biomaterials to tooth-on-a-chip strategies. *Clin. Oral Investig.* **2021**, *25*, 4749–4779. [CrossRef]

8. Ribeiro, J.S.; Münchow, E.A.; Bordini, E.A.F.; de Oliveira da Rosa, W.L.; Bottino, M.C. Antimicrobial Therapeutics in Regenerative Endodontics: A Scoping Review. *J. Endod.* **2020**, *46*, S115–S127. [CrossRef]

9. Murray, P.E. Review of guidance for the selection of regenerative endodontics, apexogenesis, apexification, pulpotomy, and other endodontic treatments for immature permanent teeth. *Int. Endod. J.* **2022**. [CrossRef]

10. Albuquerque, M.T.P.; Nagata, J.Y.; Diogenes, A.R.; Azabi, A.A.; Gregory, R.L.; Bottino, M.C. Clinical Perspective of Electrospun Nanofibers as a Drug Delivery Strategy for Regenerative Endodontics. *Curr. Oral Health. Rep.* **2016**, *3*, 209–220. [CrossRef]

11. Bottino, M.C.; Albuquerque, M.T.P.; Azabi, A.; Munchow, E.A.; Spolnik, K.J.; Nör, J.E.; Edwards, P.C. A novel patient-specific three-dimensional drug delivery construct for regenerative endodontics. *J. Biomed. Mater. Res. B Appl. Biomater.* **2019**, *107*, 1576–1586. [CrossRef] [PubMed]

12. Sharifi, S.; Sharifi, H.; Akbari, A.; Chodosh, J. Systematic optimization of visible light-induced crosslinking conditions of gelatin methacryloyl (GelMA). *Sci. Rep.* **2021**, *11*, 23276. [CrossRef] [PubMed]

13. Wei, D.; Xiao, W.; Sun, J.; Zhong, M.; Guo, L.; Fan, H.; Zhang, X. A biocompatible hydrogel with improved stiffness and hydrophilicity for modular tissue engineering assembly. *J. Mater. Chem. B* **2015**, *3*, 2753–2763. [CrossRef] [PubMed]

14. Qu, L.; Dubey, N.; Ribeiro, J.S.; Bordini, E.A.F.; Ferreira, J.A.; Xu, J.; Castilho, R.M.; Bottino, M.C. Metformin-loaded nanospheres-laden photocrosslinkable gelatin hydrogel for bone tissue engineering. *J. Mech. Behav. Biomed. Mater.* **2021**, *116*, 104293. [CrossRef]

15. Aldana, A.A.; Malatto, L.; Rehman, M.A.U.; Boccaccini, A.R.; Abraham, G.A. Fabrication of Gelatin Methacrylate (GelMA) Scaffolds with Nano- and Micro-Topographical and Morphological Features. *Nanomater. Basel* **2019**, *9*, 120. [CrossRef]

16. Ribeiro, J.S.; Bordini, E.A.F.; Ferreira, J.A.; Mei, L.; Dubey, N.; Fenno, J.C.; Piva, E.; Lund, R.G.; Schwendeman, A.; Bottino, M.C. Injectable MMP-Responsive Nanotube-Modified Gelatin Hydrogel for Dental Infection Ablation. *ACS Appl. Mater. Interfaces* **2020**, *12*, 16006–16017. [CrossRef]

17. Sato, R.; Harada, R.; Shigeta, Y. Theoretical analyses on a flipping mechanism of UV-induced DNA damage. *Biophys. Phys.* **2016**, *13*, 311–319. [CrossRef]

18. Nguyen, A.K.; Goering, P.L.; Reipa, V.; Narayan, R.J. Toxicity and photosensitizing assessment of gelatin methacryloyl-based hydrogels photoinitiated with lithium phenyl-2,4,6-trimethylbenzoylphosphinate in human primary renal proximal tubule epithelial cells. *Biointerphases* **2019**, *14*, 021007. [CrossRef]

19. Firth, A.; Prathapan, P. Azithromycin: The First Broad-spectrum Therapeutic. *Eur. J. Med. Chem.* **2020**, *207*, 112739. [CrossRef]

20. Andrada, A.C.; Azuma, M.M.; Furusho, H.; Hirai, K.; Xu, S.; White, R.R.; Sasaki, H. Immunomodulation Mediated by Azithromycin in Experimental Periapical Inflammation. *J. Endod.* **2020**, *46*, 1648–1654. [CrossRef]

21. Zhang, Z.; Zhu, Y.; Yang, X.; Li, C. Preparation of azithromycin microcapsules by a layer-by-layer self-assembly approach and release behaviors of azithromycin. *Colloids Surf. A Physicochem. Eng. Asp.* **2010**, *362*, 135–139. [CrossRef]

22. Siqueira, J.F.; Rôças, I.N. Diversity of Endodontic Microbiota Revisited. *J. Dent. Res.* **2009**, *88*, 969–981. [CrossRef] [PubMed]

23. Augustine, R.; Malik, H.N.; Singhal, D.K.; Mukherjee, A.; Malakar, D.; Kalarikkal, N.; Thomas, S. Electrospun polycaprolactone/ZnO nanocomposite membranes as biomaterials with antibacterial and cell adhesion properties. *J. Polym. Res.* **2014**, *21*, 347. [CrossRef]

24. Li, J.; Mooney, D.J. Designing hydrogels for controlled drug delivery. *Nat. Rev. Mater.* **2016**, *1*, 16071. [CrossRef] [PubMed]

25. Klotz, B.J.; Gawlitta, D.; Rosenberg, A.J.W.P.; Malda, J.; Melchels, F.P.W. Gelatin-Methacryloyl Hydrogels: Towards Biofabrication-Based Tissue Repair. *Trends Biotechnol.* **2016**, *34*, 394–407. [CrossRef] [PubMed]

26. Ribeiro, J.S.; Daghrery, A.; Dubey, N.; Li, C.; Mei, L.; Fenno, J.C.; Schwendeman, A.; Aytac, Z.; Bottino, M.C. Hybrid Antimicrobial Hydrogel as Injectable Therapeutics for Oral Infection Ablation. *Biomacromolecules* **2020**, *21*, 3945–3956. [CrossRef] [PubMed]

27. Parihar, A.S.; Das, A.C.; Sahoo, S.K.; Bhardwaj, S.S.; Babaji, P.; Varghese, J.G. Evaluation of role of periodontal pathogens in endodontic periodontal diseases. *J. Fam. Med. Prim. Care* **2020**, *9*, 239–242. [CrossRef]

28. Nagata, J.Y.; Soares, A.J.; Souza-Filho, F.J.; Zaia, A.A.; Ferraz, C.C.; Almeida, J.F.; Gomes, B.P. Microbial Evaluation of Traumatized Teeth Treated with Triple Antibiotic Paste or Calcium Hydroxide with 2% Chlorhexidine Gel in Pulp Revascularization. *J. Endod.* **2014**, *40*, 778–783. [CrossRef]

29. Sassone, L.M.; Fidel, R.; Faveri, M.; Fidel, S.; Figueiredo, L.; Feres, M. Microbiological evaluation of primary endodontic infections in teeth with and without sinus tract. *Int. Endod. J.* **2008**, *41*, 508–515. [CrossRef]

30. Kulik, E.M.; Thurnheer, T.; Karygianni, L.; Walter, C.; Sculean, A.; Eick, S. Antibiotic Susceptibility Patterns of Aggregatibacter actinomycetemcomitans and Porphyromonas gingivalis Strains from Different Decades. *Antibiot. Basel* **2019**, *8*, 253. [CrossRef]

31. Narita, M.; Shibahara, T.; Takano, N.; Fujii, R.; Okuda, K.; Ishihara, K. Antimicrobial Susceptibility of Microorganisms Isolated from Periapical Periodontitis Lesions. *Bull. Tokyo Dent. Coll.* **2016**, *57*, 133–142. [CrossRef] [PubMed]

32. Lai, P.-C.; Schibler, M.R.; Walters, J.D. Azithromycin Enhances Phagocytic Killing of *Aggregatibacter actinomycetemcomitans* Y4 by Human Neutrophils. *J. Periodontol.* **2015**, *86*, 155–161. [CrossRef] [PubMed]

33. Cosme-Silva, L.; Benetti, F.; Dal-Fabbro, R.; Gomes Filho, J.E.; Sakai, V.T.; Cintra, L.T.A.; Alvarez, N.; Ervolino, E.; Viola, N.V. Biocompatibility and biomineralization ability of Bio-C Pulpecto. A histological and immunohistochemical study. *Int. J. Paediatr. Dent.* **2019**, *29*, 352–360. [CrossRef]

34. Cigana, C.; Assael, B.M.; Melotti, P. Azithromycin Selectively Reduces Tumor Necrosis Factor Alpha Levels in Cystic Fibrosis Airway Epithelial Cells. *Antimicrob. Agents Chemother.* **2007**, *51*, 975–981. [CrossRef] [PubMed]

35. Kanoh, S.; Rubin, B.K. Mechanisms of Action and Clinical Application of Macrolides as Immunomodulatory Medications. *Clin. Microbiol. Rev.* **2010**, *23*, 590–615. [CrossRef]

36. Haydar, D.; Cory, T.J.; Birket, S.E.; Murphy, B.S.; Pennypacker, K.R.; Sinai, A.P.; Feola, D.J. Azithromycin Polarizes Macrophages to an M2 Phenotype via Inhibition of the STAT1 and NF-kappaB Signaling Pathways. *J. Immunol.* **2019**, *203*, 1021–1030. [CrossRef]

37. Letsiou, E.; Kitsiouli, E.; Nakos, G.; Lekka, M.E. Mild stretch activates cPLA2 in alveolar type II epithelial cells independently through the MEK/ERK and PI3K pathways. *Biochim. Biophys. Acta* **2011**, *1811*, 370–376. [CrossRef]

38. Ahlstrand, T.; Kovesjoki, L.; Maula, T.; Oscarsson, J.; Ihalin, R. *Aggregatibacter actinomycetemcomitans* LPS binds human interleukin-8. *J. Oral Microbiol.* **2018**, *11*, 1549931. [CrossRef]

39. Daghrery, A.; Ferreira, J.A.; de Souza Araújo, I.J.; Clarkson, B.H.; Eckert, G.J.; Bhaduri, S.B.; Malda, J.; Bottino, M.C. A Highly Ordered, Nanostructured Fluorinated CaP-Coated Melt Electrowritten Scaffold for Periodontal Tissue Regeneration. *Adv. Health Mater.* **2021**, *10*, e2101152. [CrossRef]

40. Cosme-Silva, L.; Dal-Fabbro, R.; Gonçalves, L.D.O.; Prado, A.S.D.; Plazza, F.A.; Viola, N.V.; Cintra, L.T.A.; Gomes Filho, J.E. Hypertension affects the biocompatibility and biomineralization of MTA, High-plasticity MTA, and Biodentine®. *Braz. Oral Res.* **2019**, *33*, e060. [CrossRef]

41. Lattouf, R.; Younes, R.; Lutomski, D.; Naaman, N.; Godeau, G.; Senni, K.; Changotade, S. Picrosirius red staining: A useful tool to appraise collagen networks in normal and pathological tissues. *J. Histochem. Cytochem.* **2014**, *62*, 751–758. [CrossRef] [PubMed]

International Journal of
Molecular Sciences

Article

Engineering of Injectable Antibiotic-Laden Fibrous Microparticles Gelatin Methacryloyl Hydrogel for Endodontic Infection Ablation

Juliana S. Ribeiro [1,2], Eliseu A. Münchow [3], Ester A. F. Bordini [1,4], Nathalie S. Rodrigues [1], Nileshkumar Dubey [1], Hajime Sasaki [1], John C. Fenno [5], Steven Schwendeman [6] and Marco C. Bottino [1,7,*]

1   Department of Cariology, Restorative Sciences, and Endodontics, School of Dentistry, University of Michigan, Ann Arbor, MI 48104, USA; sribeirooj@gmail.com (J.S.R.); esterbordini@gmail.com (E.A.F.B.); nathalie.sard@gmail.com (N.S.R.); nileshd@nus.edu.sg (N.D.); hajimes@umich.edu (H.S.)
2   Department of Restorative Dentistry, School of Dentistry, Federal University of Pelotas, Pelotas 96015-560, Rio Grande do Sul, Brazil
3   Department of Conservative Dentistry, School of Dentistry, Federal University of Rio Grande do Sul, Porto Alegre 90035-003, Rio Grande do Sul, Brazil; eliseumunchow@gmail.com
4   Department of Dental Materials and Prosthodontics, School of Dentistry, São Paulo State University, Araraquara 14801, São Paulo, Brazil
5   Department of Biologic and Materials Sciences & Prosthodontics, University of Michigan School of Dentistry, Ann Arbor, MI 48104, USA; fenno@umich.edu
6   Department of Pharmaceutical Sciences and the Biointerfaces Institute, University of Michigan, Ann Arbor, MI 48104, USA; schwend@med.umich.edu
7   Department of Biomedical Engineering, College of Engineering, University of Michigan, Ann Arbor, MI 48104, USA
*   Correspondence: mbottino@umich.edu; Tel.: +1-734-763-2206; Fax: +1-734-936-1597

Citation: Ribeiro, J.S.; Münchow, E.A.; Bordini, E.A.F.; Rodrigues, N.S.; Dubey, N.; Sasaki, H.; Fenno, J.C.; Schwendeman, S.; Bottino, M.C. Engineering of Injectable Antibiotic-Laden Fibrous Microparticles Gelatin Methacryloyl Hydrogel for Endodontic Infection Ablation. *Int. J. Mol. Sci.* **2022**, *23*, 971. https://doi.org/10.3390/ijms23020971

Academic Editor: Matthias Widbiller

Received: 21 December 2021
Accepted: 14 January 2022
Published: 16 January 2022

Publisher's Note: MDPI stays neutral with regard to jurisdictional claims in published maps and institutional affiliations.

Abstract: This study aimed at engineering cytocompatible and injectable antibiotic-laden fibrous microparticles gelatin methacryloyl (GelMA) hydrogels for endodontic infection ablation. Clindamycin (CLIN) or metronidazole (MET) was added to a polymer solution and electrospun into fibrous mats, which were processed via cryomilling to obtain CLIN- or MET-laden fibrous microparticles. Then, GelMA was modified with CLIN- or MET-laden microparticles or by using equal amounts of each set of fibrous microparticles. Morphological characterization of electrospun fibers and cryomilled particles was performed via scanning electron microscopy (SEM). The experimental hydrogels were further examined for swelling, degradation, and toxicity to dental stem cells, as well as antimicrobial action against endodontic pathogens (agar diffusion) and biofilm inhibition, evaluated both quantitatively (CFU/mL) and qualitatively via confocal laser scanning microscopy (CLSM) and SEM. Data were analyzed using ANOVA and Tukey's test (α = 0.05). The modification of GelMA with antibiotic-laden fibrous microparticles increased the hydrogel swelling ratio and degradation rate. Cell viability was slightly reduced, although without any significant toxicity (cell viability > 50%). All hydrogels containing antibiotic-laden fibrous microparticles displayed antibiofilm effects, with the dentin substrate showing nearly complete elimination of viable bacteria. Altogether, our findings suggest that the engineered injectable antibiotic-laden fibrous microparticles hydrogels hold clinical prospects for endodontic infection ablation.

Keywords: electrospinning; cryomilling; biodegradation; antibiotics; fibrous particles; regeneration; dentistry; endodontics

1. Introduction

The success of endodontic regenerative treatment depends on the elimination of intraradicular microorganisms and the establishment of a microenvironment favorable to the proliferation and differentiation of stem cells [1]. Over decades, calcium hydroxide was the antimicrobial agent most commonly used for the disinfection of contaminated root canals [2], but its effectiveness was revealed to be limited against some pathogens, such as *Enterococcus faecalis* (*E. faecalis*), *Actinomyces naeslundii* (*A. naeslundii*), and *Candida albicans* (*C. albicans*) [3,4]. Moreover, the overall efficacy of calcium hydroxide within the dentinal tubules seemed unreliable [5]. Thus, other intracanal medicaments (e.g., antibiotics) or the combination of different agents seemed paramount to maximizing the eradication of microorganisms [6,7]. Notwithstanding, the local administration of antibiotics may result in some negative side effects (e.g., bacterial resistance, cell toxicity, among others) [8–10], making it relevant to the continued search for an effective and biologically safe drug delivery method capable of penetrating the root canal system of infected pulp tissues.

In light of offering a biocompatible scenario for the disinfection of contaminated root canals, biodegradable drug delivery systems (e.g., scaffolds and hydrogels) laden with antimicrobial agents have gained the attention of researchers, and several studies are currently characterizing their clinical potential [11–14]. Concerning the use of fibrous scaffolds, electrospinning is a straightforward method for synthesizing antibiotic-laden polymeric nanofibers [6,9,15,16]. Notably, a tubular 3D construct based on triple antibiotic-eluting fibers has already been developed to fit within the intracanal space of infected teeth [6]; thus, allowing significant antimicrobial activity, the elimination of bacterial biofilm inside dentinal tubules, and more importantly, healing of damaged periapical tissues [6]. Nevertheless, electrospun fibers may not be capable of complete eradication of root canal infections, especially due to the complex anatomical geometry of the root canal system (e.g., the existence of lateral canals and apical ramifications), which may be difficult to access by a solid drug-releasing approach such as the 3D construct described earlier. In order to improve the bioavailability of antimicrobial agents into the entire root canal system, the incorporation of electrospun fibers into hydrogels has also been attempted, showing promising results due to more controlled degradation and drug release profile [17–19]. Despite effectiveness, the strategy of combining electrospun fibers with a hydrogel may present some drawbacks, such as inadequate dispersion of the fibers within the hydrogel matrix. Alternatively, enhanced miscibility between nanofibers and the hydrogel matrix was demonstrated when electrospun fibers were further processed via cryomilling into microspheres or small-sized particles [20,21]. Relevant to clinical dentistry, gelatin methacryloyl (GelMA) is a semi-synthetic biocompatible and biodegradable hydrogel with interesting structural characteristics, such as curing capacity, stability at physiological temperature, and cell-friendly behavior [22]. Of note, a recent study by Monteiro et al. [23] demonstrated that GelMA can be easily photopolymerized using a dental curing light, representing a promising method of placing injectable biomaterials into the root canal system following a chairside procedure. Considering that no previous attempt has been made to engineer injectable antibiotic-laden fibrous microparticles GelMA hydrogels, herein, electrospun fibers loaded with clindamycin (CLIN) or metronidazole (MET) were effectively processed into antibiotic-laden microparticles and successfully used to modify a well-known photo-curable gelatin methacryloyl hydrogel, guaranteeing excellent biological and antimicrobial/antibiofilm properties to fight endodontic infections. Altogether, our findings suggest that the engineered injectable antibiotic-laden fibrous microparticles hydrogels hold clinical prospects for endodontic infection ablation.

2. Results

2.1. Antibiotic-laden Fibers and Cryomilled Fibrous Microparticles

The SEM micrographs shown in Figure 1 demonstrate a bead-free fiber morphology for the obtained electrospun fiber mats. While the MET-based fibers (Figure 1C) were thicker ($p < 0.001$) than the PLGA fibers (Figure 1A), the CLIN-based fibers (Figure 1B) exhibited a thinner fiber diameter distribution ($p = 0.042$). All the fiber mats showed an average fiber diameter that was statistically different from each other ($p < 0.001$; Figure 1D). SEM micrographs of the cryomilled fibers showed the formation of a uniform set of GelMA-based fiber particles, with both the CLIN- (Figure 1E) and MET-laden (Figure 1F) microparticles demonstrating homogeneous dispersion within the GelMA matrix without any signs of aggregation. Even though some typical peaks of CLIN and MET overlapped with the peaks of GelMA, Figure 1G reveals that both antibiotics were successfully incorporated within the GelMA particles. The peaks at 1560 cm^{-1} (C=C), 1690 cm^{-1} (C=O), and 1965 cm^{-1} (C–H) are related to the stretching vibration of CLIN, [24] whereas the peaks at 870 cm^{-1} (C–NO$_2$), 1275 cm^{-1} (C–O), 1537 cm^{-1} (N=O), and 3097 cm^{-1} (=C–H) are related to the stretching vibration of MET [25,26].

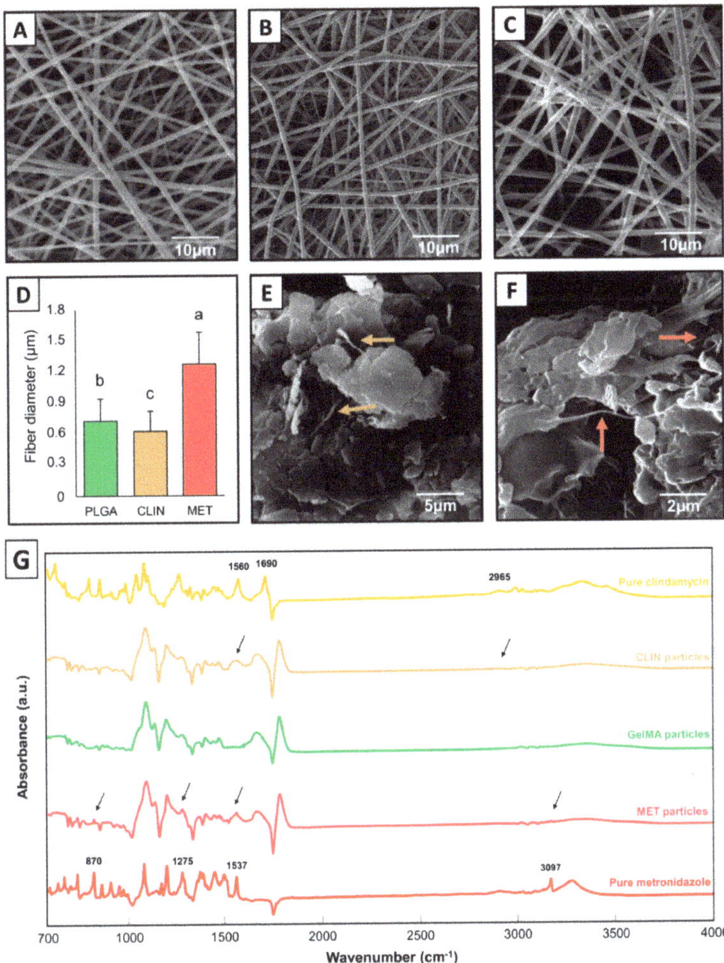

Figure 1. Representative SEM micrographs of electrospun and cryomilled fibers. (**A**) PLGA fibers (control); (**B**) CLIN-laden fibers; (**C**) MET-laden fibers; (**D**) graph showing the average fiber diameter of the electrospun fibers, with different letters above standard deviation bars, indicating statistical differences among the groups ($p < 0.05$); (**E**) fiber-based particles comprised of CLIN, yellow arrows indicate CLIN-laden fibers, and (**F**) fiber-based particles comprised of MET, red arrows indicate MET-laden fibers; and (**G**) FTIR spectra of pristine antibiotics (CLIN and MET) and GelMA particles, as well as the processed antibiotic-modified GelMA particles, black arrows indicate characteristics peaks of each antibiotic. SEM, Scanning Electron Microscope; PLGA, Poly(lactic-co-glycolic acid); CLIN, Clindamycin; MET, metronidazole; FTIR, Fourier-transform infrared spectroscopy; GelMA, gelatin methacryloyl hydrogel.

2.2. Antibiotic-laden Fibrous Microparticles in Gelatin Methacryloyl Hydrogels

2.2.1. Swelling and Biodegradation

Figure 2 shows the swelling ratio (A), degradation profile (B), and cell viability (C) of the synthesized hydrogels. Modification of GelMA with the antibiotic-laden cryomilled particles significantly increased the swelling ratio of the hydrogel ($p < 0.05$), ranging from 15.4% in the fiber-free GelMA to 37% in the CLIN+MET-based GelMA. The hydrogels incorporated with MET-laden fiber particles (MET and CLIN+MET groups) resulted in a greater swelling ratio than the CLIN-based hydrogel ($p < 0.05$). All hydrogels displayed a degradation profile starting within the first hour of the experiment, although the groups comprised of MET-based particles seemed to produce a faster and more intense degradation profile as compared with the CLIN and GelMA groups. CLIN+MET hydrogel lost nearly 50% of the initial mass after 6 h, and at the 72 h time point, it was degraded. CLIN- and MET-based hydrogels showed a slower extent of degradation than the CLIN+MET group, with complete degradation of the former occurring at the 168 h time point. The fiber-free GelMA was not completely degraded after 336 h, showing approximately 20% of the remaining mass after enzymatic degradation.

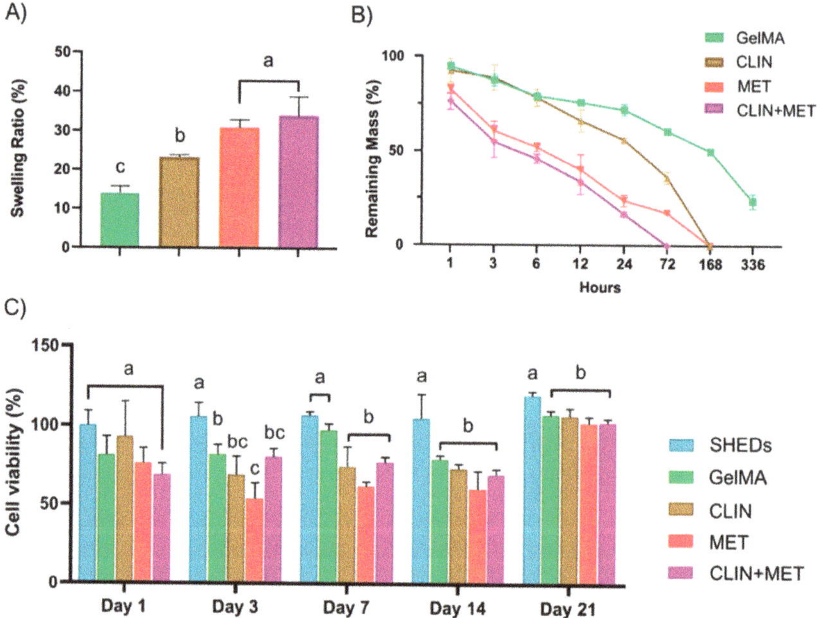

Figure 2. Graphs showing the swelling ratio (**A**), the degradation profile (**B**), and the cell viability (**C**) of the engineered hydrogels. The swelling ratio (%) results indicate the amount of water absorbed by the hydrogels within a 24 h period at 37 °C. The results for the in vitro degradation test reveal the mass loss of all hydrogels after exposure to DI water containing 1 U/mL of collagenase type I at 37 °C for 336 h. The results for the viability of stem cells from exfoliated deciduous teeth (SHEDs) were obtained indirectly using an MTS assay after 24 h of cell exposure in response to aliquots of the hydrogels at days 1, 3, 7, 14, and 21. The percentage of cell viability was normalized by the mean absorbance of SHEDs cultured at day 1 (100%). All the results are presented as mean ± SD values ($n = 4$/group). Distinct letters above the standard deviation bars indicate statistically significant differences among the groups ($p < 0.05$). GelMA, gelatin methacryloyl hydrogel; CLIN, Clindamycin; MET, metronidazole.

2.2.2. Cell Viability

Concerning cytocompatibility of the tested hydrogels, a reduction in SHEDs' viability was observed at every time point investigated, especially for the materials containing MET. However, at day 1 there was no statistical difference between the groups and the control (SHEDs) ($p > 0.05$), although, at day 3, all groups showed a cell viability potential that was statistically lower ($p < 0.05$) than the control and below the 80% level. The antibiotic-laden fibrous microparticle hydrogels sustained a lower cytocompatibility compared with the control at days 7 and 14, although without any cytotoxic behavior, ranging from 61% to 70% of cell viability. All hydrogels presented cell viability of ~100% when testing aliquots collected at day 21.

2.2.3. Antimicrobial Efficacy

Results for the antimicrobial properties (agar diffusion assay) of the antibiotic-laden fibrous microparticles hydrogels are shown in Figure 3A. All hydrogels showed antimicrobial action against the bacteria, except for MET against *E. faecalis* at all periods tested and against A. naeslundii at days 7 and 14; and the CLIN+MET group when tested against *E. faecalis* at days 1 and 3, which did not display any inhibition potential. The antibiotic-modified hydrogels showed overall lower antimicrobial activity compared with the control (CHX), except when tested against A. naeslundii, in which CLIN resulted in greater inhibition zones ($p < 0.05$). Among the hydrogel groups, CLIN demonstrated increased antimicrobial effectiveness, especially against A. naeslundii. MET was as effective as CLIN when tested against *F. nucleatum*, although its antimicrobial action against A. naeslundii was limited to the first 3 days, and against *E. faecalis*, it presented no inhibition potential. The CLIN+MET-laden hydrogel overall exhibited lesser effectiveness compared with the isolated counterparts ($p < 0.05$), and its antimicrobial activity against *E. faecalis* was time-dependent, resulting in inhibition zones only after day 7.

The antibiotic-free hydrogel (GelMA) and the negative control group (bacterial growth) neither impair biofilm formation nor reduce cells viability, as demonstrated by high colony-forming units' ($\log_{10}$ CFU/mL) values (Figure 3B). Meanwhile, all antibiotic-modified hydrogels showed antibiofilm effects, resulting in lower CFU counts than the controls ($p < 0.05$). The antibiofilm activity was similar among the modified hydrogels, regardless of the type or combination of antibiotics ($p > 0.05$). SEM micrographs shown in Figure 3 reveal a homogeneous accumulation of biofilm for the negative control (Figure 3D) and GelMA (Figure 3E) groups after 7 days of A. naeslundii culture on dentin. Conversely, dentin treated with the antibiotic-modified hydrogels showed the absence of biofilm (Figure 3F–H), with the dentin tubules apparently being empty. As verified in the CLSM micrographs related to the negative control and GelMA groups (Figure 3I,J), a dense population of viable bacteria attached to the dentin surface and penetrating dentinal tubules was identified. However, after the application of the antibiotic-modified hydrogels, the dentin substrate showed nearly complete elimination of viable bacteria (Figure 3K–M), similar to the positive control (Figure 3N).

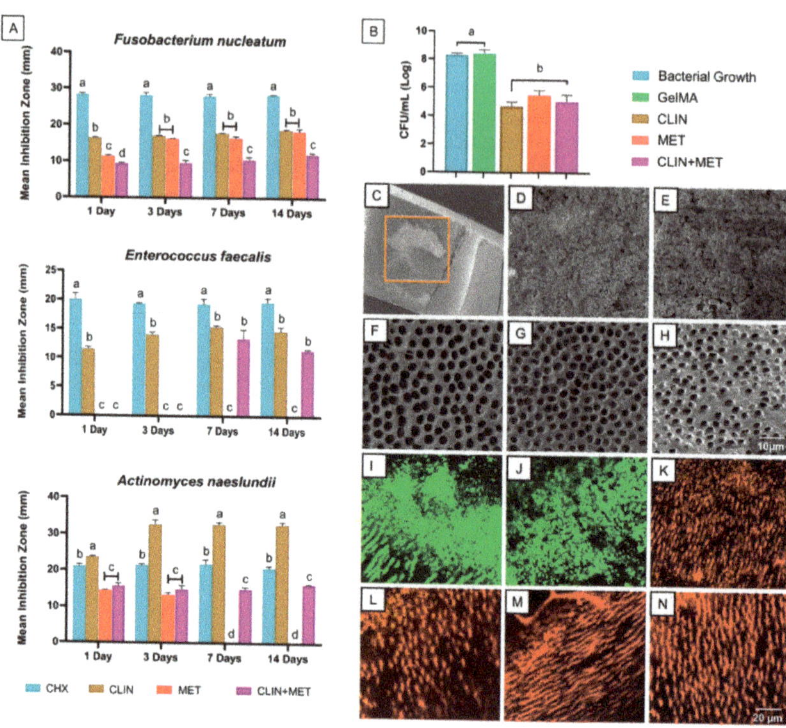

Figure 3. Results for the antimicrobial properties of the tested hydrogels. (**A**) Graphs showing the results (mean inhibition zones, in mm) from the agar diffusion assays against three bacteria at days 1, 3, 7, and 14. Chlorhexidine (CHX) served as the positive control. (**B**) Graph depicting the results (mean counts of colony-forming unit [CFU/mL]) from the *A. naeslundii* biofilm model used in the study, having a negative control group consisting of untreated bacterial growth. Distinct letters above the standard deviation bars indicate statistically significant differences among the groups ($p < 0.05$). (**C**) Representative SEM micrograph showing the evaluated areas of each sample (inner root walls of dentin slices). (**D–H**) SEM micrographs for the negative control (**D**) and groups treated with GelMA (**E**), CLIN-based hydrogel (**F**), MET-based hydrogel (**G**), and CLIN+MET-based hydrogel (**H**). (**I–N**) CLSM micrographs of 7-day *A. naeslundii* biofilm imaged from inner root canal walls. Images are related to the negative control group (**I**), antibiotic-free GelMA (**J**), CLIN-based hydrogel (**K**), MET-based hydrogel (**L**), CLIN+MET-based hydrogel (**M**), and 2.5% sodium hypochlorite (**N**), which served as the positive control. CLSM images were collected in sequential illumination mode by using 488 nm and 552 nm laser lines. Fluorescent emission was collected in 2 HyD spectral detectors with filter range set up to 500–550 nm and 590–655 nm for green (SYTO9) and red dye (PI), respectively. SEM, Scanning Electron Microscope; GelMA, gelatin methacryloyl hydrogel; CLIN, Clindamycin; MET, metronidazole; CLSM, confocal laser scanning microscopy.

3. Discussion

The drug delivery system developed in this study combined the effects of two straightforward releasing vehicles (i.e., electrospun fibers and fiber-based particles) to obtain a hybrid mechanism for the safe and sustained release of antibiotics targeting the elimination of root canal infections. More importantly, our main goal was to offer a drug delivery system in the form of an injectable and photo-curable hydrogel, aiming to establish a feasible chairside procedure of easy application with effective antimicrobial action.

The fibers synthesized via electrospinning were all morphologically adequate, showing a smooth and homogeneous fiber diameter distribution. The CLIN-based fibers presented the thinnest average fiber diameter of 628 nm ($\pm$194 nm), which was lower than the PLGA fibers (729 ± 214 nm) and the MET-based fibers (1.3 ± 0.3 μm). Here, both antibiotics (CLIN and MET) possess a hydrophilic nature [27], which may decrease the viscosity of the polymer solution, allowing for the formation of thinner fibers [28]. However, the MET-based fibers did not show such thin distribution, exhibiting ca. two-fold higher thickness values, probably due to the single hydroxyl group found in MET, which may have increased hydrogen bonding interactions and cross-linking of the polymer network, thereby reducing spinnability during electrospinning and the acquisition of thicker fibers [28]. Overall, the incorporation of antibiotics did not compromise morphological features of the electrospun fibers, which is an essential aspect for their proper functioning as a drug-releasing vehicle.

Considering that electrospun fibers may result in the burst release of drugs during the first 24 h [16,27,29], we have further processed them into small-sized particles, aiming to obtain a more controlled and sustained release of CLIN/MET. The method used here was cryomilling, which consists of cooling the material, then reducing it to smaller particles. According to some studies [29,30], this approach can maintain or even improve the therapeutic properties of the original fibers. The matrix used for the embedment of the fibers prior to cryomilling was GelMA (i.e., the same polymer used to prepare the injectable hydrogel system) so that better distribution and chemical compatibility between the fiber-based particles and the hydrogel could be expected without affecting hydrogel injectability. Considering that the incorporation of CLIN and MET into the respective hydrogels was confirmed by the identification of typical FTIR peaks (Figure 1G), we anticipated that the engineered antibiotic-laden fibrous microparticles hydrogels would demonstrate antimicrobial properties.

The release rate of any drug or therapeutic compound relies on the degradation speed of the carrying vehicle [12,22]. In the case of GelMA, the degradation speed is inversely correlated to three main factors: the degree of functionalization of the compound, the concentration of the hydrogel, and the total amount of enzymes [31]. The foregoing characteristics were all kept constant in our study. In light of verifying the degradation ability of the hydrogels, we performed two different tests: hygroscopic swelling and in vitro enzymatic degradation. As shown in Figure 2A, the antibiotic-modified hydrogels absorbed a greater amount of humidity (PBS) when compared with the neat GelMA, probably due to the hydrophilic nature of CLIN and MET [27]. Both of these antibiotics are comprised of hydroxyl groups, hence increasing polar interactions and the formation of hydrogen bonding. Nevertheless, it may be suggested that MET is more hydrophilic than CLIN due to the more heterogeneous composition of the latter (e.g., elements, such as phosphorus, chlorine, and sulfur). Thus, the presence of MET would make the GelMA matrix swell to a greater extent as compared with the presence of CLIN. No less important, the porosity of the synthesized fibers may have also played a role in the swelling ratio of the fiber-modified hydrogels, since the higher the degree of porosity of the fiber mat, the more intense its hygroscopic behavior [32,33]. Despite the fact that we did not conduct any analysis to determine the porosity level of the fiber mats, one should note that the MET-laden fibers showed the thickest morphology and the most porous fiber architecture. Thus, the total amount of GelMA matrix embedded within the fiber mat was also probably higher, turning the MET fibrous particles more prone to hydrolysis [27]. This may explain the more intense hygroscopic swelling and biodegradation patterns demonstrated by the MET and CLIN+MET hydrogels when compared with the CLIN group.

From the findings presented here, one can note that our antibiotic-modified hydrogels would work adequately since they demonstrated a one-week driven degradation profile, which is indeed desirable for a dressing medication that aims to eliminate root intracanal infection during a clinical inter-appointment period. It is the sustained release of antibiotics during that one-week interval that could effectively act in the ablation of root canal infections. However, it is paramount that the release of antibiotics does not reach cytotoxic levels; otherwise, the hydrogels would interfere with tissue healing/regeneration events, thus impairing the clinical translation of our strategy. According to the results shown in Figure 2C, the hydrogels incorporated with the antibiotic-laden fibrous particles did not exhibit any significant cytotoxic effects on SHEDs. Even so, one may note that lower cell viability was identified for the antibiotic-modified hydrogels at days 3 and 7, probably due to the release of CLIN and MET during degradation of the GelMA matrix. Even though antibiotics are usually cytotoxic when administered at high concentration levels [34], the amount of CLIN and MET used in our study was minimal, so their release would not be responsible for a significant reduction in the viability of SHEDs. Here, one would suggest that cell proliferation into the photopolymerized hydrogel may occur at a slower rate of speed, increasing once the moderately cross-linked structure of the hydrogel starts to degrade. Of note, cell viability was importantly increased after complete degradation of the hydrogels, suggesting that any byproducts originated during degradation and the release of CLIN and MET did not cause toxic effects to the cells. Last but not least, all the experimental hydrogels demonstrated cell viability values above the cytotoxicity threshold of 50%, suggesting their clinical safety and suitability.

In light of verifying the antimicrobial properties of the experimental hydrogels, we conducted two distinct antimicrobial analyses: agar diffusion and biofilm inhibition assays. Concerning the first analysis, the CLIN-based hydrogel was the only GelMA material showing antimicrobial activity against all the bacteria species at every time point investigated in the study (Figure 3A), corroborating to the findings of a previous study [35], which showed that microorganisms are usually highly susceptible to CLIN but not always to MET. Here, three bacteria (*F. nucleatum*, *E. faecalis*, and *A. naeslundii*) were considered in the agar diffusion assay due to their broad association to cases of infected immature teeth and failed endodontic treatment with persistent infection [36,37]. The foregoing bacteria are usually pathogenic and difficult to eliminate, but as verified by our findings, the strategy of using a hydrogel system incorporated with antibiotics resulted in important inhibition values, although it was dependent on the type of antibiotic(s) used, as well as the type of bacteria tested. While CLIN is typically a bacteriostatic agent, acting on the inhibition of bacterial protein synthesis [27], MET is a bactericidal antimicrobial competing with the biological electron acceptors of bacteria, disturbing their energy metabolism, and thus causing cell death [27]. Having this in mind, MET was expected to be less effective against Gram-positive (G+) bacteria, which presents a more organized cell wall structure, compared with the Gram-negative (G−) counterparts. Indeed, MET was effective when tested against *F. nucleatum* (G−), but it did not result in any inhibition potential against *E. faecalis* (G+). It seems that the concentration of MET released through the present drug-delivery strategy was insufficient for the proper growth inhibition of G+ bacteria. Conversely, CLIN was effective against all three bacteria, and especially to *A. naeslundii* (G+), which suggests that this bacterial species is a very sensitive microorganism to CLIN. Indeed, CLIN may act as a direct peptidyltransferase inhibitor in the case of sensitive microorganisms, hence affecting the process of the peptide chain initiation and stimulating dissociation of peptidyl-tRNA from ribosomes, i.e., a potent antibacterial inhibition mechanism [38].

In our study, one hydrogel was prepared by mixing equal amounts of the CLIN- and MET-laden fibrous microparticles to elucidate whether the antimicrobial activity would be potentiated upon the presence of both antibiotics into the same hydrogel. Nevertheless, the inhibition potential was not amplified, indicating the existence of a minimum inhibition concentration level for each of the drugs. One should note that the concentration of each antibiotic released from the CLIN+MET hydrogel was probably lower than the concentrations derived from the single-mix CLIN and MET groups; therefore, explaining the overall lower or lack of statistical differences in the antimicrobial results of those groups. Remarkably, the MET hydrogel did not result in consistent antimicrobial activity to all of the tested conditions, although the hydrogel modified with the CLIN-laden fibrous microparticles exhibited a steadier antimicrobial action, thus supporting further investigations in pre-clinical animal models of periapical disease. It is noteworthy that on one hand, we could increase the concentration level of MET incorporated into the fibrous particles, aiming to obtain a more significant gain in antimicrobial effect, but on the other hand, we could negatively increase the cytotoxicity of the resultant hydrogel, thus limiting the applicability of MET when using the same strategy. Last, the presence of MET fibrous particles in the CLIN+MET hydrogel formulation allowed a faster GelMA degradation, with the complete breakdown occurring at 72 h of incubation, i.e., a shorter period as compared with the one-week profile shown by the single-mix counterparts, which, as discussed earlier, would be more desirable as an intracanal dressing.

Concerning the biofilm inhibition assay evaluated in this study, SEM and CLSM analyses (Figure 3) revealed almost complete removal of the *A. naeslundii* biofilm upon exposure to our antibiotic-laden fibrous microparticles GelMA, indicating that this hybrid strategy is effective in the ablation of even complex structures such as a highly-organized biofilm. Different from the negative control (bacterial growth) and antibiotic-free GelMA groups, the hydrogels containing CLIN- and MET-laden microparticles significantly decreased the counts of viable bacteria in our *A. naeslundii* biofilm model, showing that the fiber-particle vehicle proposed here would provide a satisfactory diffusion of antibiotics through dentinal tubules, as well as to the entire extent of the root canal system (e.g., lateral canals and apical ramifications). Even though our study used a standard sample of human radicular dentin, rather than a full root canal model, the present methodology is commonly used due to its ability to mimic the clinical scenario as it facilitates the continuous formation of biofilms and penetration of bacteria within dentinal tubules [39].

The great novelty of the present study relates to the fact that the antimicrobial mechanisms of small-sized particles are still complex to understand due to the variety of mechanisms involved, making it difficult for bacterial cells to become resistant [40]. This highlights the importance of our strategy, which combines the beneficial effects of small-sized fibers and particles to that of the minimum use of antibiotics to combat bacterial infection, consisting of a biocompatible and therapeutic effective drug delivery approach with minimal possibility of causing bacterial resistance. Moreover, the release of small-sized compounds, such as the antibiotic-loaded fibrous particles synthesized here, may improve the availability of antibiotics at sites of difficult access within the root canal system, perhaps contributing to a more efficacious treatment. To the best of our knowledge, this is the first study that has loaded PLGA electrospun fibers with CLIN and MET and has further processed the fibers via cryomilling in order to obtain microparticles with the drug-releasing ability and the capability of being encapsulated into a photo-curable GelMA hydrogel.

4. Materials and Methods

4.1. Reagents

Poly(DL-lactide-co-glycolide) (PLGA [75:25], Mw = 97,100, [η] = 0.55–0.75 dL/g, Lactel Absorbable Polymers, Birmingham, AL, USA) pellets, metronidazole (MET), type A gelatin (300 bloom from porcine skin), and methacrylic anhydride were bought from Sigma-Aldrich (St. Louis, MO, USA). Clindamycin phosphate 97.0%+ (CLIN) and lithium phenyl-2,4,6-trimethylbenzoylphosphinate (LAP) were obtained from TCI America Inc. (Portland, OR, USA). Chloroform and methanol were procured from Thermo Fisher Scientific (Waltham, MA, USA), whereas Dulbecco's phosphate-buffered saline (DPBS) was purchased from Gibco Invitrogen Corporation (Grand Island, NY, USA). All the reagents were used without further purification.

4.2. Synthesis of Antibiotic-Releasing Fibers

Three stock polymer solutions were prepared by dissolving PLGA in chloroform to produce an 18 wt.% solution, which was incorporated with different antibiotic mixtures (CLIN or MET dissolved in methanol) at a 15 wt.% concentration relative to the total polymer weight. One stock solution was not incorporated with antibiotics, serving as the control (PLGA). The solutions were stirred overnight, loaded into 5 mL plastic syringes (Becton, Dickson and Company, Franklin Lakes, NJ, USA) fitted with a 27 G metallic needle (Small Parts, Inc., Miami, FL, USA), and then processed via electrospinning using the following parameters: rotating mandrel (120 rpm of speed), a flow rate of 1–2 mL/h, a spinning distance of 20 cm, and 18 kV. The obtained fiber mats were dried overnight at room temperature to remove any residual solvent and stored at 4 °C until use. The morphology and architecture of the fiber mats were analyzed via scanning electron microscopy (SEM; JSM-6390, JEOL, Tokyo, Japan). Samples obtained from each group were mounted on Al stubs and sputter-coated with gold before imaging. The average fiber diameter of 100 single fibers was measured with ImageJ (National Institutes of Health, Bethesda, MD, USA) and expressed as mean $\pm$ SD values (in μm).

4.3. Preparation of Fiber-Based Particles and Morpho-Chemical Characterizations

The antibiotic-laden fiber mats were processed into fine, small particles via cryomilling [29]. First, the fibrous mats were gently soaked in GelMA, which was synthesized as described elsewhere [22,41]. Briefly, gelatin was solubilized in DPBS at 50 °C to produce a 15 wt.% solution, followed by the dropwise addition of 8 mL of methacrylic anhydride. After 2 h of stirring, 8 mL of DPBS was added at 40 °C to interrupt the reaction, followed by dialysis in deionized water for 1 week to remove salts and unreacted monomers. After hydrogel (GelMA) synthesis, each electrospun fiber mat was cut into small pieces and completely soaked in GelMA. A photocrosslinker (LAP) was added to the mixture at the 0.05% level and stirring (240 rpm) was conducted at 50 °C; crosslinking was achieved for 60 s using a light-emitting diode (LED) curing unit (Bluephase; Ivoclar-Vivadent, Amherst, NY, USA) at 385 nm, resulting in GelMA/fibrous mat samples with a 50 wt.% fiber content. The resin-fiber samples were left to dry in a fume hood overnight, then placed in appropriate metallic milling vials, precooled for 2 min in liquid nitrogen, and milled for 15 min utilizing a cryogenic impact mill (model SPEX CertiPrep 6750, SPEX CertiPrep, Metuchen, NJ, USA). The cryomilling process included alternating 1 min milling cycles separated by 1 min cooling intervals, respectively. The obtained fiber-based particles were stored in a desiccator containing silica at room temperature until further use. SEM imaging was done to verify the morphology of the obtained particles. Fourier-transform infrared spectroscopy (FTIR; Thermo Fischer Scientific, Inc.) was utilized in attenuated total reflection mode ranging between 700–4000 cm^{-1} at a resolution of 4 cm^{-1} to confirm the chemical characteristics of the fiber-based particles incorporated with CLIN and MET.

4.4. Fabrication of the Antibiotic-Laden Fibrous Microparticles Gelatin Methacryloyl Hydrogel

The antibiotic-laden fiber particles were sieved (45 μm) before their incorporation into 4 mL of hydrogel solution (15% GelMA), i.e., the same hydrogel synthesized earlier for the GelMA/fiber mats soaking process. Photocrosslinker was added, as described before, followed by the addition of the fiber-based particles at a 5% (*w/v*) level. Four groups were prepared: GelMA—fiber-free GelMA (control); CLIN—GelMA comprised of CLIN-laden fiber particles; MET—GelMA comprised of MET-laden fiber particles; and CLIN+MET—GelMA comprised of equal amounts of CLIN- and MET-laden fiber particles.

4.5. Characterization Analyses

Fiber-modified and fiber-free GelMA mixtures were prepared to obtain distinct hydrogel-based samples for the analyses described below. To that end, 100 μL of each mixture was placed into an elastomeric (CutterSil Putty PLUS; Kulzer Dental North America, South Bend, IN, USA) mold of varying dimensions (depending on the test) and irradiated (photopolymerized) for 15 s with the LED at 385 nm.

4.5.1. Swelling and Biodegradation

Cylindrical specimens ($n = 3$/group) were immersed in PBS to allow for swelling for 24 h at 37 °C. Next, the specimens were weighed to establish their wet weight (Ww), then they were lyophilized and weighed again to establish their dry weight (Wd). The volumetric swelling ratio (in %) of each hydrogel group was calculated using the following equation:

$$\text{Swelling Ratio} = (\text{Ww} - \text{Wd})/\text{Wd} \times 100 \tag{1}$$

The in vitro biodegradation of the hydrogels was carried out by incubating cylindrical specimens (6 mm diameter × 2 mm thick) of each hydrogel in an enzymatic solution comprised of DPBS and collagenase type I (1 U/mL, Roche Holding AG, Basel, Switzerland). The specimens (n = 4/group) were weighed at baseline (W0) before their incubation in 5 mL of enzymatic solution for 3 weeks at 37 °C; then, the solution was renewed every 3 days with fresh solution to maintain constant enzyme activity. At present time points, the specimens were removed from the solution, washed twice with sterile DI water, blotted dry with low-lint wipes, and weighed again (Wt). The degradation (in %) of each hydrogel group was calculated using the following equation:

$$\text{Degradation} = (\text{Wt}/\text{W0}) \times 100 \tag{2}$$

4.5.2. Cell Viability

To determine whether modifying the GelMA hydrogel with antibiotic-laden fiber particles would result in cell toxicity, hydrogel specimens (6 mm in diameter × 2-mm thick) were prepared for an in vitro assay following the International Standards Organization guidelines (ISO, 10993-5). Initially, the specimens (n = 5/group) were UV-treated for 30 min on each side to disinfect them. Then, they were individually placed in sterile scintillation glass vials (VWR International, LLC, Radnor, PA, USA) that contained 5 mL of alpha-modified Eagle's Medium (α-MEM; Gibco Invitrogen Corporation, Grand Island, NY, USA) supplemented with 10% fetal bovine serum (FBS; Gibco), L-glutamine (Sigma), 1% penicillin-streptomycin (Gibco), and 1 U/mL of collagenase type I (Roche Holding AG). Next, they were incubated at 37 °C for up to 21 days, with 500 µL aliquots collected at different time points (after 1, 3, 7, 14, and 21 days of storage) to investigate the potential cytotoxic effects of leachable hydrogels' byproducts (e.g., antibiotics) over time. Equal amounts of storage medium were added back to each vial to maintain a constant extraction volume. Finally, prior to cell exposure, the collected aliquots were filtered through a 0.22 µm membrane (MilliporeSigma, Burlington, MA, USA).

Stem cells derived from human exfoliated deciduous teeth (SHEDs; Lonza, Walkersville, MD, USA) were cultured in an incubator at 37 °C with 5% CO_2 in α-MEM supplemented with 10% FBS, 1% L-glutamine, and 1% penicillin–streptomycin. Cells at passages 4 to 7 were utilized. SHEDs were seeded at a density of 2.5×10^3 cells/well and were allowed to adhere in the wells of 96-well plates (Corning Incorporated, Corning, NY, USA). After 24 h, the media were subsequently replaced by collected extracts (100 µL) taken from GelMA-based hydrogels. For 24 h, the aliquots were kept in contact with the cells. An amount of 30 µL of CellTiter 96 AQueous One Solution Reagent (Promega Corporation, Madison, WI, USA) was then added to the test wells and allowed to react at 37 °C in a humidified 5% CO_2 atmosphere. The incorporated dye was measured by reading the absorbance at 490 nm (SpectraMax iD3; Molecular Devices LLC, San Jose, CA, USA) and comparing it with a blank column. SHEDs cultured in complete α-MEM were used as the positive control. Absorbance values were converted to percentages and compared with the test groups' values.

4.5.3. Antimicrobial Efficacy

The antimicrobial efficacy of the antibiotic-laden fiber-modified hydrogels was verified against endodontic pathogens (agar diffusion) and biofilm inhibition using an infected dentin *A. naeslundii* biofilm model evaluated both quantitatively (CFU/mL) and qualitatively via confocal laser scanning microscopy (CLSM) and SEM. For the agar diffusion assay, the hydrogels were tested against *Actinomyces naeslundii* (*A. naeslundii*, ATCC 12104), *Fusobacterium nucleatum* (*F. nucleatum*, ATCC 25586), and *Enterococcus faecalis* (*E. faecalis*, ATCC 19433) bacteria. Cylindrical-shaped (6 mm diameter × 2 mm thick) specimens were prepared (n = 3/group) and disinfected by UV-irradiation (30 min/side). *A. naeslundii* and *F. nucleatum* were anaerobically cultured for 24 h in 5 mL of brain and heart infusion (BHI) broth. *E. faecalis* was aerobically cultured in 5 mL of BHI broth for 24 h. Each bacterial suspension was spectrophotometrically (405 nm) adjusted to obtain 3×10^8 CFU/mL. An amount of 100 µL of each broth was swabbed onto BHI agar plates to form a bacterial lawn. To evaluate the antimicrobial properties over time, GelMA-based hydrogel specimens of the same dimensions were prepared ($n = 3$/group); the specimens were individually incubated in glass vials with 5 mL of sterile PBS for 3 weeks at 37 °C. At predetermined time intervals (1, 3, 7, and 14 days), 500 µL aliquots were drawn and replaced with equivalent amounts of fresh DPBS. The retrieved aliquots were stored at −20 °C until further use. The agar plate was divided into zones: 10 µL of 2% chlorhexidine digluconate (CHX; positive control), 10 µL of DI water (negative control), and 20 µL of GelMA-based aliquots. After incubating for 48 h (*F. nucleatum* and *A. naeslundii*), diameters (in mm) of the clear zones of growth inhibition were measured.

For the anti-biofilm assay, the experiment was conducted only after approval by the local Institutional Review Board (IRB protocol no. 1407656657; University of Michigan). Fifty-four recently extracted, single-rooted human teeth were cleaned and stored in 0.1% thymol until use. The teeth were cut to obtain dentin slices (2 mm thick) so that the crown portion was sectioned 2 mm above the cementum–enamel junction and was cut along the buccolingual plane. The specimens were then wet-finished with SiC papers (600–1200 grit) to obtain both standardized and smooth surfaces, followed by immersion in 2.5% NaOCl and 17% ethylenediaminetetraacetic acid (EDTA) solutions for 5 and 3 min, respectively, under the ultrasonic bath (L&R 2014 Ultrasonic Cleaning System, L&R Ultrasonics, Kearny, NJ, USA). Then, the specimens were rinsed in sterile saline solution for 10 min and autoclaved at 121 °C for 20 min.

A. naeslundii (ATCC 12104) was cultivated in an anaerobic chamber for 48 h. The bacterial suspension was adjusted for approximately 7.5×10^7 colony-forming units per milliliter in BHI broth. The sterile dentin slices were placed into 24-well plates that contained 1.8 mL of BHI broth and 0.2 mL of the inoculum and were incubated in an anaerobic chamber for 7 days to allow for the formation of biofilm. The broth was renewed every 2 days. Infected dentin slices ($n = 6$/group) were randomly divided into six groups: GelMA, CLIN, MET, and CLIN+MET hydrogels, as well as Ca(OH)$_2$ paste (positive control) and an untreated 7-day-old biofilm (negative control). After 7 days, non-adherent bacteria were removed by gently rinsing the samples in PBS. Next, 50 µL of each material was placed above the biofilm formed and crosslinked for 15 s with the LED. The samples were then incubated for 3 days in the anaerobic chamber and divided for colony forming units (CFU/mL; $n = 4$), scanning electron microscopy (SEM; $n = 2$), and confocal laser scanning microscopy (CLSM; $n = 3$) analyses.

For CFU/mL, the incubated samples were cautiously removed from the wells and placed in Eppendorf tubes containing 1 mL of saline solution. The tubes were sonicated at 30 W for 30 s to detach the biofilms formed on the dentin slices. After that, 100 µL aliquots were collected and subjected to serial dilution, which was carried out in BHI blood agar plates. The plates were then incubated at 37 °C for 24 h in an anaerobic chamber, with the CFU/mL being counted. To prepare them for SEM evaluation, the samples were gently washed in PBS and fixed overnight in 2.5% glutaraldehyde. Next, they were dehydrated in increased concentrations of alcohol/water solutions, treated with increased concentrations of HMDS solutions, and sputter-coated with Au-Pd prior to imaging. For CLSM evaluation, after each treatment was applied, the samples were gently washed with DI water and stained using the fluorescent LIVE/DEAD BacLight Bacterial Viability Kit L-7012 (Molecular Probes, Inc., Eugene, OR, USA). Three areas of each sample were analyzed utilizing 3D reconstruction. They were selected randomly, always starting from the root canal space to the cementum side. A 40× lens (Leica SP2 CL5Mt; Leica Microsystems GmbH, Wetzlar, HE, Germany) was used. The sequence of segments through tissue depth (Z-stacks) was collected using optimal step-size settings (0.35 µm), with the images being composed of 512×512 pixels. The excitation-emission maxima for the dyes, respectively, were approximately 480/500 nm for SYTO 9 and 490/635 nm for PI. The images were reconstructed using ImageJ and the percentages of live/dead bacteria were compared with the positive [Ca(OH)$_2$] and negative (no treatment) controls to establish statistical significance among the groups.

4.6. Statistical Analysis

The obtained data were statistically analyzed (SigmaPlot version 12; Systat Software, Inc., Chicago, IL, USA) using analysis of variance and Tukey's test for multiple comparisons at the $\alpha = 5\%$ level of significance.

5. Conclusions

In this work, we successfully designed a photo-curable injectable hydrogel loaded with antibiotic-laden fibrous microparticles, which demonstrated effective antimicrobial activity and non-cytotoxic behavior. Moreover, based on the collected data, the proposed hydrogel holds clinical promise for bacterial infection ablation before regenerative endodontics procedures. Further pre-clinical animal studies (e.g., periapical disease model in rodents) are paramount to investigating the antimicrobial efficacy of this new injectable system.

Author Contributions: Conceptualization, M.C.B. and J.S.R.; methodology, J.S.R., N.S.R., N.D. and E.A.F.B.; software, J.S.R.; validation, N.S.R. and N.D.; formal analysis, N.D.; investigation, J.S.R.; resources, J.C.F. and M.C.B.; data curation, E.A.M.; writing—original draft preparation, J.S.R. and N.S.R.; writing—review and editing, S.S., J.C.F. and M.C.B.; visualization, H.S.; supervision, J.C.F.; project administration, M.C.B.; funding acquisition, M.C.B. All authors have read and agreed to the published version of the manuscript.

Funding: M.C.B. acknowledges the National Institutes of Health (NIH, National Institute of Dental and Craniofacial Research/NIDCR, grants K08DE023552 and R01DE026578) and the International Association for Dental Research (IADR-GSK Innovation in Oral Care Award). E.A.F.B and J.S.R. acknowledge the scholarship support from the São Paulo State Research Foundation (FAPESP) (grant numbers 2016/15674-5 and 2018/14257-7) and CAPES Foundation (Brazil), respectively.

Institutional Review Board Statement: The study was conducted in accordance with the Declaration of Helsinki, and approved by the Institutional Review Board (IRB) of the University of Michigan (protocol code: HUM00164678 and date of approval: 6/24/2019) for studies involving humans.

Informed Consent Statement: Not applicable. The proposed study does not fit the definition of research involving human subjects (45CFR46.102) because the researchers intending to contribute to generalizable knowledge do not interact with human subjects, nor obtain identifiable private information or identifiable biospecimens.

Data Availability Statement: The data presented in this study are available on request from the corresponding author.

Acknowledgments: The authors are thankful for the assistance provided by Jennifer Walker with the cryomilling experiments.

Conflicts of Interest: The authors declare no conflict of interest.

Abbreviations

SEM, Scanning Electron Microscope; PLGA, Poly(lactic-co-glycolic acid); CLIN, Clindamycin; MET, metronidazole; FTIR, Fourier-transform infrared spectroscopy; GelMA, gelatin methacryloyl hydrogel; SHEDs, stem cells from exfoliated deciduous teeth; CLSM, confocal laser scanning microscopy; *A. naeslundii*, *Actinomyces naeslundii*; *F. nucleatum*, *Fusobacterium nucleatum*; *E. faecalis*, *Enterococcus faecalis*.

References

1. Diogenes, A.; Henry, M.A.; Teixeira, F.B.; Hargreaves, K.M. An update on clinical regenerative endodontics. *Endod. Top.* **2013**, *28*, 2–23. [CrossRef]
2. Mohammadi, Z.; Dummer, P.M. Properties and applications of calcium hydroxide in endodontics and dental traumatology. *Int. Endod. J.* **2011**, *44*, 697–730. [CrossRef] [PubMed]

3.   Kayaoglu, G.; Erten, H.; Bodrumlu, E.; Orstavik, D. The resistance of collagen-associated, planktonic cells of enterococcus faecalis to calcium hydroxide. *J. Endod.* **2009**, *35*, 46–49. [CrossRef]

4.   Mohammadi, Z.; Shalavi, S.; Yazdizadeh, M. Antimicrobial activity of calcium hydroxide in endodontics: A review. *Chonnam Med. J.* **2012**, *48*, 133–140. [CrossRef]

5.   Estrela, C.; Pimenta, F.C.; Ito, I.Y.; Bammann, L.L. Antimicrobial evaluation of calcium hydroxide in infected dentinal tubules. *J. Endod.* **1999**, *25*, 416–418. [CrossRef]

6.   Bottino, M.C.; Albuquerque, M.T.P.; Azabi, A.; Münchow, E.A.; Spolnik, K.J.; Nör, J.E.; Edwards, P.C. A novel patient-specific three-dimensional drug delivery construct for regenerative endodontics. *J. Biomed. Mater. Res. B Appl. Biomater.* **2019**, *107*, 1576–1586. [CrossRef] [PubMed]

7.   Chuensombat, S.; Khemaleelakul, S.; Chattipakorn, S.; Srisuwan, T. Cytotoxic effects and antibacterial efficacy of a 3-antibiotic combination: An in vitro study. *J. Endod.* **2013**, *39*, 813–819. [CrossRef]

8.   Bansal, R.; Jain, A.; Goyal, M.; Singh, T.; Sood, H.; Malviya, H.S. Antibiotic abuse during endodontic treatment: A contributing factor to antibiotic resistance. *J. Fam. Med. Prim. Care* **2019**, *8*, 3518–3524. [CrossRef] [PubMed]

9.   Karczewski, A.; Feitosa, S.A.; Hamer, E.I.; Pankajakshan, D.; Gregory, R.L.; Spolnik, K.J.; Bottino, M.C. Clindamycin-modified triple antibiotic nanofibers: A stain-free antimicrobial intracanal drug delivery system. *J. Endod.* **2018**, *44*, 155–162. [CrossRef] [PubMed]

10.   Porter, M.L.; Munchow, E.A.; Albuquerque, M.T.; Spolnik, K.J.; Hara, A.T.; Bottino, M.C. Effects of novel 3-dimensional antibiotic-containing electrospun scaffolds on dentin discoloration. *J. Endod.* **2016**, *42*, 106–112. [CrossRef]

11.   Moreira, M.S.; Sarra, G.; Carvalho, G.L.; Gonçalves, F.; Caballero-Flores, H.V.; Pedroni, A.C.F.; Lascala, C.A.; Catalani, L.H.; Marques, M.M. Physical and biological properties of a chitosan hydrogel scaffold associated to photobiomodulation therapy for dental pulp regeneration: An in vitro and in vivo study. *BioMed Res. Int.* **2021**, *2021*, 6684667. [CrossRef] [PubMed]

12.   Ribeiro, J.S.; Bordini, E.A.F.; Ferreira, J.A.; Mei, L.; Dubey, N.; Fenno, J.C.; Piva, E.; Lund, R.G.; Schwendeman, A.; Bottino, M.C. Injectable mmp-responsive nanotube-modified gelatin hydrogel for dental infection ablation. *ACS Appl. Mater. Interfaces* **2020**, *12*, 16006–16017. [CrossRef] [PubMed]

13.   AlSaeed, T.; Nosrat, A.; Melo, M.A.; Wang, P.; Romberg, E.; Xu, H.; Fouad, A.F. Antibacterial efficacy and discoloration potential of endodontic topical antibiotics. *J. Endod.* **2018**, *44*, 1110–1114. [CrossRef] [PubMed]

14.   Bekhouche, M.; Bolon, M.; Charriaud, F.; Lamrayah, M.; Da Costa, D.; Primard, C.; Costantini, A.; Pasdeloup, M.; Gobert, S.; Mallein-Gerin, F.; et al. Development of an antibacterial nanocomposite hydrogel for human dental pulp engineering. *J. Mater. Chem. B* **2020**, *8*, 8422–8432. [CrossRef] [PubMed]

15.   Albuquerque, M.T.; Evans, J.D.; Gregory, R.L.; Valera, M.C.; Bottino, M.C. Antibacterial tap-mimic electrospun polymer scaffold: Effects on p. Gingivalis-infected dentin biofilm. *Clin. Oral Investig.* **2016**, *20*, 387–393. [CrossRef] [PubMed]

16.   Albuquerque, M.T.; Ryan, S.J.; Munchow, E.A.; Kamocka, M.M.; Gregory, R.L.; Valera, M.C.; Bottino, M.C. Antimicrobial effects of novel triple antibiotic paste-mimic scaffolds on actinomyces naeslundii biofilm. *J. Endod.* **2015**, *41*, 1337–1343. [CrossRef]

17.   Bruggeman, K.F.; Wang, Y.; Maclean, F.L.; Parish, C.L.; Williams, R.J.; Nisbet, D.R. Temporally controlled growth factor delivery from a self-assembling peptide hydrogel and electrospun nanofibre composite scaffold. *Nanoscale* **2017**, *9*, 13661–13669. [CrossRef] [PubMed]

18.   Li, W. Supramolecular nanofiber-reinforced puerarin hydrogels as drug carriers with synergistic controlled release and antibacterial properties. *J. Mater. Sci.* **2020**, *55*, 6669–6677. [CrossRef]

19.   Wu, R.; Niamat, R.A.; Sansbury, B.; Borjigin, M. Fabrication and evaluation of multilayer nanofiber-hydrogel meshes with a controlled release property. *Fibers* **2015**, *3*, 296–308. [CrossRef]

20.   Buchanan, F.; Gallagher, L.; Jack, V.; Dunne, N. Short-fibre reinforcement of calcium phosphate bone cement. *Proc. Inst. Mech. Eng. H* **2007**, *221*, 203–211. [CrossRef]

21.   Knotek, P.; Pouzar, M.; Buzgo, M.; Krizkova, B.; Vlcek, M.; Mickova, A.; Plencner, M.; Navesnik, J.; Amler, E.; Belina, P. Cryogenic grinding of electrospun poly-epsilon-caprolactone mesh submerged in liquid media. *Mater. Sci. Eng. C Mater. Biol. Appl.* **2012**, *32*, 1366–1374. [CrossRef] [PubMed]

22.   Nichol, J.W.; Koshy, S.T.; Bae, H.; Hwang, C.M.; Yamanlar, S.; Khademhosseini, A. Cell-laden microengineered gelatin methacrylate hydrogels. *Biomaterials* **2010**, *31*, 5536–5544. [CrossRef]

23.   Monteiro, N.; Thrivikraman, G.; Athirasala, A.; Tahayeri, A.; França, C.M.; Ferracane, J.L.; Bertassoni, L.E. Photopolymerization of cell-laden gelatin methacryloyl hydrogels using a dental curing light for regenerative dentistry. *Dent. Mater.* **2018**, *34*, 389–399. [CrossRef] [PubMed]

24.   Sangnim, T.; Limmatvapirat, S.; Nunthanid, J.; Sriamornsak, P.; Wannachaiyasit, S.; Sittikijyothin, W.; Huanbutta, K. Design and characterization of clindamycin-loaded nanofiber patches composed of polyvinyl alcohol and tamarind seed gum and fabricated by electrohydrodynamic atomization. *Asian J. Pharm. Sci.* **2018**, *13*, 450–458. [CrossRef] [PubMed]

25.   Kumar, M.; Awasthi, R. Development of metronidazole-loaded colon-targeted microparticulate drug delivery system. *Polim. Med.* **2015**, *45*, 57–65. [PubMed]

26.   Trivedi, M.K.; Patil, S.; Shettigar, H.; Bairwa, K.; Jana, S. Spectroscopic characterization of biofield treated metronidazole and tinidazole. *Med. Chem.* **2015**, *5*. [CrossRef]

27. Carnaval, T.G.; Goncalves, F.; Romano, M.M.; Catalani, L.H.; Mayer, M.A.P.; Arana-Chávez, V.E.; Nishida, A.C.; Lage, T.C.; Francci, C.E.; Adde, C.A. In vitro analysis of a local polymeric device as an alternative for systemic antibiotics in dentistry. *Braz. Oral Res.* **2017**, *31*, e92. [CrossRef] [PubMed]

28. Dargaville, B.L.; Vaquette, C.; Rasoul, F.; Cooper-White, J.J.; Campbell, J.H.; Whittaker, A.K. Electrospinning and crosslinking of low-molecular-weight poly(trimethylene carbonate-co-(l)-lactide) as an elastomeric scaffold for vascular engineering. *Acta Biomater.* **2013**, *9*, 6885–6897. [CrossRef] [PubMed]

29. Munchow, E.A.; da Silva, A.F.; Piva, E.; de Albuquerque, M.T.P.; Pinal, R.; Gregory, R.L.; Breschi, L.; Bottino, M.C. Development of an antibacterial and anti-metalloproteinase dental adhesive for long-lasting resin composite restorations. *J. Mater. Chem. B* **2020**, *8*, 10797–10811. [CrossRef] [PubMed]

30. Ribeiro, J.S.; Daghrery, A.; Dubey, N.; Li, C.; Ling, M.; Fenno, J.C.; Schwendeman, A.; Aytac, Z.; Bottino, M.C. Hybrid antimicrobial hydrogel as injectable therapeutics for oral infection ablation. *Biomacromolecules* **2020**, *21*, 3945–3956. [CrossRef] [PubMed]

31. Pepelanova, I.; Kruppa, K.; Scheper, T.; Lavrentieva, A. Gelatin-methacryloyl (gelma) hydrogels with defined degree of functionalization as a versatile toolkit for 3d cell culture and extrusion bioprinting. *Bioengineering* **2018**, *18*, 55. [CrossRef] [PubMed]

32. Gokturk, H.; Ozkocak, I.; Buyukgebiz, F.; Demir, O. An in vitro evaluation of various irrigation techniques for the removal of double antibiotic paste from root canal surfaces. *J. Appl. Oral Sci.* **2016**, *24*, 568–574. [CrossRef]

33. McCann, J.T.; Marquez, M.; Xia, Y. Highly porous fibers by electrospinning into a cryogenic liquid. *J. Am. Chem. Soc.* **2006**, *128*, 1436–1437. [CrossRef] [PubMed]

34. Ribeiro, J.S.; Munchow, E.A.; Bordini, E.A.F.; da Rosa, W.L.O.; Bottino, M.C. Antimicrobial therapeutics in regenerative endodontics: A scoping review. *J. Endod.* **2020**, *46*, S115–S127. [CrossRef] [PubMed]

35. Sousa, E.L.; Gomes, B.P.; Jacinto, R.C.; Zaia, A.A.; Ferraz, C.C.R. Microbiological profile and antimicrobial susceptibility pattern of infected root canals associated with periapical abscesses. *Eur. J. Clin. Microbiol. Infect. Dis.* **2013**, *32*, 573–580. [CrossRef] [PubMed]

36. Nagata, J.Y.; Soares, A.J.; Souza-Filho, F.J.; Zaia, A.A.; Ferraz, C.C.R.; Almeida, J.F.A.; Gomes, B.P.F.A. Microbial evaluation of traumatized teeth treated with triple antibiotic paste or calcium hydroxide with 2% chlorhexidine gel in pulp revascularization. *J. Endod.* **2014**, *40*, 778–783. [CrossRef] [PubMed]

37. Sassone, L.M.; Fidel, R.; Faveri, M.; Fidel, S.; Figueiredo, L.; Fereset, M. Microbiological evaluation of primary endodontic infections in teeth with and without sinus tract. *Int. Endod. J.* **2008**, *41*, 508–515. [CrossRef]

38. Spizek, J.; Rezanka, T. Lincosamides Chemical structure, biosynthesis, mechanism of action, resistance, and applications. *Biochem. Pharmacol.* **2017**, *133*, 20–28. [CrossRef] [PubMed]

39. Alyas, S.M.; Fischer, B.I.; Ehrlich, Y.; Spolnik, K.; Gregory, R.G.; Yassen, G.H. Direct and indirect antibacterial effects of various concentrations of triple antibiotic pastes loaded in a methylcellulose system. *J. Oral Sci.* **2016**, *58*, 575–582. [CrossRef]

40. Wang, L.; Hu, C.; Shao, L. The antimicrobial activity of nanoparticles: Present situation and prospects for the future. *Int. J. Nanomed.* **2017**, *12*, 1227–1249. [CrossRef] [PubMed]

41. Van Den Bulcke, A.I.; Bogdanov, B.; De Rooze, N.; Schacht, E.H.; Cornelissen, M.; Berghmanset, H. Structural and rheological properties of methacrylamide modified gelatin hydrogels. *Biomacromolecules* **2000**, *1*, 31–38. [CrossRef] [PubMed]

International Journal of
Molecular Sciences

Article

Kappa-Carrageenan/Chitosan/Gelatin Scaffolds Provide a Biomimetic Microenvironment for Dentin-Pulp Regeneration

Konstantinos Loukelis [1], Foteini Machla [2], Athina Bakopoulou [2,*] and Maria Chatzinikolaidou [1,3,*]

[1] Department of Materials Science and Technology, University of Crete, 70013 Heraklion, Greece
[2] Department of Prosthodontics, School of Dentistry, Faculty of Health Sciences, Aristotle University of Thessaloniki, 54124 Thessaloniki, Greece
[3] Foundation for Research and Technology Hellas-Institute of Electronic Structure and Laser (FORTH-IESL), 70013 Heraklion, Greece
* Correspondence: abakopoulou@dent.auth.gr (A.B.); mchatzin@materials.uoc.gr (M.C.)

Abstract: This study aims to investigate the impact of kappa-carrageenan on dental pulp stem cells (DPSCs) behavior in terms of biocompatibility and odontogenic differentiation potential when it is utilized as a component for the production of 3D sponge-like scaffolds. For this purpose, we prepared three types of scaffolds by freeze-drying (i) kappa-carrageenan/chitosan/gelatin enriched with KCl (KCG-KCl) as a physical crosslinker for the sulfate groups of kappa-carrageenan, (ii) kappa-carrageenan/chitosan/gelatin (KCG) and (iii) chitosan/gelatin (CG) scaffolds as a control. The mechanical analysis illustrated a significantly higher elastic modulus of the cell-laden scaffolds compared to the cell-free ones after 14 and 28 days with values ranging from 25 to 40 kPa, showing an increase of 27–36%, with the KCG-KCl scaffolds indicating the highest and CG the lowest values. Cell viability data showed a significant increase from days 3 to 7 and up to day 14 for all scaffold compositions. Significantly increasing alkaline phosphatase (ALP) activity has been observed over time in all three scaffold compositions, while the KCG-KCl scaffolds indicated significantly higher calcium production after 21 and 28 days compared to the CG control. The gene expression analysis of the odontogenic markers DSPP, ALP and RunX2 revealed a two-fold higher upregulation of DSPP in KCG-KCl scaffolds at day 14 compared to the other two compositions. A significant increase of the RunX2 expression between days 7 and 14 was observed for all scaffolds, with a significantly higher increase of at least twelve-fold for the kappa-carrageenan containing scaffolds, which exhibited an earlier ALP gene expression compared to the CG. Our results demonstrate that the integration of kappa-carrageenan in scaffolds significantly enhanced the odontogenic potential of DPSCs and supports dentin-pulp regeneration.

Keywords: biocompatibility; dental pulp stem cells; dental tissue engineering; odontogenic differentiation; scaffolds; tissue regeneration

Citation: Loukelis, K.; Machla, F.; Bakopoulou, A.; Chatzinikolaidou, M. Kappa-Carrageenan/Chitosan/Gelatin Scaffolds Provide a Biomimetic Microenvironment for Dentin-Pulp Regeneration. *Int. J. Mol. Sci.* **2023**, 24, 6465. https://doi.org/10.3390/ijms24076465

Academic Editors: Kerstin M. Galler and Matthias Widbiller

Received: 9 February 2023
Revised: 23 March 2023
Accepted: 27 March 2023
Published: 30 March 2023

1. Introduction

Progress in dental tissue engineering and regenerative dentistry has been tremendous over the past decades, with recent milestones on translational research having led to the development of innovative concepts in tissue engineering of hard and soft dental tissues, among which is the dentin-pulp complex [1]. Since the discovery of dental pulp stem cells (DPSCs), many biomaterials and signaling molecules have been investigated for their odontogenic response [2]. For example, in cases of direct and indirect pulp capping, various biocompatible platforms containing calcium and silicate-based materials have been utilized, aiming to reconstitute the exposed area and induce the odontogenic differentiation of the cells and the formation of a biomimetic mineralized barrier (tertiary dentinogenesis) [3].

Adult stem cells are considered to be a versatile model system and can be isolated from multiple organs of the human body. Their main advantage is their inert trait of self-renewal and their capability towards various differentiation lineages in the presence of appropriate stimulants [4,5]. DPSCs are a particular category of ectoderm-derived mesenchymal stem cells originating from migrating neural crest cells [6,7]. They reside in the pulp of permanent teeth and their main role is the production of odontoblasts and dentin. Moreover, the accessibility and the relatively easy process of isolation compared to other stem cell types, as well as their excellent behavior under in vitro conditions, make them a very promising cell type for experimentation in the regenerative medicine area [2,4,8]. For these reasons, DPSCs have been used in dental tissue engineering to evaluate their potential for dental reconstruction [9,10].

Scaffold-based therapies have been employed promisingly to achieve suitable matrices able to accommodate the regeneration of adjacent tissues [11]. Chitosan, a natural biopolymer that can be extracted from chitin, the main ingredient of the exoskeleton of different arthropod species such as shrimps, is one of the most commonly utilized biomaterials in tissue engineering [12]. Chitosan possesses alternating glucosamine and N-glucosamine groups to which it owes its excellent biocompatibility and antibacterial properties [12,13]. Based on its abundance and easy processability, chitosan has found applicability in various research fields, from the food industry [14] to the construction of tissue engineering scaffolds [15]. Additionally, it has been used as a promising scaffold for dentin and pulp tissue engineering [16]. One of the hurdles be overcome in the case of any potential regenerative medicine device is its ability to avoid an excessive inflammatory response, which can in turn lead to the rejection of the inserted transplant [17]. Gelatin is a natural biomaterial derived from hydrolyzed collagen comprising an alternating sequence of the arginine, glycine and aspartate (RGD) tripeptide motif, which promotes cell adhesion and a very low exhibition of immune responses [18]. Since gelatin is quite inexpensive while retaining most of the biological attributes of collagen, it is frequently employed in tissue engineering either in its native [19] or methacrylated form [20]. Due to their great biochemical response, chitosan and gelatin have been combined and studied, especially in the field of bone and dental tissue engineering [21–23]. For example, Georgopoulou et al. [15] reported on the preparation of chitosan/gelatin scaffolds, which were chemically crosslinked with a low concentration of either glutaraldehyde or genipin solution, to showcase the bone differentiation capacity of pre-osteoblasts and bone marrow derived mesenchymal stem cells (MSCs) in vitro and in vivo. Scaffolds depicted an excellent biocompatibility and upregulation of the examined osteogenic markers RUNX2, ALP and OSC up to day 14, while their insertion into a mouse femur illustrated the establishment of a well-spread extracellular matrix without adverse reactions. Capitalizing on the antibacterial properties of chitosan, Pereda et al. [24] reported on chitosan/gelatin edible films for long term food item preservation. Their results showed great antimicrobial activity, while the membranes were impenetrable to water, a crucial parameter for microbial growth inhibition.

Carrageenans are a family of various polysaccharides deriving from red seaweeds. Among them, kappa and iota-carrageenan have attracted great attention due to their gelling, thickening, stability and non-toxicity properties in food and commodity industries, but also as ingredients for regenerative medicine constructs due to their ability to promote cell growth, proliferation and differentiation [25–27]. The biological properties of kappa-carrageenan stem from its three hydroxyl groups per disaccharide repeating unit, which makes it highly hydrophilic, and its one negatively charged sulfate group, which enables chemical reactions [28]. Apart from the biocompatibility, tissue engineered constructs should possess sufficient mechanical stiffness similar to the tissue type targeted to regenerate to be successful [29]. In the presence of cations, especially monovalent ones such as potassium, kappa-carrageenan formation turns from a coil shape to a helicoidal structure, leading to increased gelling strength [30,31] and making it a prime candidate for tissue engineering applications [32].

Chitosan/gelatin scaffolds of various compositions have been previously examined for their odontogenic capacity, with promising in vitro and in vivo results [20,21]. In a previous study, we showed that the integration of kappa-carrageenan in chitosan/gelatin scaffolds positively influenced the mechanical properties and the osteogenic potential of pre-osteoblasts within the scaffolds [33]. In the context of dental applications, kappa-carrageenan has been described as a potent anti-human papillomavirus agent in cells and tissues of the oral cavity and as a hemostatic sponge when combined with gelatin, without any cytotoxicity in L929 fibroblasts [34,35]. However, there are no reports in the literature on tissue engineered carrageenan-based scaffolds for odontogenic differentiation. Therefore, in this study we aim to evaluate the impact of kappa-carrageenan in chitosan/gelatin scaffolds as a biomimetic microenvironment for dentin-pulp regeneration. In the present study, kappa-carrageenan/chitosan/gelatin (KCG) and kappa-carrageenan/chitosan/gelatin enriched with KCl (KCG-KCl) scaffolds have been produced, and the biological responses of DPSCs seeded onto them, including viability, proliferation, morphology and differentiation capacity towards the formation of mature odontoblasts, have been examined. Moreover, the mechanical strength of cell-loaded scaffolds was compared to that of cell-free scaffolds to deduce if the viscoelastic nature of DPSCs contributes to the reinforcement of the total robustness of the constructs, as an essential biomechanical attribute that controls cellular responses. To investigate the odontogenic capacity of the scaffolds, alkaline phosphatase (ALP) activity and calcium secretion were determined, and the gene expression levels of dental-related markers were evaluated by means of quantitative polymerase chain reaction (qPCR).

2. Results

2.1. Immunophenotypic Characterization of DPSC Cultures

The analysis of flow cytometry results revealed the high expression of MSC markers of the cell population, as illustrated in Figure 1. Specifically, CD90 was present at 98% and CD73 at 96% of the total cells compared to the control, and CD146 and STRO-1 at 73 and 18%, respectively. The endothelial cell marker CD105 was observed at 91%, while the hematopoietic markers CD34 and CD45 had minor expression (<10%). All expressed surface markers indicated a highly enriched DPSCs population.

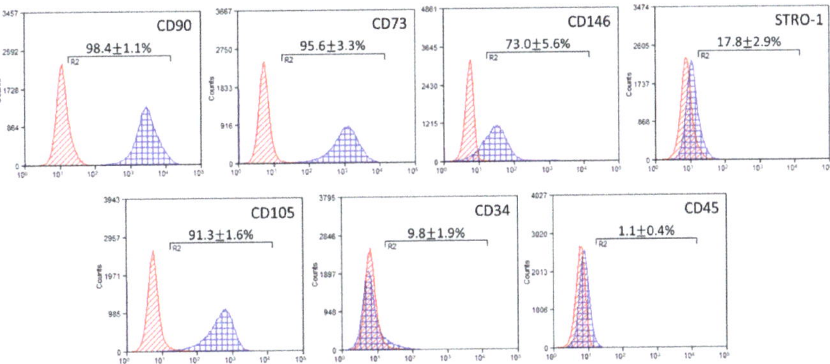

Figure 1. Representative histograms of immunophenotypic characterization by flow cytometry results on DPSCs surface markers: mesenchymal (CD90, CD73, CD146, STRO1), endothelial (CD105) and hematopoietic (CD34, CD45). The red area illustrates the expression of the control (unstained cells), and the blue area the marker of interest. The arithmetical values are the mean expression ± SD (*n* = 3).

2.2. DPSCs Contribution to the Mechanical Integrity of the Scaffolds

The Young's modulus was measured at a strain of 5–20% and velocity of 1 mm/s (Figure 2) to monitor if the presence of cells affected the mechanical strength of the scaffolds. The measurements were performed under wet conditions, and the wet cell-free scaffolds were compared to the wet DPSCs laden scaffolds after 14 and 28 days in culture. Prior to the measurements, the excess of culture medium was removed from the scaffolds. Among the cell-free scaffolds, the KCG-KCl ones demonstrated the highest value, close to 28 kPa, while the KCG and control CG retained comparable values, approximately at 23 kPa. After 14 days in culture, all scaffolds showed a slight increase in the Young's modulus, with the KCG-KCl indicating the highest value of 32 kPa, followed by the KCG at 25 kPa and the CG at 24 kPa. At day 28, all scaffold types exhibited the highest Young's modulus values, with the KCG-KCl at 39 kPa and the other two at 30 kPa, demonstrating a 36% and 33% increase between day 0 and day 28, respectively. These findings strongly support the idea that the infiltration and proliferation of the cells inside the scaffolds led to mechanically enhanced constructs.

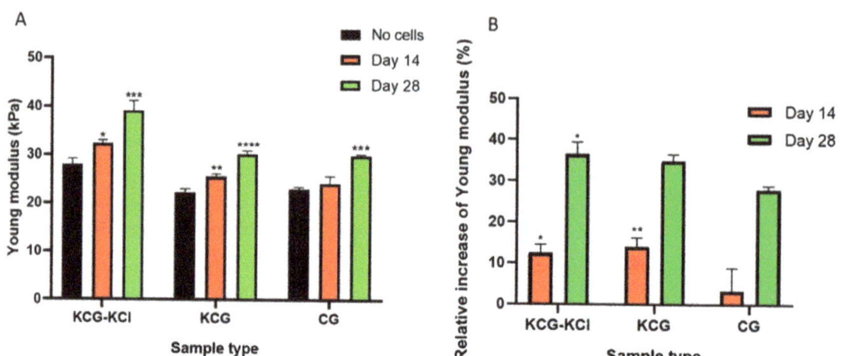

Figure 2. (**A**) Evaluation of the Young's modulus at 5–20% strain and 1 mm/s velocity for the three scaffold compositions. (**B**) Relative expression of Young's modulus increase of cell-loaded scaffolds between days 0 and 14 (red bars) and between days 0 and 28 (green bars) compared to cell-free scaffolds. Each bar represents the mean ± SD of $n = 6$ (* $p < 0.05$, ** $p < 0.01$, *** $p < 0.001$, **** $p < 0.0001$ compared to cell-free scaffolds (no cells)).

2.3. SEM Analysis of DPSCs Morphology and Adhesion

Scanning electron microscopy enables the illustration of the morphology of the scaffolds, the cell adhesion on their surface and the cell infiltration into the pores of the scaffolds. Figure 3 depicts representative SEM images of the three types of scaffolds, KCG-KCl, KCG and CG, either cell-free or loaded with DPSCs after 2 and 10 days of culture. Cell-free images are displayed in Figure 3 (upper panel) and exhibit the initially empty pores of the scaffolds that had been completely filled with cells by day 10. At day 2, adhered DPSCs could be detected within the pores, and the cell nuclei of healthy cells are also visible (white arrows point to the them). The morphology of attached cells on the scaffolds did not show any differences among the various compositions. At day 10, the pores of all three scaffold compositions were filled with elongated dense intercellular formations of DPSCs, signifying tissue growth.

2.4. DPSCs' Growth and Proliferation within the Various Scaffold Types

All scaffold types demonstrated great biocompatible character, but no significant differences in their cytocompatibility levels could be detected (Figure 4). In detail, at day 3, all samples depicted similar absorbance values and, by day 7, almost a two-fold increase was evident for every composition compared to the previous time point, with the KCG-KCl one exceeding the rest by a small margin. At day 14, the same motif of growth was observed, with an almost three-fold increase in cell number compared to day 7 and with the KCG-KCl retaining a slightly higher value than the KCG and the CG control.

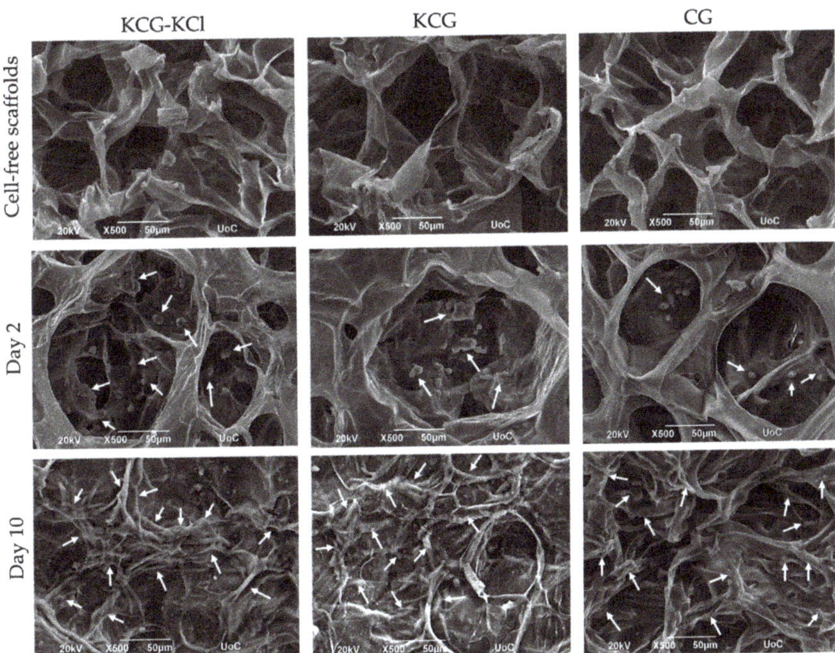

Figure 3. Representative SEM images from cell-free (upper panel) and cell-laden scaffolds after 2 days (middle panel) and 10 days (lower panel) in culture. KCG-KCl scaffolds (left column), KCG scaffolds (middle column) and CG control scaffolds (right column) are visualized. White arrows in the middle panel point to adhered cells and their nuclei on day 2, while in the lower panel the arrows point to dense elongated multicellular formations on day 10. Magnifications are $500\times$ and scale bars represent 50 µm.

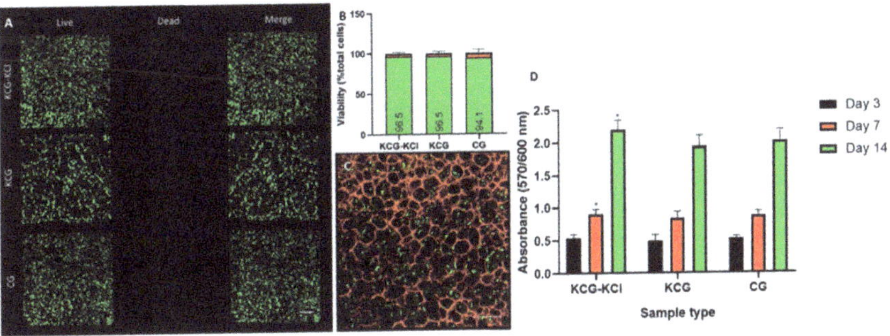

Figure 4. Cell viability assessment by CLSM live/dead fluorescent staining at 72 h and by the metabolic PrestoBlue™ cell viability assay of DPSCs seeded on KCG-KCl, KCG and CG scaffolds at days 3, 7 and 14. (**A**) Representative CLSM (live and dead cells) images. (**B**) Mean percentage (%) of live to the total number of cells. Error bars indicate the standard deviation ($n = 3$). No statistically significant differences were observed among the groups. (**C**) CLSM image of the autofluorescent CG scaffold and the fluorescent cells, illustrating the homogenous distribution of the cells on the scaffold. Scale bars indicate 100 µm. (**D**) Cell viability assessment of DPSCs in the presence of the three scaffold types on days 3, 7 and 14. Each bar represents the mean $\pm$ SD of triplicates of two independent experiments (* $p < 0.05$, compared to the CG control scaffold at the corresponding time point).

CLSM imaging confirmed that cells attached firmly and were evenly distributed on the scaffolds 72 h after seeding. Presentative CLSM photos of each scaffold composition are shown in Figure 4A. The green-appearing viable cells (as indicated by the enzymatic digestion of the fluorescent dye calcein AM) were predominant. The presence of red-appearing dead cells (stained by the ethidium bromide homodimer III) was limited in all groups. The percentage of live cells to the total cell number (live and dead) of each group is depicted in Figure 4B. Games Howell's post hoc test revealed no statistically significant differences among the tested groups ($p > 0.05$). DPSCs attached firmly in the CG scaffolds, as shown in Figure 3F.

2.5. Odontogenic Differentiation of DPSCs in the Presence of the Scaffolds

2.5.1. Determination of the ALP Activity of DPSCs

Since alkaline phosphatase activity is considered an early marker of odontogenesis, day 3, day 7 and day 14 were selected as the time points for its investigation (Figure 5A). At day 3, only a mild expression of the enzyme's activity was evident, with all scaffold types having identical values. Between day 3 and day 7, a steep increase was observable for all compositions of approximately quadruple magnitude, with the KCG-KCl and CG scaffolds exhibiting slightly higher values than the KCG ones. Finally, at day 14, all scaffolds showed a two-fold increase, with no profound variations between them.

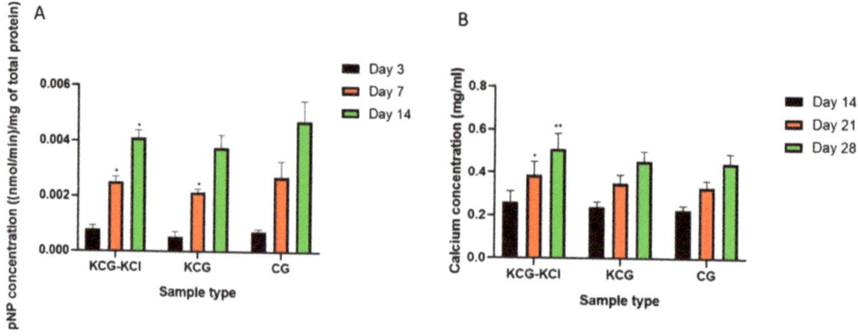

Figure 5. (**A**) Normalized alkaline phosphatase activity of DPSCs in the presence of the three scaffold types on days 3, 7 and 14. Each bar represents the mean ± SD of triplicates of two independent experiments (* $p < 0.05$ compared to the CG control scaffold at each time point). (**B**) Calcium concentration measured in the collected supernatants on days 14, 21 and 28 in culture from the three scaffold compositions. Each bar represents the mean ± SD of triplicates of two independent experiments (* $p < 0.05$, ** $p < 0.01$ compared to the CG control scaffold at each time point).

2.5.2. Evaluation of the Secreted Calcium in Supernatants

The concentration of secreted calcium secreted by the DPSCs was quantified on days 14, 21 and 28 (Figure 5B). All samples showed not only similar values with regard to calcium concentration, but also an identical trend in the way it increased. Specifically, at day 14, the KCG-KCl scaffolds slightly surpassed the other two compositions, with the same motif being repeated for the other two time points as well. KCG and CG samples retained similar values throughout the 28 days period, at all different time points. The increase of calcium concentration was linear between the subsequent time points, with an almost 50% increase between the previous and the next time point.

2.5.3. Real-Time PCR Odontogenic Markers

The gene expression of odontogenic markers was upregulated at different levels in all three tested scaffolds, as assessed by real-time PCR at 3, 7 and 14 days. The KCG-KCl scaffold showed the highest odontogenic effect on DPSCs. Gene expressions normalized to the two HKGs (*SDHA* and *B2M*) are displayed as fold change and illustrated in Figure 6.

More specifically, the expression of *DSPP*, which is a late marker of odontogenic differentiation, was not expressed until day 14 in any of three examined DPSC-seeded scaffolds. CG and KCG compositions revealed an approximately two-fold upregulation, compared to baseline control, while a statistically higher (five-fold) value was identified at day 14 of treatment inside the KCG-KCl scaffolds ($p < 0.001$). Similarly, the expression of the *RunX2* gene was upregulated only at day 14, for all groups, with an approximately three-fold increase, while the *ALP* marker expression levels of KCG-KCl and KCG scaffolds depicted a seven-fold and eight-fold increase at day 7, respectively, followed by a decline at day 14. Conversely, CG scaffolds showed a continuous upregulation profile, up to day 14, which peaked at a seven-fold increase at the last time point.

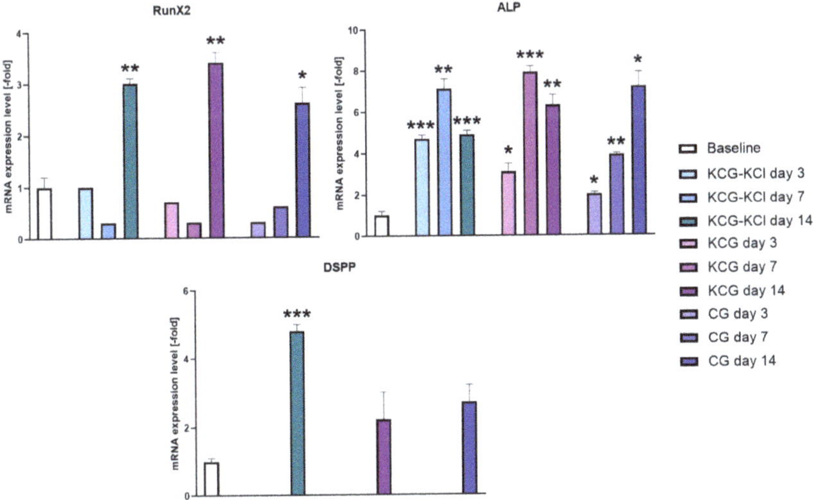

Figure 6. Odontogenic differentiation of DPSCs after 3, 7 and 14 days in contact with the tested scaffolds. The mRNA expressions of *RunX2*, *ALP* and *DSPP* are expressed as fold change compared to the two housekeeping genes (*B2M* and *SDHA*). DPSCs that were scaffold-free acted as the baseline. Asterisks indicate the statistically significant differences compared to the negative control (* $p \leq 0.05$, ** $p \leq 0.01$, *** $p \leq 0.001$) ($n = 3$).

3. Discussion

Oral-cavity-related diseases belong to the most prevalent categories of health problems, encompassing a wide range of pathophysiologies from the more typical tooth and periodontal damage to life threatening cancers [34]. DPSCs are stem cells that can be isolated from the dental pulp of permanent teeth and are capable of being guided towards different lineage pathways, making them an ideal cell line to work with for the investigation of regenerative model systems [4]. One of their most popular applications is related to their directed differentiation towards odontoblast-like cells to enable the recuperation of dental tissue [23]. Research in tissue regeneration strives to propel current medicinal standards further. To this end, researchers have investigated the development of novel treatment methods that can accommodate the restoration of tissue defects caused by exogenous trauma or health issues resulting from genetic abnormalities [35]. Dentin and pulp tissue engineering would significantly benefit dental clinical practice, and therefore recent studies have focused on the discovery of new bioactive materials to facilitate this purpose. Indirect and direct pulp capping materials have been utilized throughout the years to accommodate induced odontogenesis at the site of pulp exposure or proximity [2].

Biomaterials have been developed over the last decades as a source for the construction of cell friendly platforms that interact with the human body and enable gradual regeneration in a desired, controllable way. They can be broadly classified into two categories based on their origin and the method of their production, natural and synthetic, and have been used in various combinations to create tissue engineering scaffolds that enhance the growth of specific tissue types. Natural biomaterials are similar to structural biomolecules that compose native tissue, and thus their biological effect is usually more potent than that of the synthetic ones [36]. However, one of the main downsides of natural biopolymers is their low mechanical strength, making it pivotal to either crosslink or mix them with other, firmer compounds. Chitosan and gelatin have been extensively studied in tissue engineering applications, due to their superb biological response towards different cell lines [15,23]. Chitosan scaffolds, in combination with various bioactive substances, have been investigated for their dentin and pulp tissues regeneration ability with advantageous results [11]. Kappa-carrageenan, a natural biocompatible polysaccharide, possesses a negatively charged sulfate group which provides the capability to form firm gels at room temperature in the presence of a cation such as sodium or potassium [30]. Moreover, the spare sulfate groups that do not interact with these cations are free to form ionic binding with other positively charged groups. This property can be exploited when kappa-carrageenan and chitosan are combined to form polyelectrolyte complexes that stem from the ionic interactions occurring between the positively charged amino groups of chitosan and the negatively charged sulfate groups of kappa-carrageenan, leading to an amplification of the scaffolds' mechanical strength [37]. In this study, we fabricated two types of kappa-carrageenan-containing scaffolds with and without the addition of potassium chloride based on their capacity to promote the osteogenic responses of pre-osteoblastic cells [33]. We attempted to illustrate their contribution to induce the odontogenic differentiation of DPSCs towards odontoblast-like cells. Chitosan/gelatin scaffolds served as the control group.

Under wet conditions, all three types of scaffolds behaved as sponge-like materials, able to absorb large quantities of water, with Young's modulus values ranging between 22.5 and 28.5 kPa. The KCG-KCl indicated the highest and the KCG and CG had almost identical values, underlying the impact of potassium chloride in the overall mechanical robustness of the scaffolds. Such structures of spongy nature have been proven to allow and promote bone differentiation [38]. Although the produced scaffolds provided an excellent biocompatible 3D environment for in vitro biological evaluation, their elastic modulus values were mechanically inferior compared to the native dental pulp tissue [39], which may limit their use as substitutes. Regarding degradable materials, when cells are seeded onto scaffolds, two antagonistic procedures constantly take place: the degradation of the scaffold, and the de novo synthesis of the extracellular matrix by the cells. The degradation rate of the three scaffolds compositions, as well as the role of kappa-carrageenan and the crosslinker KCl in reducing the degradation rate of the chitosan/gelatin scaffolds, has been thoroughly investigated and reported on previously [33]. Different research groups have shed light on the viscoelastic attributes of various cell lines, which derive from the intricate complexes that their different structural biomolecules formulate [40,41]. To evaluate whether the cells contribute to the stiffness of the fabricated scaffolds, measurements of the Young's modulus were conducted after 14 and 28 days of cell culture. Between days 0 and 14, a slight increase of approximately 12% was observed for the two kappa-carrageenan containing scaffolds, with the KCG-KCl exceeding KCG and the CG scaffolds showing a 4% increase. However, at day 28, a significant rise to 39 kPa was detected for the KCG-KCl, followed by the KCG and CG scaffolds at 30 kPa. All scaffolds showed increased elastic modulus values by a range of 27–36% compared to cell-free scaffolds. These significant differences reinforced our initial hypothesis: that the presence of DPSCs have a positive effect on the elastic modulus of the cell laden scaffolds. These results are in alignment with other research works reporting that cells can show elastic modulus values of several kPa [42,43].

All scaffolds showed excellent biocompatibility, with the living cell numbers gradually climbing towards higher values over time. The differences in the absorbance values were almost negligible, with the KCG-KCl scaffolds surpassing the other types at all time points. Scaffold porosity is a crucial parameter for a tissue engineering construct, as it can directly affect cell migration and infiltration after the saturation of the surface [44]. In a previous study, we showed that these scaffold types displayed a porosity of at least 80% [33]. This architecture allowed for a deep infiltration of the pores by the DPSCs after surface saturation, resulting in a cell number increase up to day 14 [45]. SEM images at day 2 illustrated that the cells had adhered to the material surface, and by day 10 dense elongated intercellular formations covering the pores of the scaffolds were visible. These findings indicated tissue growth formation, and they were confirmed by the live/dead staining results, revealing excellent biocompatibility. The living cells were distributed within the scaffolds, as illustrated by the CLSM images. These findings are in accordance with reports from the literature on chitosan scaffolds evaluated in direct contact with DPSCs [22,46].

Alkaline phosphatase activity is a crucial marker of odontogenesis since the enzyme is able to cleave phosphate groups from different molecules and make them available for the later stage of mineralization. As such, the enzyme activity is expected to have high expression during the early odontogenesis phase, and to decline afterwards [47]. All three scaffold types showed almost a two-fold increase of the ALP activity between days 3 and 7, with the same motif being evident for the period between days 7 and 14. The CG control had the highest values at all time points; however, this was not significant compared to the other two compositions. Calcium secretion is representative of the final stage of odontogenesis and is essential for the formation of hydroxyapatite, the main inorganic component of dental tissue. It is considered a middle-to-late marker of odontogenesis [48]; thus, it was monitored at days 14, 21 and 28. At day 14, all scaffolds depicted similar calcium secretion levels. Between days 14 and 21, a rise in calcium production was detected for all scaffold compositions, and similarly up to day 28. The KCG-KCl scaffolds indicated significantly increased calcium production on days 21 and 28 compared to the CG control. The results from both odontogenesis assays coincided with the findings from our previous work, investigating the pre-osteoblasts maturation, as well as with other studies focusing on the differentiation capability of kappa-carrageenan-containing scaffolds [33,47,49].

The odontogenic effect of all three scaffold compositions was investigated by means of qPCR (Figure 6). The KCG-KCl scaffolds showed the optimal odontogenesis induction effect, as indicated by the *DSPP* expression. *DSPP* belongs to the family of Small Integrin-Binding Ligand N-linked Glycoproteins (SIBLINGs), and is an odontogenic specific marker [50]. This protein participates in the mineralization of the dentin matrix and has been found to be involved in bone mineralization as well, but on a lesser scale [51]. In our study, *DSPP* expression was upregulated at day 14 in all three scaffold groups, with only the KCG-KCl scaffolds displaying a statistically significant increase. Even though *RunX2* and *ALP* are not specific markers for odontogenic differentiation, their effect on the mineralization of the dentin matrix is essential [52]. *RunX2* expression increased at day 14 in all three investigated scaffolds, while the expression of *ALP* was statistically higher than the baseline at all timepoints. Interestingly, the ALP expression was more upregulated at days 3 and 7 in the kappa-carrageenan-containing scaffolds, compared to the CG control, with a decline on day 14, while the CG scaffolds retained their highest values at this time point, implying that the odontogenic process started earlier in the carrageenan containing scaffolds. It has been shown that when dexamethasone, a general induction factor, was applied on DPSCs, the expression of *RunX2* was not upregulated [53], while in this study all tested scaffold compositions caused increased *RunX2* expression. The upregulation of the expression of the transcription factor *RunX2* is essential for the odontogenic differentiation of DPSCs to odontoblast-like cells [54]. It has also been shown that bioactive pulp-capping materials that are commonly used in dental clinical praxis, such as silicate cements and mineral trioxide aggregate (MTA), do not affect the *ALP* gene expression, while CG, KCG and KCG-KCl significantly upregulated this gene [55]. In previous works of our group revolving around the odontogenic potential of chitosan/gelatin-based scaffolds, we have shown the upregulation of these markers in a similar 3D environment [23,56]. The abovementioned findings strongly indicate the odontogenic stimuli of the CG, KCG and KCG-KCl scaffolds as early as day 3. The sponge-like architecture of the produced scaffolds favors the physical absorption of high amounts of water, thus rendering them excellent candidates for enhanced functional platforms for the absorption and subsequent release of drugs, growth factors and various biomolecules, as well as bioactive inorganic compounds that are dissolvable in biological environments.

4. Materials and Methods

4.1. Materials

Chitosan (low molecular weight), gelatin from bovine skin, kappa-carrageenan, potassium chloride, magnesium chloride, glutaraldehyde, Triton X-100, p-nitro-phenyl-phosphate (pNPP), L-ascorbic acid 2-phosphate, β-glycerophosphate and dexamethasone were purchased from Sigma-Aldrich (St. Louis, MO, USA). The PrestoBlue™ reagent for cell viability and proliferation was purchased from Invitrogen Life Technologies (Carlsbad, CA, USA). Trypsin/EDTA (0.25%), phosphate-buffered saline (PBS), amphotericin-B (fungizone), L-glutamine and penicillin/streptomycin (P/S) were all purchased from Gibco ThermoFisher Scientific (Waltham, MA, USA). Minimum essential Eagle's medium (alpha-MEM) and fetal bovine serum (FBS) were purchased from PAN-Biotech (Aidenbach, Germany). O-cresol phthalein complexone reagent kit (CPC) was purchased from Biolabo (Les Hautes Rives Maizy, France). Bradford reagent was purchased from AppliChem GmbH (Darmstadt, Germany). Collagenase I was purchased from Abiel Biotech (Palermo, Italy), and dispase II from Roche (Basel, Switzerland). The viability assay kit for live/dead staining (calcein AM and EthD-III) was purchased from Biotium (Fremont, CA, USA), the NucleoSpin RNA isolation kit from Macherey-Nagel (Duren, Germany), the cDNA Synthesis kit PrimeScript first-strand from Takara (Shiga, Japan), and the SYBR PCR Master Mix from Applied Biosystems (Foster City, CA, USA). The fluorochrome-conjugated antibodies for cell surface marker characterization were purchased from BioLegend (Fell, Germany).

4.2. Scaffolds Preparation

Three types of scaffolds were produced following a protocol based on our previous work [33]: (i) kappa-carrageenan/chitosan/gelatin (KCG), (ii) kappa-carrageenan/chitosan/gelatin enriched with potassium chloride (KCG-KCl) and (iii) chitosan/gelatin (CG) acting as a control. For the fabrication of the scaffolds, an established protocol was followed based on our previous work [33]. Briefly, for the KCG scaffolds a 2.5% w/v chitosan in 1% v/v acetic acid solution was allowed to mix at 50 °C for 1 h, while another 5% w/v gelatin solution was prepared in deionized water for 20 min at 50 °C. The two solutions were mixed in a volume ratio of 2:1 chitosan/gelatin, and afterwards the final solution was mixed with a 1% w/v kappa-carrageenan water solution. The KCG-KCl mixtures were prepared in exactly the same way, but with the addition of 0.25 M KCl. Subsequently, the resulting solutions were left under mild stirring for 4 h at 60 °C to homogenize completely. From each of them, 10 mL were transferred into a 15 mL tube and, after the addition of 50 μL of 0.25% v/v glutaraldehyde, 800 μL of the blend was cast onto the wells of a 24 well-plate and freeze-dried for 24 h to create porous 3D scaffolds. Prior to cell seeding, the scaffolds were UV treated at 262 nm to disinfect them for 30 min. Table 1 presents the scaffold designation and their chemical composition, while Scheme 1 summarizes the production process.

Table 1. Acronyms of the various scaffold compositions.

Type	Scaffold Acronym	Composition
(i)	KCG-KCl	1% w/v kappa-carrageenan, 1.67% w/v chitosan, 1.67% w/v gelatin, 0.25 M KCl
(ii)	KCG	1% w/v kappa-carrageenan, 1.67% w/v chitosan, 1.67% w/v gelatin
(iii)	CG	1.67% w/v chitosan, 1.67% w/v gelatin

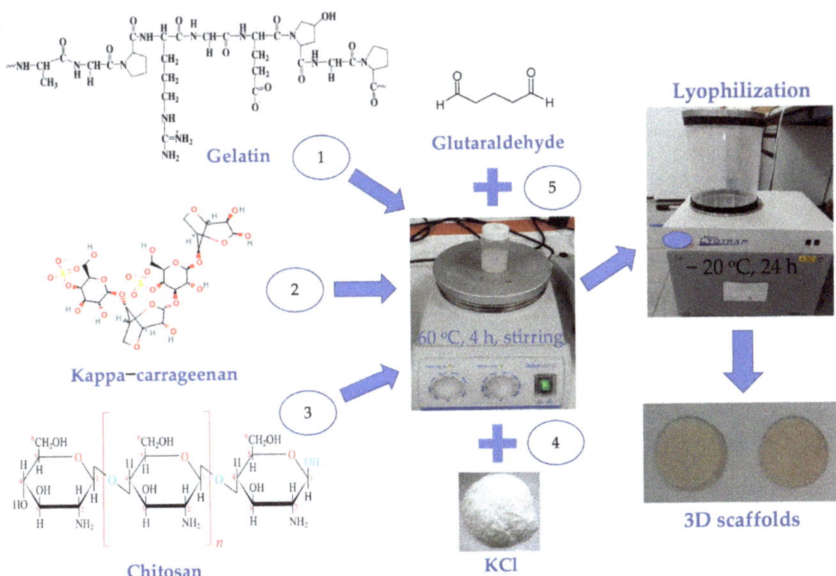

Scheme 1. Graphical representation of the scaffold production. The numbers 1–5 in the scheme show the sequential addition of each compound (1 of gelatin, 2 of kappa carrageenan, 3 of chitosan, 4 of KCl and 5 of glutaraldehyde).

4.3. Establishment and Culture of DPSCs

The protocol of this study received approval from the Institutional Ethics Board Committee (number 46/20-3-2019). Three patients donated their healthy third molars, which were extracted for orthodontic/preventive reasons. Dental pulp stem cell pulled cultures were established using the enzymatic dissociation method, as described previously [57]. Briefly, immediately after tooth extraction teeth were cleaned from soft remnant tissues and disinfected with iodine. The pulp chamber was exposed after a cut was performed through the cemento-enamel junction of each tooth using a high-speed handpiece instrument. The pulp tissue was transported in serum-free cell culture medium to sterile conditions. The pulp was then mechanically minced and enzymatically dissociated for 45 min at 37 °C in a buffer solution of 3 mg/mL collagenase I and 4 mg/mL dispase II.

DPSCs were cultured in complete culture medium (CCM), consisting of a minimum essential medium supplemented with 15% fetal bovine serum (FBS), 0.1 mM L-ascorbic acid and a triple antibiotic solution of 100 U/mL penicillin, 0.1 mg/mL streptomycin and 0.25 µg/mL amphotericin B. Cells' incubation conditions were 37 °C, 5% CO_2 and 100% relative humidity (RH) (Heal Force, Shanghai, China) until 70–80% confluency was reached. Cell detachment was performed by 0.25% Trypsin/1 mM EDTA, and passages of 2–6 were used for all experiments. For the differentiation experiments, the medium was enriched with 50 µg/mL l-ascorbic acid, 10 mM β-glycerophosphate and 10 nM dexamethasone.

4.4. Immunophenotypic Characterization of DPSCs

DPSCs were characterized for specific surface markers by flow cytometry. Cells were suspended, washed with flow cytometry staining buffer (FCSB) containing 1% BSA and 0.1% NaN3 in PBS, and stained for the relevant markers. The mesenchymal (STRO1, CD146, CD90, CD73), endothelial (CD105), embryonic (SSEA4) and hematopoietic (CD34, CD45) surface markers on single-cell suspensions were labelled with the following fluorochrome-conjugated antibodies: CD90-FITC, CD34-APC, CD45-PE, SSEA4-FITC, STRO1-APC, CD146-PE, CD105-APC and CD73-PE. Data were acquired from 50.000 events per sample using a Guava EasyCyte 8HT Benchtop flow cytometer (Merck Millipore,

Darmstadt, Germany) and analyzed with the Summit 5.1 software (Summit Software Technologies, Fort Wayne, IN, USA) ($n = 3$).

4.5. Cell Viability Assessment

4.5.1. Cell Viability Assessment with the PrestoBlue™ Metabolic Assay

A 10 µL cell suspension containing 8×10^4 cells was seeded onto each scaffold and 400 µL of fresh medium was added to a final volume of 400 µL for each well of a 24 well-plate. The cell viability of the scaffolds was investigated by employing PrestoBlue™ assay, whose base ingredient is resazurin, a reductive agent that changes color from blue to purple according to cell metabolism levels. Briefly, based on an established protocol [33], at three different time points, days 3, 7 and 14, the old medium for each scaffold was replaced with a mixture of 40 µL of PrestoBlue™ and 360 µL of fresh medium and the 24 well-plate was incubated for 1 h at 37 °C. Subsequently, 100 µL of this mixture was transferred to a 96 well-plate and were photometrically measured with a spectrophotometer at 570 and 600 nm (Synergy HTX Multi-Mode Micro-plate Reader, BioTek, Bad Friedrichshall, Germany). All samples were analyzed in triplicates of two independent experiments ($n = 6$).

4.5.2. Cell Viability Assessment by Live/Dead Staining

Scaffolds containing 24-well plates were seeded with 10^5 cells/well in 1 mL CCM/well. The cell-seeded scaffolds were cultured further for 3 days, and then cell viability was assessed by live/dead double-staining using 2 mM calcein AM ((λex/λem(max): 494/517 nm)/ 4 mM ethidium homodimer III ((λex/λem(max): 532/625 nm) for up to 45 min. Calcein AM was cleaved by esterases of live cells to produce the green fluorescent dye calcein, while EthD-III stained the nucleus of dead cells with bright red fluorescence. The seeded HTFD scaffolds were observed by confocal laser scanning microscopy (CLSM) (Leica Microsystems, Wetzlar, Germany). Approximately 35 steps of 10 µm-step size z-stacked images were generated. Two images were taken per scaffold. Cells seeded on a glass surface acted as a negative control. The numbers of live and dead cells were measured with the ImageJ software (version 1.8.0, NIH, Bethesda, MD, USA). The % percentage of live cells to the total number of cells was calculated ($n = 3$ per condition). When the gains of the infra-red laser were increased, the structure of the scaffold was observed and cell morphology, attachment and distribution were observed.

4.6. Mechanical Properties of the DPSCs Seeded Scaffolds

The various scaffold compositions were examined mechanically by determining the differences of the Young's modulus values between cell seeded and cell-free scaffolds. The elastic modulus value of the cells can play an important role to the reinforcement of the mechanical integrity of the final construct, and thus it is an important parameter to take into account [58]. Compression testing was conducted by means of a compression and tensile strength test unit (UniVert, CellScale, Waterloo, ON, Canada) with a maximum cell load of 50 N. The measurements were conducted under wet conditions ($n = 6$). The control cell-free scaffolds were compared to DSPCs laden scaffolds after 14 and 28 days of culture. The Young's modulus was calculated at a strain of 5–20% and velocity of 1 mm/s using the following formula:

$$Young\ modulus: \frac{F \times L}{A \times \Delta l}$$

where F is the perpendicular force applied to the surface of the scaffolds, A is the surface, Δl is the height strain of the scaffolds after the force application and L is the initial height.

4.7. Scanning Electron Microscopy (SEM)

The morphology of both the cell-free and cell laden scaffolds was investigated by means of scanning electron microscopy (JEOL JSM-6390 LV) at days 2 and 10. The cell-laden scaffolds were rinsed twice with PBS and fixed using a PFA 4% solution. The scaffolds were then dehydrated by employing increasing ethanol concentrations, from 30 to 100%

pure ethanol. For the cell-free scaffolds, the same procedure was followed but without fixation. At the last stage, the scaffolds were dried in a critical point drier (Baltec CPD 030), sputter-coated with an 80 nm thick layer of gold (Baltec SCD 050) and observed under a scanning electron microscope at an accelerating voltage of 20 kV (JEOL JSM-6390 LV).

4.8. Odontogenic Differentiation

4.8.1. Alkaline Phosphatase Activity

The ALP activity is indicative of the early stages of odontogenesis. For these experiments, 1×10^5 DPSCs were seeded onto the different scaffold compositions and the enzyme activity was evaluated at days 3, 7 and 14. The scaffolds were rinsed with PBS multiple times, and then incubated in a solution comprising 0.1% Triton X-100 and 50 mM Tris-HCl, at pH 10.5, so that cell lysis can take place. After three freezing/thawing cycles between room temperature and $-20\,°C$, a 100 µL suspension from each well was mixed with 100 µL of a 2 mg/mL p-nitro-phenyl-phosphate (pNPP) solution diluted in a buffer containing 50 mM Tris-HCl and 2 mM $MgCl_2$. The resulting mixtures were allowed to incubate at $37\,°C$, and then 100 µL from each well was transferred to a 96-well plate and photometrically measured at 405 nm in a spectrophotometer (Synergy HTX Multi-Mode Micro-plate Reader, BioTek, Bad Friedrichshall, Germany) [59]. The enzymatic activity was calculated using the equation [units = nmol p-nitrophenol/min] and normalized to total cellular protein in lysates determined using the Bradford protein concentration assay (AppliChem GmbH, Darmstadt, Germany). All samples were analyzed in triplicates of two independent experiments ($n = 6$).

4.8.2. Calcium Secretion Levels Determination

The production of calcium is connected with the formation of hydroxyapatite, one of the basic and most crucial components of bone and dental tissue. The O-cresol phthalein complexone (CPC) method was used to determine the calcium concentration in the supernatants that were collected every three days of cell culture, up to day 28. Briefly, 10 µL of culture medium from each sample was mixed with 100 µL of calcium buffer and 100 µL of calcium dye, as previously described [33], and transferred to a 96-well plate for photometrical measurement at 550 nm in a spectrophotometer (Synergy HTX Multi-Mode Micro-plate Reader, BioTek, Bad Friedrichshall, Germany) [60]. All samples were analyzed in triplicates of two independent experiments ($n = 6$).

4.8.3. Gene Expression of DPSCs

Scaffolds were placed in 24-well plates and were seeded with 6×10^5 cells/well, in 1 mL CCM/well for 24 h. The effect of odontogenic differentiation was evaluated at 3, 7 and 14 days by quantitative real-time polymerase chain reaction (qPCR), as described previously [23]. A group consisting of cells alone, without the presence of scaffolds, acted as a negative control (baseline expression group). Cells were lysed and the RNA was isolated using the NucleoSpin RNA isolation kit. The cDNA was then synthesized using the PrimeScript first-strand cDNA Synthesis Kit. Finally, qPCR was performed in a thermal cycler (Step One Plus, Applied Biosystems, Waltham, MA, USA) using the SYBR PCR Master Mix. Two incubation steps at $50\,°C$ for 2 min and at $95\,°C$ for 2 min were followed by 40 cycles of PCR, comprising denaturation for 15 s at $95\,°C$ and annealing/extension for 1 min at $60\,°C$. The relevant primers (Table 1) were designed with the Primer-Blast software (NCBI, NIH, Bethesda, MD, USA). Dehydrogenase complex, subunit A, flavoprotein (*SDHA*) and beta-2-microglobulin (*B2M*) were used as house-keeping genes (HKGs). The genes of interest (GOI) were the following: dentin sialophosphoprotein (*DSPP*), alkaline phosphatase (*ALP*) and the transcription factor Runt-Related Transcription Factor-2 (*RUNX2*) (Table 2). The results were adjusted by amplification efficiency with the LinRegPCR software (Amsterdam, The Netherlands) and normalized to the HKGs. The quality of amplification and specificity were checked by the amplification plot and the standard melting curves. The expression of

GOI was normalized to the HKGs, displayed as a fold expression and statistically analyzed ($n = 3$).

Table 2. Primers design and amplicon size for the genes of interest and housekeeping genes used in the qPCR analysis.

Gene Symbol	Forward (5′–3′)	Reverse (5′–3′)	Amplicon Size (bp)
DSPP	GCTGGCCTGGATAATTCCGA	CTCCTGGCCCTTGCTGTTAT	135
ALP	CCGTGGCAACTCTATCTTTGG	CAGGCCCATTGCCATACAG	89
RUNX2	CCACCGAGACCAACAGAGTC	TCACTGTGCTGAAGAGGCTG	118
B2M	TGTCTTTCAGCAAGGACTGGT	ACATGTCTCGATCCCACTTAAC	138
SDHA	GCATGCCAGGGAAGACTACA	GCCAACGTCCACATAGGACA	127

4.9. Statistical Analysis

Statistical analysis for the assessment of cell viability, ALP activity and calcium production was performed using the one-way ANOVA Dunnett's multi-comparison test in GraphPad Prism version 8 software (GraphPad Software, San Diego, CA, USA), comparing each kappa-carrageenan-containing scaffold with the CG control at each experimental time point. A p-value < 0.05 was considered significant. The results of live/dead staining were analyzed with two-way ANOVA followed by Tukey's post hoc test. PCR results for the relevant gene expression were analyzed by Welch's ANOVA (Kolmogorov–Smirnov $p > 0.05$), followed by Games Howell's post hoc test. Data were expressed as means $\pm$ standard deviation (SD).

5. Conclusions

This study investigated the role of kappa-carrageenan in odontogenesis promoting when incorporated in a chitosan/gelatin biomimetic matrix, which is widely regarded as one of the most favorable scaffolds for mineralized tissue engineering. Both kappa-carrageenan containing scaffolds exhibited great biocompatible and odontogenic differentiation capabilities. Real-time PCR analysis revealed an earlier expression of the ALP marker for the two scaffolds, compared to the control, while KCG-KCl samples had the highest upregulation of the odontogenesis specific DSPP marker at day 14. Moreover, all scaffold compositions showed an increase in Young's modulus values with progressing culture time, validating our initial hypothesis of the cell contribution to the reinforcement of the scaffolds' total mechanical robustness. These results underline that kappa-carrageenan scaffolds offer an excellent 3D environment for DPSCs, with a remarkable potential for dentin-pulp reconstitution. However, more specific biological markers could be evaluated in the future to gain a better insight into the regeneration capacity of the complex dentin-pulp tissue using kappa-carrageenan scaffolds. In addition, the use of carrageenans as dental pulp implants is still weighed down by their relatively low mechanical strength, giving room for further future experimentation.

Author Contributions: Conceptualization, A.B. and M.C.; methodology, K.L. and F.M.; investigation, K.L. and F.M.; writing—original draft preparation, K.L. and F.M.; writing—review and editing, A.B. and M.C.; supervision, A.B. and M.C.; funding acquisition, A.B. and M.C. All authors have read and agreed to the published version of the manuscript.

Funding: This research was funded by the Hellenic Foundation for Research and Innovation (H.F.R.I.) under the "1st Call for H.F.R.I. Research Projects to support Faculty members and Researchers and the procurement of high-cost research equipment grant" (project number HFRI-FM17-1999) and "2nd Call for H.F.R.I. Research Projects to support Faculty members and Researchers" (project number HFRI 3549).

Institutional Review Board Statement: Not applicable.

Informed Consent Statement: Informed consent was obtained from all subjects involved in the study, as described in Section 4.3 'The protocol of this study received approval by the Institutional Ethics Board Committee (number 46/20-3-2019)'.

Data Availability Statement: Data can be made available upon request.

Conflicts of Interest: The authors declare no conflict of interest.

References

1. Yelick, P.C.; Sharpe, P.T. Tooth Bioengineering and Regenerative Dentistry. *J. Dent. Res.* **2019**, *98*, 1173–1182. [CrossRef] [PubMed]
2. Machla, F.; Angelopoulos, I.; Epple, M.; Chatzinikolaidou, M.; Bakopoulou, A. Biomolecule-Mediated Therapeutics of the Dentin–Pulp Complex: A Systematic Review. *Biomolecules* **2022**, *12*, 285. [CrossRef] [PubMed]
3. Giraud, T.; Jeanneau, C.; Rombouts, C.; Bakhtiar, H.; Laurent, P.; About, I. Pulp capping materials modulate the balance between inflammation and regeneration. *Dent. Mater.* **2019**, *35*, 24–35. [CrossRef] [PubMed]
4. Egusa, H.; Sonoyama, W.; Nishimura, M.; Atsuta, I.; Akiyama, K. Stem cells in dentistry—Part I: Stem cell sources. *J. Prosthodont. Res.* **2012**, *56*, 151–165. [CrossRef]
5. Feng, R.; Lengner, C. Application of Stem Cell Technology in Dental Regenerative Medicine. *Adv. Wound Care* **2013**, *2*, 296–305. [CrossRef]
6. Gronthos, S.; Mankani, M.; Brahim, J.; Robey, P.; Shi, S. Postnatal human dental pulp stem cells (DPSCs) In Vitro and In Vivo. *Proc. Natl. Acad. Sci. USA* **2001**, *97*, 13625–13630. [CrossRef]
7. La Noce, M.; Stellavato, A.; Vassallo, V.; Cammarota, M.; Laino, L.; Desiderio, V.; Del Vecchio, V.; Nicoletti, G.F.; Tirino, V.; Papaccio, G.; et al. Hyaluronan-Based Gel Promotes Human Dental Pulp Stem Cells Bone Differentiation by Activating YAP/TAZ Pathway. *Cells* **2021**, *10*, 2899. [CrossRef]
8. Casagrande, L.; Cordeiro, M.M.; Nor, S.A.; Nor, J.E. Dental pulp stem cells in regenerative dentistry. *Odontology* **2011**, *99*, 1–7. [CrossRef]
9. Demarco, F.F.; Conde, M.C.M.; Cavalcanti, B.N.; Casagrande, L.; Sakai, V.T.; Nör, J.E. Dental pulp tissue engineering. *Braz. Dent. J.* **2011**, *22*, 3–13. [CrossRef]
10. Lymperi, S.; Ligoudistianou, C.; Taraslia, V.; Kontakiotis, E.; Anastasiadou, E. Dental stem cells and their applications in dental tissue engineering. *Open Dent. J.* **2013**, *7*, 76. [CrossRef]
11. Bordini, E.; Cassiano, F.; Bronze-Uhle, E.; Alamo, L.; Hebling, J.; Costa, C.; Soares, D. Chitosan in association with osteogenic factors as a cell-homing platform for dentin regeneration: Analysis in a pulp-in-a-chip model. *Dent. Mater.* **2022**, *38*, 655–669. [CrossRef] [PubMed]
12. Di Martino, A.; Sittinger, M.; Risbud, M.V. Chitosan: A versatile biopolymer for orthopaedic tissue-engineering. *Biomaterials* **2005**, *26*, 5983–5990. [CrossRef]
13. Hsu, S.-h.; Chang, Y.-B.; Tsai, C.-L.; Fu, K.-Y.; Wang, S.-H.; Tseng, H.-J. Characterization and biocompatibility of chitosan nanocomposites. *Colloids Surf. B Biointerfaces* **2011**, *85*, 198–206. [CrossRef] [PubMed]
14. Aider, M. Chitosan application for active bio-based films production and potential in the food industry: Review. *LWT—Food Sci. Technol.* **2010**, *43*, 837–842. [CrossRef]
15. Georgopoulou, A.; Papadogiannis, F.; Batsali, A.; Marakis, J.; Alpantaki, K.; Eliopoulos, A.G.; Pontikoglou, C.; Chatzinikolaidou, M. Chitosan/gelatin scaffolds support bone regeneration. *J. Mater. Sci. Mater. Med.* **2018**, *29*, 59. [CrossRef]
16. Aguilar, A.; Zein, N.; Harmouch, E.; Hafdi, B.; Bornert, F.; Offner, D.; Clauss, F.; Fioretti, F.; Huck, O.; Benkirane-Jessel, N.; et al. Application of Chitosan in Bone and Dental Engineering. *Molecules* **2019**, *24*, 3009. [CrossRef] [PubMed]
17. Kiernan, C.H.; Wolvius, E.B.; Brama, P.A.J.; Farrell, E. The Immune Response to Allogeneic Differentiated Mesenchymal Stem Cells in the Context of Bone Tissue Engineering. *Tissue Eng. Part B Rev.* **2017**, *24*, 75–83. [CrossRef]
18. Santoro, M.; Tatara, A.; Mikos, A. Gelatin carriers for drug and cell delivery in tissue engineering. *J. Control. Release* **2014**, *190*, 210–218. [CrossRef] [PubMed]
19. Liu, X.; Smith, L.A.; Hu, J.; Ma, P.X. Biomimetic nanofibrous gelatin/apatite composite scaffolds for bone tissue engineering. *Biomaterials* **2009**, *30*, 2252–2258. [CrossRef]
20. Jiang, G.; Li, S.; Yu, K.; He, B.; Hong, J.; Xu, T.; Meng, J.; Ye, C.; Chen, Y.; Shi, Z.; et al. A 3D-printed PRP-GelMA hydrogel promotes osteochondral regeneration through M2 macrophage polarization in a rabbit model. *Acta Biomater.* **2021**, *128*, 150–162. [CrossRef]
21. Peter, M.; Ganesh, N.; Selvamurugan, N.; Nair, S.V.; Furuike, T.; Tamura, H.; Jayakumar, R. Preparation and characterization of chitosan–gelatin/nanohydroxyapatite composite scaffolds for tissue engineering applications. *Carbohydr. Polym.* **2010**, *80*, 687–694. [CrossRef]
22. Bakopoulou, A.; Georgopoulou, A.; Grivas, I.; Bekiari, C.; Prymak, O.; Loza, K.; Epple, M.; Papadopoulos, G.C.; Koidis, P.; Chatzinikolaidou, M. Dental pulp stem cells in chitosan/gelatin scaffolds for enhanced orofacial bone regeneration. *Dent. Mater.* **2019**, *35*, 310–327. [CrossRef] [PubMed]
23. Vagropoulou, G.; Trentsiou, M.; Anthie, G.; Papachristou, E.; Prymak, O.; Kritis, A.; Epple, M.; Chatzinikolaidou, M.; Bakopoulou, A.; Koidis, P. Hybrid chitosan/gelatin/nanohydroxyapatite scaffolds promote odontogenic differentiation of dental pulp stem cells and in vitro biomineralization. *Dent. Mater.* **2020**, *37*, e23–e36. [CrossRef] [PubMed]

24. Pereda, M.; Ponce, A.G.; Marcovich, N.E.; Ruseckaite, R.A.; Martucci, J.F. Chitosan-gelatin composites and bi-layer films with potential antimicrobial activity. *Food Hydrocoll.* **2011**, *25*, 1372–1381. [CrossRef]

25. Roshanfar, F.; Hesaraki, S.; Dolatshahi-Pirouz, A. Electrospun Silk Fibroin/kappa-Carrageenan Hybrid Nanofibers with Enhanced Osteogenic Properties for Bone Regeneration Applications. *Biology* **2022**, *11*, 751. [CrossRef]

26. Rode, M.; Angulski, A.; Gomes, F.; Silva, M.; Jeremias, T.; Carvalho, R.; Vieira, D.; De Oliveira, L.F.; Maia, L.; Trentin, A.; et al. Carrageenan hydrogel as a scaffold for skin-derived multipotent stromal cells delivery. *J. Biomater. Appl.* **2018**, *33*, 422–434. [CrossRef]

27. Gashti, M.P.; Stir, M.; Hulliger, J. Synthesis of bone-like micro-porous calcium phosphate/iota-carrageenan composites by gel diffusion. *Colloids Surf. B Biointerfaces* **2013**, *110*, 426–433. [CrossRef]

28. Campo, V.L.; Kawano, D.F.; Silva, D.B.d.; Carvalho, I. Carrageenans: Biological properties, chemical modifications and structural analysis—A review. *Carbohydr. Polym.* **2009**, *77*, 167–180. [CrossRef]

29. Nukavarapu, S.; Dorcemus, D. Osteochondral Tissue Engineering: Current Strategies and Challenges. *Biotechnol. Adv.* **2012**, *31*, 706–721. [CrossRef]

30. Hermansson, A.M.; Eriksson, E.; Jordansson, E. Effects of potassium, sodium and calcium on the microstructure and rheological behaviour of kappa-carrageenan gels. *Carbohydr. Polym.* **1991**, *16*, 297–320. [CrossRef]

31. Tavakoli, S.; Kharaziha, M.; Kermanpur, A.; Mokhtari, H. Sprayable and injectable visible-light Kappa-carrageenan hydrogel for in-situ soft tissue engineering. *Int. J. Biol. Macromol.* **2019**, *138*, 590–601. [CrossRef] [PubMed]

32. Yegappan, R.; Selvaprithiviraj, V.; Amirthalingam, S.; Jayakumar, R. Carrageenan based hydrogels for drug delivery, tissue engineering and wound healing. *Carbohydr. Polym.* **2018**, *198*, 385–400. [CrossRef]

33. Loukelis, K.; Papadogianni, D.; Chatzinikolaidou, M. Kappa-carrageenan/chitosan/gelatin scaffolds enriched with potassium chloride for bone tissue engineering. *Int. J. Biol. Macromol.* **2022**, *209*, 1720–1730. [CrossRef] [PubMed]

34. Peres, M.A.; Macpherson, L.M.D.; Weyant, R.J.; Daly, B.; Venturelli, R.; Mathur, M.R.; Listl, S.; Celeste, R.K.; Guarnizo-Herreño, C.C.; Kearns, C.; et al. Oral diseases: A global public health challenge. *Lancet* **2019**, *394*, 249–260. [CrossRef] [PubMed]

35. Dzobo, K.; Thomford, N.E.; Senthebane, D.; Shipanga, H.; Rowe, A.; Dandara, C.; Pillay, M.; Motaung, S. Advances in Regenerative Medicine and Tissue Engineering: Innovation and Transformation of Medicine. *Stem Cells Int.* **2018**, *2018*, 2495848. [CrossRef]

36. Kim, B.-S.; Baez, C.E.; Atala, A. Biomaterials for tissue engineering. *World J. Urol.* **2000**, *18*, 2–9. [CrossRef]

37. Araujo, J.V.; Davidenko, N.; Danner, M.; Cameron, R.E.; Best, S. Novel Porous scaffolds of pH Responsive Chitosan/Carrageenan-based Polyelectrolyte Complexes for Tissue Engineering. *J. Biomed. Mater. Res. Part A* **2014**, *102*, 4415–4426. [CrossRef]

38. Tziveleka, L.A.; Sapalidis, A.; Kikionis, S.; Aggelidou, E.; Demiri, E.; Kritis, A.; Ioannou, E.; Roussis, V. Hybrid Sponge-Like Scaffolds Based on Ulvan and Gelatin: Design, Characterization and Evaluation of Their Potential Use in Bone Tissue Engineering. *Materials* **2020**, *13*, 1763. [CrossRef]

39. Lin, S.-L.; Lee, S.-Y.; Lin, Y.-C.; Huang, Y.-H.; Yang, J.-C.; Huang, H.-M. Evaluation of mechanical and histological properties of cryopreserved human premolars under short-term preservation: A preliminary study. *J. Dent. Sci.* **2014**, *9*, 244–248. [CrossRef]

40. Yamada, S.; Wirtz, D.; Kuo, S.C. Mechanics of Living Cells Measured by Laser Tracking Microrheology. *Biophys. J.* **2000**, *78*, 1736–1747. [CrossRef]

41. Bausch, A.R.; Ziemann, F.; Boulbitch, A.A.; Jacobson, K.; Sackmann, E. Local Measurements of Viscoelastic Parameters of Adherent Cell Surfaces by Magnetic Bead Microrheometry. *Biophys. J.* **1998**, *75*, 2038–2049. [CrossRef] [PubMed]

42. Guz, N.; Dokukin, M.; Kalaparthi, V.; Sokolov, I. If Cell Mechanics Can Be Described by Elastic Modulus: Study of Different Models and Probes Used in Indentation Experiments. *Biophys. J.* **2014**, *107*, 564–575. [CrossRef] [PubMed]

43. Kim, Y.; Kim, M.; Shin, J.H.; Kim, J. Characterization of cellular elastic modulus using structure based double layer model. *Med. Biol. Eng. Comput.* **2011**, *49*, 453–462. [CrossRef] [PubMed]

44. Prakoso, A.T.; Basri, H.; Adanta, D.; Yani, I.; Ammarullah, M.I.; Akbar, I.; Ghazali, F.A.; Syahrom, A.; Kamarul, T. The Effect of Tortuosity on Permeability of Porous Scaffold. *Biomedicines* **2023**, *11*, 427. [CrossRef] [PubMed]

45. Samourides, A.; Browning, L.; Hearnden, V.; Chen, B. The effect of porous structure on the cell proliferation, tissue ingrowth and angiogenic properties of poly(glycerol sebacate) urethane) scaffolds. *Mater. Sci. Eng. C* **2020**, *108*, 110384. [CrossRef] [PubMed]

46. Zheng, K.; Feng, G.; Zhang, J.; Xing, J.; Huang, D.; Lian, M.; Zhang, W.; Wu, W.; Hu, Y.; Lu, X.; et al. Basic Fibroblast Growth Factor Promotes Human Dental Pulp Stem Cells cultured in 3D Porous Chitosan Scaffolds to Neural Differentiation. *Int. J. Neurosci.* **2020**, *131*, 625–633. [CrossRef]

47. González Ocampo, J.I.; Machado de Paula, M.M.; Bassous, N.J.; Lobo, A.O.; Ossa Orozco, C.P.; Webster, T.J. Osteoblast responses to injectable bone substitutes of kappa-carrageenan and nano hydroxyapatite. *Acta Biomater.* **2019**, *83*, 425–434. [CrossRef]

48. Abuarqoub, D.; Awidi, A.; Abuharfeil, N. Comparison of osteo/odontogenic differentiation of human adult dental pulp stem cells and stem cells from apical papilla in the presence of platelet lysate. *Arch. Oral Biol.* **2015**, *60*, 1545–1553. [CrossRef]

49. Yegappan, R.; Selvaprithiviraj, V.; Amirthalingam, S.; Mohandas, A.; Hwang, N.; Rangasamy, J. Injectable angiogenic and osteogenic carrageenan nanocomposite hydrogel for bone tissue engineering. *Int. J. Biol. Macromol.* **2018**, *122*, 320–328. [CrossRef]

50. Yamakoshi, Y. Dentinogenesis and Dentin Sialophosphoprotein (DSPP). *J. Oral Biosci.* **2009**, *51*, 134–142. [CrossRef]

51. Qin, C.; Brunn, J.C.; Cadena, E.; Ridall, A.; Tsujigiwa, H.; Nagatsuka, H.; Nagai, N.; Butler, W.T. The Expression of Dentin Sialophosphoprotein Gene in Bone. *J. Dent. Res.* **2002**, *81*, 392–394. [CrossRef] [PubMed]

52. Lei, S.; Liu, X.-M.; Liu, Y.; Bi, J.; Zhu, S.; Chen, X. Lipopolysaccharide Downregulates the Osteo-/Odontogenic Differentiation of Stem Cells From Apical Papilla by Inducing Autophagy. *J. Endod.* **2020**, *46*, 502–508. [CrossRef] [PubMed]

53. Machla, F.; Sokolova, V.; Platania, V.; Prymak, O.; Kostka, K.; Kruse, B.; Agrymakis, M.; Pasadaki, S.; Kritis, A.; Alpantaki, K.; et al. Tissue engineering at the dentin-pulp interface using human treated dentin scaffolds conditioned with DMP1 or BMP2 plasmid DNA-carrying calcium phosphate nanoparticles. *Acta Biomater.* **2023**, *159*, 156–172. [CrossRef] [PubMed]
54. Camilleri, S.; McDonald, F. Runx2 and dental development. *Eur. J. Oral Sci.* **2006**, *114*, 361–373. [CrossRef] [PubMed]
55. Widbiller, M.; Lindner, S.R.; Buchalla, W.; Eidt, A.; Hiller, K.A.; Schmalz, G.; Galler, K.M. Three-dimensional culture of dental pulp stem cells in direct contact to tricalcium silicate cements. *Clin. Oral Investig.* **2016**, *20*, 237–246. [CrossRef] [PubMed]
56. Papadogiannhs, F.; Batsali, A.; Klontzas, M.; Karabela, M.; Anthie, G.; Mantalaris, A.; Zafeiropoulos, N.; Chatzinikolaidou, M.; Pontikoglou, C. Osteogenic differentiation of bone marrow mesenchymal stem cells on chitosan/gelatin scaffolds: Gene expression profile and mechanical analysis. *Biomed. Mater.* **2020**, *15*, 064101. [CrossRef]
57. Bakopoulou, A.; Leyhausen, G.; Volk, J.; Tsiftsoglou, A.; Garefis, P.; Koidis, P.; Geurtsen, W. Assessment of the Impact of Two Different Isolation Methods on the Osteo/Odontogenic Differentiation Potential of Human Dental Stem Cells Derived from Deciduous Teeth. *Calcif. Tissue Int.* **2011**, *88*, 130–141. [CrossRef]
58. Ding, Y.; Xu, G.; Wang, G. On the determination of elastic moduli of cells by AFM based indentation. *Sci. Rep.* **2017**, *2017*, 45575. [CrossRef]
59. Hadjicharalambous, C.; Kozlova, D.; Sokolova, V.; Epple, M.; Chatzinikolaidou, M. Calcium phosphate nanoparticles carrying BMP-7 plasmid DNA induce an osteogenic response in MC3T3-E1 pre-osteoblasts. *J. Biomed. Mater. Res. Part A* **2015**, *103*, 3834–3842. [CrossRef]
60. Wang, X.; Wenk, E.; Zhang, X.; Meinel, L.; Vunjak-Novakovic, G.; Kaplan, D.L. Growth factor gradients via microsphere delivery in biopolymer scaffolds for osteochondral tissue engineering. *J. Control. Release* **2009**, *134*, 81–90. [CrossRef]

MDPI

St. Alban-Anlage 66

4052 Basel

Switzerland

www.mdpi.com

International Journal of Molecular Sciences Editorial Office

E-mail: ijms@mdpi.com

www.mdpi.com/journal/ijms